Valuing Agroforestry Systems

Advances in Agroforestry

Volume 2

Series Editor:

P.K.R. Nair
School of Forest Resources and Conservation,
University of Florida, Gainesville, Florida, U.S.A.

Aims and Scope

Agroforestry, the purposeful growing of trees and crops in interacting combinations, began to attain prominence in the late 1970s, when the international scientific community embraced its potentials in the tropics and recognized it as a practice in search of science. During the 1990s, the relevance of agroforestry for solving problems related to deterioration of family farms, increased soil erosion, surface and ground water pollution, and decreased biodiversity was recognized in the industrialized nations too. Thus, agroforestry is now receiving increasing attention as a sustainable land-management option the world over because of its ecological, economic, and social attributes. Consequently, the knowledge-base of agroforestry is being expanded at a rapid rate as illustrated by the increasing number and quality of scientific publications of various forms on different aspects of agroforestry.

Making full and efficient use of this upsurge in scientific agroforestry is both a challenge and an opportunity to the agroforestry scientific community. In order to help prepare themselves better for facing the challenge and seizing the opportunity, agoroforestry scientists need access to synthesized information on multi-dimensional aspects of scientific agroforesty.

The aim of this new book-series, *Advances in Agroforestry*, is to offer state-of-the art synthesis of research results and evaluations relating to different aspects of agroforestry. Its scope is broad enough to encompass any and all aspects of agroforestry research and development. Contributions are welcome as well as solicited from competent authors on any aspect of agroforestry. Volumes in the series will consist of reference books, subject-specific monographs, peer-reviewed publications out of conferences, comprehensive evaluations of specific projects, and other book-length compilations of scientific and professional merit and relevance to the science and practice of agroforestry worldwide.

Valuing Agroforestry Systems

Methods and Applications

by

JANAKI R.R. ALAVALAPATI

School of Forest Resources and Conservation,
University of Florida, Gainesville, Florida, U.S.A.

and

D. EVAN MERCER

U.S. Forest Service,
Southern Research Station,
Research Triangle Park, North Carolina, U.S.A.

KLUWER ACADEMIC PUBLISHERS
DORDRECHT / BOSTON / LONDON

A C.I.P. Catalogue record for this book is available from the Library of Congress.

ISBN 1-4020-2412-6 (HB)
ISBN 1-4020-2413-4 (e-book)

Published by Kluwer Academic Publishers,
P.O. Box 17, 3300 AA Dordrecht, The Netherlands.

Sold and distributed in North, Central and South America
by Kluwer Academic Publishers,
101 Philip Drive, Norwell, MA 02061, U.S.A.

In all other countries, sold and distributed
by Kluwer Academic Publishers,
P.O. Box 322, 3300 AH Dordrecht, The Netherlands.

Cover photograph by D. Evan Mercer.

Printed on acid-free paper

Printed in the Netherlands.

TABLE OF CONTENTS

CONTRIBUTORS

Janaki R.R. Alavalapati
School of Forest Resources and Conservation, University of Florida, Gainesville, Florida, United States.

Peter Blandon
Department of Business Studies, London Metropolitan University, London, United Kingdom.

Geoff Bright
School of Agricultural and Forest Sciences, University of Wales, Bangor, United Kingdom.

Oscar Cacho
School of Economics, University of New England, Armidale, Australia.
Cooperative Research Centre for Plant-based Management of Dryland Salinity.

Ben H. J. De Jong
El Colegio de la Frontera Sur, Unidad Villahermosa, Villahermosa, Tabasco, México.

Brooks M. Depro
RTI International, Research Triangle Park, North Carolina, United States.

W. Mark Drew
School of Forest Resources and Conservation, University of Florida, Gainesville, Florida, United States.

Elsa Esquivel-Bazán
Ambio, San Cristóbal de las Casas, Chiapas, México.

Steven Franzel
World Agroforestry Centre (ICRAF), Nairobi, Kenya.

Stephen C. Grado
Department of Forestry, College of Forest Resources, Mississippi State University, Mississippi, United States.

Robyn Hean
Agroforestry Unit, New South Wales Agriculture, Tamworth, Australia.
Cooperative Research Centre for Plant-based Management of Dryland Salinity.

Peter E. Hildebrand
Department of Food and Resource Economics, University of Florida, Gainesville, Florida, United States.

Amanda L. Husak
Department of Forestry, College of Forest Resources, Mississippi State University, Mississippi, United States.

Shashi Kant
Faculty of Forestry, University of Toronto, Ontario, Canada.

Emma Lehrer
Faculty of Forestry, University of Toronto, Ontario, Canada.

D. Evan Mercer
United States Department of Agriculture – Forest Service, Southern Research Station, Research Triangle Park, North Carolina, United States.

Jensen R. Montambault
School of Natural Resources and Environment, University of Florida, Gainesville, Florida, United States.

Maxwell Mudhara
Department of Agricultural Economics and Extension, University of Zimbabwe, Harare, Zimbabwe.

P. K. R. Nair
School of Forest Resources and Conservation, University of Florida, Gainesville, Florida, United States.

Susana Ochoa-Gaona
El Colegio de la Frontera Sur, Unidad Villahermosa, Villahermosa, Tabasco, México.

Nicolás Peréz-Hernández
Ambio, San Cristóbal de las Casas, Chiapas, México.

Subhrendu K. Pattanayak
RTI International, Research Triangle Park, North Carolina, United States.
North Carolina State University, Raleigh, North Carolina, United States.

Sotero Quechulpa-Montalvo
Ambio, San Cristóbal de las Casas, Chiapas, México.

Ram K. Shrestha
School of Forest Resources and Conservation, University of Florida, Gainesville, Florida.

Ann Snook
Universidad de Quintana Roo, Chetumal, Quitana Roo, Mexico.

Paul H. Thangata
World Agroforestry Centre (ICRAF), Southern Africa Regional Programme, Harare, Zimbabwe.

Runsheng Yin
Department of Forestry, Michigan State University, East Lansing, Michigan, United States.

PREFACE

The primary objective of this book is to offer practical means for strengthening the economics and policy dimension of the agroforestry discipline. This book, written by the leading experts in economics and agroforestry, encompasses case studies from Australia, China, Kenya, India, Indonesia, Malawi, Mexico, Micronesia, Tanzania, United Kingdom, United States, Zambia, and Zimbabwe. The applied economic methodologies encompass a wide variety of case studies including enterprise/farm budget models through Faustmann models, Policy Analysis Matrix, production function approach, risk assessment models, dynamic programming, linear programming, meta-modeling, contingent valuation, attribute-based choice experiments, econometric modeling, and institutional economic analysis. It is our belief that these methodologies help agroforestry students and professionals conduct rigorous assessment of economic and policy aspects of agroforestry systems and to produce less biased and more credible information.

Furthermore, the economic and policy issues explored in the book – profitability, environmental benefits, risk reduction, household constraints, rural development, and institutional arrangements – are central to further agroforestry adoption in both tropical and temperate regions.

All of the chapters in this volume were subject to rigorous peer review by at least one other contributing author and one external reviewer. We would like to acknowledge the indispensable collaboration of those who provided careful external reviews: Ken Andrasko, Chris Andrew, Peter Boxall, Norman Breuer, Bill Hyde, Tom Holmes, Sherry Larkin, Jagannadharao Matta, Venkatrao Nagubadi, Roz Naylor, Thomas Randolph, Gerald Shively, Changyou Sun, Bo Jellesmark Thorsen, and Yaoqi Zhang. All reviews were coordinated by the book editors.

We would like to take this opportunity to express our gratitude to all the authors and co-authors of each chapter for their valuable contribution and timely response. Special thanks are extended to Jensen Montambault for making the production of this volume on a tight time schedule possible through editing, technical, and formatting assistance. We are also grateful for additional formatting and indexing assistance supplied by Terri Mashour, Troy Timko, and Fauzia Zamir. Generous in-kind and direct support were provided by the University of Florida and the USDA Forest Service. Finally, we would like to express our appreciation for the guidance of the editors at Kluwer Academic Publishing, Helen Buitenkamp, Sandra Oomkes, and Amber Tanghe-Neely.

– The Editors

JANAKI R. R. ALAVALAPATI, D. EVAN MERCER, AND
JENSEN R. MONTAMBAULT

AGROFORESTRY SYSTEMS AND VALUATION METHODOLOGIES

An Overview

1. INTRODUCTION

Agroforestry, the deliberate integration of trees with agricultural crops and/or livestock either simultaneously or sequentially on the same unit of land, has been an established practice for centuries. Throughout the tropics and, to some extent, temperate zones, farmers have a long tradition of retaining trees on their fields and pastures, as well as growing crops or raising domestic animals in tree stands or forests (Alavalapati & Nair, 2001; Gordon & Newman, 1997; Nair, 1989). In the late 1970s, agroforestry attracted the attention of the international scientific and development communities due to its potential for improving the environment and livelihood of rural tropical communities. The agroforestry prospective increased further during the 1990s as scientists and policy makers recognized the potential for applying agroforestry systems (AFS) to problems such as soil erosion, rising salinity, surface and ground water pollution, increasing greenhouse gases, and biodiversity losses in temperate zones and developed economies. Financial viability and attractiveness has also proven AFS an important land use alternative in various settings throughout the world (Garrett, 1997), generating increased interest in this sustainable land-use management practice with potential environmental and socioeconomic benefits.

Research over the past two decades has focused on exploring the biophysical and ecological aspects of agroforestry with a limited emphasis on social aspects of agroforestry, especially economics, policy analysis, and valuation of associated environmental services (Mercer & Miller, 1998). Concern over adoption rates has highlighted the importance of integrating socioeconomic elements into traditional biophysical agroforestry research (Nair, 1998; Rochelau, 1998). As a result, there is a growing interest and need for enhancing economic and policy research among

J. R. R. Alavalapati & D. E. Mercer (Eds.), Valuing Agroforestry Systems, 1–8.

agroforestry professionals. Montambault and Alavalapati (2003) conducted an extensive review and analysis of socioeconomic research in agroforestry literature between 1992 and 2002. Results showed a clear increasing trend in publications with more complex analyses, such as econometrics and optimization. The development of more sophisticated economic models creates applications that give more realistic and useful results for agroforestry practitioners. Indeed, the first World Agroforestry Congress (June 2004, Orlando, Florida) identified economics and policy as one of the key areas for enhancing the impacts of agroforestry. As an emerging facet of an interdisciplinary science, no single reference book prior to this publication has provided adequate coverage of applied economic and policy analysis methodologies for agroforestry professionals. By addressing this need, the present text offers practical means for strengthening the economics and policy elements of the agroforestry discipline.

2. DIVERSE AGROFORESTRY SYSTEMS AND ECONOMIC METHODOLOIGIES

Small-scale AFS range from slash-and-burn and taungya systems to traditional, yet complex, homegardens. More recent innovations include alley cropping and improved fallows and have been expanded to larger-scale production. As shown in Tables 1A and 1B, the nature, complexity, and objectives of AFS vary greatly between the tropics and the temperate zone.

Table 1A. Major agroforestry practices in tropical systems.

Agroforestry practice	*Brief description*
Taungya	Agricultural crops grown during the early stages of forest plantation establishment.
Homegardens	Intimate, multistory combinations of a variety of trees and crops in homestead gardens; livestock may or may not be present.
Improved fallow	Fast-growing, preferably leguminous woody species planted during the fallow phase of shifting cultivation; the woody species improve soil fertility and may yield economic products.
Multipurpose trees	Fruit and other trees randomly or systematically planted in cropland or pasture for the purpose of providing fruit, fuelwood, fodder, and timber, among other services, on farms and rangelands.
Plantation-crop combinations	Integrated multistory mixtures of tree crops (such as coconut, cacao, coffee, and rubber), shade trees, and/or herbaceous crops.
Silvopasture	Combining trees with forage and livestock production, such as grazing in existing forests; using trees to create live fences around pasture; or to provide shade and erosion control.

(Table 1A, cont.)

Agroforestry practice	*Brief description*
Shelterbelts and windbreaks	Rows of trees around farms and fields planted and managed as part of crop or livestock operations to protect crops, animals, and soil from natural hazards including wind, excessive rain, seawater, or floods.
Alley cropping	Fast-growing, preferably leguminous woody species in single or grouped rows in agricultural fields. Prunings from the woody species are applied as mulch into the agricultural production alleys to increase organic matter and nutrients and/or are removed from the field for other purposes such as animal fodder.

Table 1B. Major agroforestry practices in temperate systems.

Agroforestry practice	*Brief description*
Alley cropping	Trees planted in single or grouped rows within agricultural or horticultural fields with crops grown in the wide alleys between the tree rows.
Forest farming	Forested areas used for production or harvest of natural standing specialty crops for medicinal, ornamental, or culinary uses (e.g., ginseng, ferns, shiitake mushrooms).
Riparian buffer strips	Strips of perennial vegetation (tree/shrub/grass) planted between croplands/pastures and water sources such as streams, lakes, wetlands, and ponds to protect water quality.
Silvopasture	Combining tress with forage and livestock production, such as: growing trees on ranchlands; grazing in existing forests; providing shade and erosion control or environmental services.
Shelterbelts and Windbreaks	Rows of trees around farms and fields planted and managed as part of crop or livestock operations to protect crops, animals, and soil from natural hazards including wind, excessive rain, seawater, or floods.

Sources for Tables 1A, 1B: Association for Temperate Agroforestry [AFTA], 1997; Alavalapati & Nair, 2001; Nair, 1994.

A variety of economic and policy issues such as profitability, household benefits, equity, sustainability, soil conservation, environmental services, markets for inputs and outputs, gender, and institutions (property rights, for example) influence the nature and magnitude of AFS adoption (Alavalapati & Nair, 2001; Mercer & Hyde, 1991). A range of economic methodologies is required to systematically investigate these issues and produce objective and unbiased information to assist land managers and policy makers with AFS related decision-making.

Economic methodologies help characterize the mental calculus of a decision maker, whether a private landowner or a policy maker. As such, these models can be viewed as abstract representations of the real world useful for hypothesis generation, forecasting, policy analysis, and decision-making (Buongiorno & Gilles, 2003). These methodologies are diverse in terms of their focus, scale (temporal and spatial), and scope (Table 2). Some economic methodologies are designed to assess simple cost and benefits of outputs and inputs for which markets are fairly established while others may be limited only by scientists' capabilities and imagination. Methodologies are also available for assessing a variety of environmental advantages and challenges (e.g., carbon sequestration, biodiversity, and soil erosion) for which there are no established markets. While some methodologies are appropriate for assessing AFS at the individual farm or household level, others are applicable at regional and national scales. Partial equilibrium models are used to assess impacts on particular economic sectors by assuming that changes in AFS only affect certain sectors of the economy. Broader impacts can be analyzed with general equilibrium models that include intersectoral linkages capturing the multiplier and/or trade impacts of changes in AFS on other sectors of the economy. Although these models and methods have been extensively applied in agricultural and forest economics literature, AFS applications are relatively rare.

Table 2. Economic methodologies common in agricultural and/or forest economics literature.

Economic methodology	*Brief description*
Enterprise/farm budget models	Estimate the profitability of a farm or enterprise by deriving indicators such as net present values (NPV), benefit-cost ratio (BCR), and internal rate of return (IRR).
Policy analysis matrix models (PAM)	Similar to farm budget models, but also include market failures, assessing their impact on profitability at a farm or regional level from both the individual and society perspectives.
Risk assessment models	Incorporate probabilities of events occurring and estimate the expected profitability of AFS enterprises.
Dynamic optimization models	Estimate optimum values (e.g., timber rotation age and tree cover) under limited, terminating time periods or perpetual scenarios.
Liner and non-linear programming models	Estimate optimum resources use/allocation subject to various constraints faced by the decision maker.
Econometric models	Estimate the relationships among variables under investigation for forecasting, policy analysis, and decision-making.
Non-market valuation models	Hedonic and contingent valuation models, for example, estimate values for environmental goods and services such as reducing soil erosion, improving water quality, and carbon sequestration.
Regional economic models	Generally used to estimate changes in income, employment, and price levels at regional or national scales, in response to a policy or programmatic change by incorporating intersectoral linkages.

As each methodology has its own strengths and weaknesses, it would be erroneous to base conclusions on the scope, scale, or complexity of the models. Model choice depends primarily upon the nature of the research problem, data availability, and the skills and training of the analyst. A state-of-the-art Computable General Equilibrium (CGE) approach (Das & Alavalapati, 2003), for example, may be inappropriate and not very useful for assessing the profitability of an improved fallow system from a private landowner's perspective. This book, written by the leading experts in the field, encompasses 16 chapters arranged under 5 subsections and consists of 14 case studies covering all the continents of the world. The countries covered include Australia, China, Kenya, India, Indonesia, Malawi, Mexico, Micronesia, Tanzania, United Kingdom, United States, Zambia, and Zimbabwe. Each case study focuses on a specific type of economic methodology, illustrating its application to an AFS.

3. ORGANIZATION OF THE BOOK

One of the key factors influencing AFS adoption is its relative profitability compared with alternative land-use practices. Therefore, assessing the profitability of an AFS from a landowner perspective is of paramount importance. Chapters 2-6, in the second section, present a variety of methods for analyzing the profitability of AFS under different settings. In particular, Chapter 2 examines the profitability of fodder shrubs in Kenya, woodlots in Tanzania, and improved fallows in Zambia using an enterprise budget methodology. Chapter 3 extends the profitability analysis by applying a Land Expectation Value (LEV) approach (often referred to as the Faustmann methodology in the forest economics literature) using a silvopastoral system in the southern United States. This chapter estimates the present value of land under silvopasture system compared to alternative investment or management strategies. Chapter 4 is devoted to analyzing the private and social profitability of AFS in Pohnpei, Federated States of Micronesia using the Policy Analysis Matrix (PAM). In addition to quantifying profitability, the effect of distortions associated with policy or market failures (comparing private prices to social/efficiency prices) are assessed in this chapter. Chapter 5 develops a theoretical framework for analyzing the product-product relationship, the Production Possibility Frontier (PPF), and then applies the PPF to construct a simulation model of a wheat-maize-unpruned leucaena (*Leucaena leucocephala)* system in the Himalayan foothills of India. The model takes diminishing returns, time and interest, tree growth over time, complementarities and extra competition into account in assessing economic productivity. Chapter 6 analyzes risk in AFS through a portfolio approach applied to British and other European silvopasture practices, showing how AFS can help reduce risk and stabilize farmers' income.

As mentioned previously, AFS provide a mix of market goods such as food, wood products, and fodder, and non-market goods and services including soil conservation, water and air quality improvement, biodiversity conservation, and scenic beauty (Alavalapati, Shrestha, & Stainback, in press; Shrestha & Alavalapati, in press). The exclusion or inclusion of non-market goods and services, often referred to as externalities, largely differentiates private and social profitability. The third section of this book (Chapters 7-10) offers several environmental economic methodologies to value both market and non-market benefits of AFS. Chapter 7 examines the cost of carbon mitigation by means of agroforestry systems using a case study of farmers' participating in the Scolel Té project, Chiapas, Mexico. The methodology includes fixed and variable costs of implementing new AFS and the opportunity cost to farmers of diverting land from current land use, in addition to the cost of monitoring and internal verification of project performance. Chapter 8 deals with the estimation of external costs of dryland salinity emergence and the environmental and monetary benefits of tree planting in Australia. Using a dynamic programming model, the optimal area for forest on agricultural land is determined by explicitly considering the interactions between trees and crops. Chapter 9 assesses key environmental services such as conservation of on-farm soils and reduction of pressure on public forests through the adoption of AFS. Household production theory is used to conceptualize environmental services and policy levers and to frame testable hypotheses. Drawing from household survey data on agroforestry-based soil and forest conservation in the Manggarai region in Indonesia, the authors use an econometric model to test the hypotheses concerning soil erosion and AFS. Chapter 10 models an important externality problem, Florida ranchers' willingness to accept (WTA) for adopting silvopasture and generating environmental services, using a dichotomous choice, contingent valuation approach. In this chapter, a price premium is used as a payment vehicle to reflect the environmental services generated through silvopasture.

Since the mid-1990s, agroforestry adoption research has increased, largely motivated by perceived discrepancy between advances in agroforestry science and low adoption rates. The fourth section (Chapters 11-13) is devoted to the issue of AFS adoption and the myriad of factors influencing the adoption decision. Using a five-year linear programming (LP) model, Chapter 11 conducts an economic assessment of household constraints to the adoption of improved fallows in Mangwende Communal Area, northeastern Zimbabwe. Chapter 12 extends the previous model by conducting a meta-analysis of factors determining agroforestry adoption and farmers' decision-making in Malawi. In this chapter, information produced from LP models is used as the basis for conducting meta-regression analyses. Chapter 13 provides another perspective by describing an alternative econometric-based method for *ex-ante* analysis of AFS adoption potential. In particular, an attribute-based choice experiment (ACE), a subset of conjoint analysis, is applied to develop information for improving the adoption potential of agroforestry projects in southeast Mexico.

Although information generated through microeconomic analyses, profitability analysis and environmental economic analysis is essential for making agroforestry adoption decisions, information about the effect of AFS on regional income and employment plays a critical role in policy making. The fifth section (Chapters 14-15) focuses on the role of AFS in rural development and institutional arrangements required to further AFS adoption. Chapter 14 assesses the economic effects of agroforestry development in Northern China. Using state-of-the-art econometric time series techniques, the effect of agroforestry on agricultural productivity, and the spatial and temporal relationships between trees and annual crops are estimated. Chapter 15 provides an institutional economics perspective of AFS. In particular, the framework presented in this chapter provides an analytical approach to institutional analysis of agroforestry systems. The framework is applied to analyze market institutions as well as non-market institutions such as land tenure, tree-harvesting rights, transportation rights, tree-processing rights, loan arrangements, and technical support systems relating to Indian agroforestry.

Finally, Chapter 16 summarizes the main results and discusses the status of economic research and modeling in agroforestry. Drawing on the issues addressed in the book, gaps are identified as opportunities for further research in economic and policy of agroforestry.

4. SUMMARY

This book presents technical discussions of various AFS, economic theories, and methodologies applied to assess these systems in order to provide insight for policy and management. In doing so, the book covers 13 countries from all five continents of the world. Although the results presented in each chapter are based on specific case study data, they can be applied broadly because they are derived through appropriate rigorous quantitative approaches. This volume is primarily intended for upper division undergraduate and graduate students, as well as agroforestry and rural development professionals across the world. In addition, this book can be a significant new reference tool for resource economists, rural sociologists, and other social scientists interested in rigorous, quantitative analysis of agroforestry systems. Finally, this text is intended to provide valuable insights for policy makers and representatives of government and non-government agencies dealing with agroforestry practices in both developing and developed countries.

5. REFERENCES

Alavalapati, J. R. R., & Nair, P. K. R. (2001). Socioeconomics and institutional perspectives of Agroforestry. In M. Palo & J. Uusivuori (Eds.), *World Forests, Society and Environment: Markets and Policies* (pp 71-81). Dordrecht, The Netherlands: Kluwer Academic Publishers.

Alavalapati, J.R.R., Shrestha, R.K., & Stainback, G.A. (in press). Agroforestry development: An environmental economic perspective. In P. K. R. Nair, M. R. Rao, & L. E. Buck (Eds.), *New Vistas in Agroforestry*. Dordrecht, The Netherlands: Kluwer Academic Publishers.

Association for Temperate Agroforestry (AFTA). (1997). *The status, opportunities, and needs for agroforestry in the United States.* Columbia, Missouri: AFTA, University of Missouri.

Buongiorno, J., & Gilles, K. (2003). *Decision methods for forest resource management.* San Diego, California: Kluwer Academic Publishers.

Das, G., & Alavalapati, J. R. R. (2003). Trade mediated biotechnology transfer and its effective absorption: An application to the U.S. forestry sector. *Technological Forecasting and Social Change: An International Journal, 70*(6), 545-562.

Garrett, H. E. G. (1997). *Agroforestry: An integrated land-use management system for production and farmland conservation.* [United States Department of Agriculture Soil Conservation Service [USDA-SCS] Report 68-3A7S-3-134]. Fort Collins, CO: USDA-SCS.

Gordon, A. M., & Newman, S. M. (Eds.) (1997). *Temperate Agroforestry Systems*. New York: CAB International.

Mercer, D. E. & Hyde, W. F. (1991). The economics of agroforestry. In W. R. Burch, Jr. & J. K. Parker (Eds.). *Social Science Applications in Asian Agroforestry* (pp. 111-143). New Delhi, India: Oxford & IBH Publishing.

Mercer, D. E., & Miller, R. P. (1998). Socioeconomic research in agroforestry: progress, prospects, priorities. *Agroforestry Systems, 38*, 177-193.

Montambault, J. R., & Alavalapati, J. R. R. (2003). Socioeconomic research in agroforestry: A decade in review. Manuscript submitted for publication.

Nair, P. K. R. (Ed.). (1989). *Agroforestry systems in the tropics*. Dordrecht, The Netherlands: Kluwer Academic Publishers.

Nair, P. K. R. (1994). *Agroforestry. Encyclopedia of agricultural sciences* (Vol. 1, pp. 13-25). New York: Academic Press.

Nair, P. K. R. (1998). Directions in tropical agroforestry research: past, present, and future. *Agroforestry Systems 38*, 223-245.

Rochelau, D. (1998). Confronting complexity, dealing with differences: Social context, content, and practice in agroforestry. In L. E. Buck, J. P. Lassoie, E. C. M. Fernandes (Eds.), *Agroforestry in Sustainable Agricultural Systems* (pp. 191-235). Boca Raton, Florida: CRC Press.

Shrestha, R.K. & Alavalapati, J. R. R. (in press). Valuing environmental benefits of silvopasture practices: A case study of the Lake Okeechobee watershed in Florida. *Ecological Economics*.

STEVEN FRANZEL

FINANCIAL ANALYSIS OF AGROFORESTRY PRACTICES

Fodder shrubs in Kenya, woodlots in Tanzania, and improved fallows in Zambia

1. INTRODUCTION

Over the last two decades, researchers and farmers in east and southern Africa have combined their expertise and knowledge to develop improved agroforestry practices that improve livelihoods and provide important environmental services. Much of the research has focused on increasing biophysical productivity (Sanchez, 1996; Cooper, Leakey, Rao, & Reynolds, 1996), but, during the last 10 years, there has been greater emphasis on social and economic considerations. For example, much work has been done to assess the profitability of these practices and their feasibility and acceptability to farmers (Franzel, Coe, Cooper, Place & Scherr, 2001; Place, Franzel, DeWolf, Rommelse, Kwesiga, Niang et al., 2002).

Analyzing the economics of agroforestry practices is more complicated than that of annual crops for two main reasons. First, agroforestry practices are complex because they involve both trees and crops. Devising field trials to assess agroforestry practices and compare them with other practices is extremely difficult, requiring large plots and, at times, large spaces between the treatments. Second, there is usually a period of several years between the time the trees are established and the impact of agroforestry practices can be measured. Conducting trials and surveys with farmers over several years is expensive and problematic. For example, the greater the length of the trial, the more likely that individual farmers will want to change trial parameters in response to changing circumstances or preferences. The more changes that each farmer makes, the less likely it is that treatments can be compared across farms (Coe, 1998; Franzel et al., 2001).

The objective of this chapter is to assess the financial[1] returns to farmers of three practices: fodder shrubs in Kenya, rotational woodlots in Tanzania, and improved fallows in Zambia. Each practice has a different objective for farmers: fodder shrubs

J. R. R. Alavalapati & D. E. Mercer (Eds.), Valuing Agroforestry Systems, 9–37.

are for increasing milk production, rotational woodlots provide firewood, and improved fallows are for improving soil fertility. In each case, the implications of the analyses for researchers, extensionists, and policy makers are discussed. Finally, conclusions are drawn concerning the attractiveness of agroforestry practices for farmers and research challenges for enhancing their profitability.

2. DESCRIPTION OF THE AGROFORESTRY PRACTICES ANALYZED

2.1. Fodder shrubs, Kenya

The low quality and quantity of feed resources is a major constraint to dairy farming in central Kenya, where farm size averages 1-2 hectares (ha) and about 80% of households have stall-fed dairy cows, averaging 1.7 cows per family. The dairy zone ranges in altitude from 1300 meters (m) – 2000 m and rainfall occurs in two seasons, averaging 1200 millimeters (mm) – 1500 mm annually. Soils, primarily Nitosols, are deep and of moderate to high fertility. The main crops are coffee, produced for cash, and maize and beans, produced for food. Most farmers also grow Napier grass (*Pennisetum purpureum)* for cutting and feeding to their cows. But Napier grass is insufficient in protein so milk yields are low, about 6 kilograms (kg) per cow per day (Murithi, 1998). Commercial dairy meal is available, but farmers consider it expensive and most do not use it (Wambugu, Franzel, Tuwei, & Karanja, 2001; Franzel, Wambugu, & Tuwei, 2003).

Researchers and farmers tested several fodder shrubs around Embu, Kenya in the early 1990s and *Calliandra calothyrsus* emerged as the best performing and most preferred by farmers. The research was led by the National Agroforestry Research Project, a collaborative effort of the Kenya Agricultural Research Institute, the Kenya Forestry Research Institute, and the World Agroforestry Centre. Farmers plant the shrubs in hedges along internal and external boundaries, around the homestead, along the contour for controlling soil erosion, or intercropped with Napier grass. When pruned at a height of 1 m, the shrubs do not compete with adjacent crops. Farmers are easily able to plant 500 shrubs, at a spacing of 50 centimeters (cm), around their farms, and are able to begin pruning them within a year after planting. Five hundred shrubs are required to provide a cow throughout the year with 2 kg dry matter per day, adding about 0.6 kg crude protein. On-farm feeding trials confirmed that the farmers could use the shrubs as a substitute for dairy meal or as a supplement to increase their milk production. Dissemination began in earnest in 1999 and by 2003, about 23,000 farmers had planted *calliandra* or three other recommended species of fodder shrubs (Wambugu et al., 2001; Franzel et al., 2003).

2.2. Rotational woodlots, Tanzania

Tabora Region, western Tanzania, is an area of undulating plains and an average annual rainfall of 880 mm, falling over 5-6 months. Soils are 800-900 g (grams) per kg sand, low in organic carbon, nitrogen, and available phosphorus (Otsyina, Minae, & Cooper, 1996a). Land is a public commodity but farmers have secure user rights to the land they use. Farm size averages about 20 ha, most of which is uncultivated (Otsyina et al., 1996a). Farmers use hand hoes for cultivation. They make extensive use of hired laborers, who migrate to Tabora during the cropping season. Livestock are few, only about 5% of the farmers own cows. (Otsyina, Msangi, Gama, Ramadhani, Nyadzi, & Shirma, 1997). Tobacco is farmers' main cash crop; other crops grown for both food and cash include maize, the main food crop, groundnuts, rice, and sorghum. About 60% of the farmers grow tobacco, averaging 1.0 ha per farm. Firewood for tobacco curing is scarce; most farmers hire trucks and cut and transport firewood themselves from the forest. Farmers do not grow trees traditionally because, until recently, wood was plentiful and because they lack information on tree growing and planting material. Both policy makers and farmers are concerned about the rapid deforestation because an important natural resource is being destroyed and because the cost of collecting firewood is increasing as the distance to sources increases (Ramadhani, Otsyina, & Franzel, 2002).

Research on woodlots in Tabora began in 1993/94 at the Agricultural Research and Training Institute, Tumbi (ARTI-Tumbi). In the rotational woodlot system, farmers intercrop food crops with leguminous trees during the first 2-3 years, to maximize returns to their scarce labor. Then they leave the trees to grow, harvest them in about the fifth year, and replant food crops (Otsyina, Msangi, Gama, Ramadhani, Madulu, & Mapunda, 1996b). The most promising species tested by the farmers, in terms of growth, is *Acacia crassicarpa*, a legume. The food crops grown following the tree harvest benefit from the increase in organic matter, nutrient recycling, and nitrogen fixed by the leguminous trees (Ramadhani et al., 2002). Dissemination began in 1997 and by 2000, 961 farmers had planted woodlots.

2.3. Improved tree fallows, Zambia

The plateau area of eastern Zambia is characterized by a flat to gently rolling landscape and altitudes ranging from 900 to 1200 m. Rainfall averages about 1000 mm per year with about 85% falling in 4 months, December-March. The main soil types are loamy sand or sand Alfisols interspersed with clay and loam Luvisols. About half of the farmers practice ox cultivation, the others cultivate by hand hoe. Average cropped land per farm is 1-1.6 ha for hand hoe cultivators and 2-4 ha for ox cultivators. Maize is the most important crop accounting for 60% of cultivated area; other crops include sunflower, groundnuts, cotton, and tobacco. Surveys in the late 1980s identified soil fertility as the farmers' main problem; fertilizer use had been common during the 1980s but the collapse of the parastatal marketing system and the cessation of subsidies caused fertilizer use to decline by 70% between 1987 and 1995. Farmers had a strong felt need for fertilizer but lacked cash for purchasing it (Peterson, 1999; Franzel, Phiri, & Kwesiga, 2002b).

In 1987, the Zambia/ICRAF Agroforestry Research Project began on-station research on improved fallows, using *Sesbania sesban.* Results were encouraging and on-farm trials began in 1992. By 1995, several hundred farmers were involved in a range of different trials, testing and comparing different options. In researcher-led trials, farmers chose among 3 different species and 2 different management options (intercropping with maize vs. growing the trees in pure stands) and compared their improved fallows with plots of continuously cropped maize with and without fertilizer. In farmer-led trials, farmers planted and managed the improved fallows as they wished. Most farmers opted for a 2-year fallow and planted their main food crop, maize, for 2 to 3 seasons following the fallow. Extension activities began in 1996 and by 2001; over 20,000 farmers in eastern Zambia had planted improved fallows (Kwesiga, Franzel, Mafongoya, Ajayi, Phiri, & Katanga, in press).

3. METHODOLOGY FOR ASSESSING PROFITABILITY

3.1. General methods

Farmers using new agroforestry practices obtain increased financial benefits, relative to their existing practices, either through increased biophysical productivity or through reduced input costs. Both are important in all three of the practices examined in this paper. Researchers assessed biophysical productivity and financial net benefits by comparing results on treatment plots in on-farm trials with those on control plots, which represented farmers' existing practices. In all three cases, the trials were designed by researchers, in consultation with farmers, and they were managed by farmers. Researcher-designed trials are more suitable than farmer-designed ones because plot sizes are standardized, facilitating the collection of labor data, and practices are more uniform, permitting comparisons across farms. Farmer-managed trials are preferred to research-managed ones because data on costs and returns will more accurately reflect what farmers experience. The returns to agroforestry practices are highly sensitive to the timing and quality of certain practices, such as pruning. Thus, farmer management helps ensure that the outcomes of these trials are representative of what farmers can obtain on their own (Franzel et al., 2001).

Financial analyses were based on the costs and returns that farmers faced. The analyses did not use time series data taken from trial farmers because the time between planting and harvesting benefits was too long, 5 years in the case of woodlots and improved fallows. Rather farmers at different stages of a practice were monitored in the same year and composite farm budgets were constructed. Enterprise budgets were used for assessing the financial benefits and costs of improved fallows and woodlots, because these practices involved major changes in the maize enterprises they were being compared to. In enterprise budgets, all costs and returns of an enterprise are assessed. On the other hand, partial budgets were drawn up in the case of fodder trees because the practice had limited impacts on the costs and returns of dairy enterprise. A partial budget is a technique for assessing the benefits and costs of a practice relative to not using the practice. It thus takes into account only those changes in costs and returns that result directly from using a new practice (Upton, 1987).

Detailed information on labor use among participating farm households was collected using two main methods: including farmers' recall just after a task was completed and monitoring of work rates through observation. Prices were collected from farmers and from local markets.

Financial analyses often calculate returns to only one resource, land, ignoring the fact that labor and capital are far greater constraints than land in many farming systems. Therefore, we calculated the net returns to land, which was relevant for farmers whose most scarce resource was land and the net returns to labor, relevant for those who lacked household labor. In calculating returns to land, land was not valued but household labor was valued at its opportunity cost as estimated by hired labor prices. Returns are expressed on a per-hectare basis. For returns to labor, household labor was not valued and returns were expressed per unit of labor, that is, per workday. Net returns to capital for agroforestry practices are often extremely high or infinite because little or no capital is used in implementing them. This finding explained the attractiveness of many of the options because the alternatives, for example, fertilizer to improve crop yields or dairy meal concentrate to increase milk yields, were very expensive for farmers.

Data for a single period are usually inadequate for evaluating the performance of an agroforestry practice. Therefore, cost-benefit analyses, also called investment appraisals (Upton, 1987), were developed for estimating costs and benefits over the lifetime of an investment. Average values for costs and returns across a sample of farmers were used to compute net present values. Also, in the case of improved fallows, net present values were calculated for each individual farm based on its particular costs and returns. This latter method allowed a better understanding of the variation in returns and thus the risk of the practices.

Whereas cost-benefit analyses are useful for determining the net present value of an enterprise that has costs and returns over many years, they do not show the increase in annual income generated. To assess increases in annual income, farm models were developed in which the farm was partitioned, to contain specified portions of land devoted to each phase (corresponding to a season or year) of the practice. For example, in the model of improved fallows in Zambia, the farm was assumed to have equal portions of area in each of the practice's four phases: planting of the improved fallow (year 1), maturing of the fallow (year 2), the first post-fallow maize crop (year 3), and the second post-fallow maize crop (year 4). The net returns of this farm were compared to two other farms having the same amount of labor (the main constraining resource): one planting fertilized maize and the other planting unfertilized maize, both continuously without fallow. The model was thus useful for estimating the impact of improved fallows on annual net farm income and maize production (Franzel et al., 2002b).

3.2. Fodder shrubs

The data on the planting and management of the shrubs are from on-farm trials conducted in the early- and mid- 1990s and are described in Franzel, Arimi, and Murithi (2002a). In these trials, farmers planted and managed the shrubs as they wished; researchers monitored farmers' experiences. The trials could thus be described as farmer-designed and farmer managed. On the other hand, the feeding trials for determining milk yields were researcher-designed and farmer-managed, that is, researchers designed the treatments, in consultation with farmers, and the farmers managed the trials. These trials were conducted in 1994 and 1995 and are described in Patterson, Roothaert, Nyaata, Akyeampong, and Hove (1996).

Partial budgets were drawn up to show the effects of using fodder shrubs on farmers' net income under two scenarios: using *calliandra* 1) as a supplement to the normal diet and 2) as a substitute for purchased dairy meal. The base analysis assumes a farm with 500 trees and 1 zero-grazed dairy cow and covers a 10-year period. In fact the productive life of the tree appears to be longer, farmers who have had their trees for 10-12 years have not yet noticed any reduction in productivity. The benefits included in the analysis are the effect of *calliandra* on milk production (in the supplementation case) and the cash saved by not purchasing dairy meal and interest on cash freed up (in the substitution case). Costs are those for producing the seedlings and labor for planting, cutting, and feeding *calliandra* in 2001. Estimates of these costs were made by interviewing farmers shortly after they had completed the tasks. All costs for producing the seedlings are for labor, except for the cost of hand tools, which are used for other enterprises as well, and for seeds, which are valued at the market rate but which many farmers obtain for free from their own trees, those of neighbors, or from organizations. Therefore, in most cases, no cash expenditures are required for producing fodder shrubs. It is assumed that dairy meal and *calliandra* are fed 365 days per year as is recommended, whether the cow is in lactation or not.

Coefficients, prices, and sources of data used in the economic analysis are shown in Appendix A. Milk output per day per unit of *calliandra* or dairy meal is likely to be higher during the rainy season than during the dry season because there is more available basal feed during the rainy season. As the feeding trials were conducted during the dry season, the milk yields and profits that farmers can get from using *calliandra* or dairy meal may be lower in this study than what farmers can actually get on an average annual basis. The variability of financial returns could not be statistically assessed because a complete set of input-output data was not available for each individual farm. However, sensitivity analysis was conducted to determine the effects of changes in key parameters on profitability.

3.3. Rotational woodlots

For the on-farm trial, tobacco farmers were chosen randomly from 3 tobacco-growing villages in 3 districts, using lists of farmers available at village offices. The selected farmers were then visited to see if they were interested in hosting the on-farm trial. Five farmers planted in 1993/94 (the planting season extends from December to February), 10 in 1994/95, 8 in 1995/96, and 37 in 1996/97. The trial involved three tree species but only the best performing one, *Acacia crassicarpa,* is included in the economic analysis. Seedlings were raised in a nursery and transported to farmers' fields. The trial was researcher-designed and farmer-managed; researchers marked out plots and advised on management but farmers conducted all operations. The trial included 3 plots, 1 for each species, planted at a spacing of 4 m by 4 m (625 trees/ha). Plot size ranged from 0.07 ha to 0.16 ha depending on the land the farmer had available; thus each farmer planted about 44 to 100 trees of each species. Farmers planted maize between the newly planted trees during the first 2 years after the trees were planted. They were also advised to weed, dig micro-catchments around each tree, and apply compound fertilizer, which is recommended for maize. In fact, weeding and applying fertilizer to maize are common practices in the area. Farmers were also trained on how to prune the trees. Wood yield was measured from 4 of the 15 farmers who planted in 1993/94 and 1994/95; only 1 other farmer had harvested their trees. Otsyina et al. (1996b) and Ramadhani et al. (2002) provide more details on the trial.

The profitability of rotational woodlots was assessed by comparing it with a maize-fallow rotation, because farmers planted woodlots on fields that they indicated would have been used for growing maize for 2 years followed by a 3-year fallow. Enterprise budgets for both rotational woodlots and maize-fallow rotations were drawn up over a 5-year period, using data on inputs, outputs, and prices obtained from the farmers and other key informants (Appendix B). The analysis assumes that farmers harvest the woodlots in the fifth year. Wood prices were valued at the price farmers pay to have wood trucked in from the forest for curing their tobacco. Labor inputs and wage rates were obtained from a formal survey of 30 trial farmers in 1997. Maize seed and harvest prices were averages of market prices over the period 1995/96-1996/97. Maize yields with and without trees were not measured, but the trees were estimated to have no effects on maize yields in the first year and to reduce maize yields by 40% in the second year, based on results from an on-station trial and observations (Otsyina et al., 1996b).

A farm-level model was drawn up to assess the impact of rotational woodlots on farm profitability. In the first scenario of the model, the farmer uses 75 workdays year^{-1} to grow 1.33 ha of rotational woodlots, planting one-fifth of this amount, 0.27 ha, each year, the area needed to provide sufficient firewood each year for domestic use and for curing 1 hectare of tobacco. In the second scenario, the farmer uses the same amount of labor to cultivate maize. As in the case of fodder trees, sensitivity analysis was used to assess how changes in key parameters affected profitability.

3.4. Improved fallows

During 1996-98, data were collected on costs and returns from 12 selected farmers planting *sesbania* improved fallows in researcher-designed, farmer-managed trials.

All trials included an improved fallow plot and plots continuously cropped with and without fertilizer. Data from these trials were supplemented by data from other farmers, local markets, and secondary sources. The 12 were the only ones who had complete sets of yield response data from the improved fallow trials during 1995/96 and 1996/97. Enterprise costs and returns were drawn up for the 12 farms and used to calculate net present values per hectare to assess returns to land and net returns to labor. The analysis covered a period of 5 years: 2 years of fallow and the 3 subsequent years for which it is assumed that maize yields would be affected. Maize yields following *sesbania* fallows were available for 5 farmers for 1996 and 7 farmers for 1997. Average data on costs were used in each individual farmer's budget; maize yields from different treatments were measured on each farm and were thus specific to each farm. Where costs were a function of yield, as in the case of harvesting labor, they were adjusted in relation to yield. Sensitivity analysis was conducted to show the effects of changes in parameters on the results of the economic analysis.

Farm models were drawn up to assess the impact of adopting improved fallows on annual income, as mentioned above. Models were drawn up for the same three scenarios as for the enterprise budgets: farms that adopt improved fallows (planting a portion of their maize area to improved fallows each year, so that each portion is in a different phase of improved fallows), farms that cultivate unfertilized maize, and those with fertilized maize.

4. RESULTS AND DISCUSSION

4.1. Fodder shrubs

Partial budgets for *calliandra* as a supplement to farmers' basal feed and as a substitute for dairy meal in 2001 are shown in Tables 1-2. Tree establishment costs (including the costs of producing bare-rooted seedlings[2] in a nursery and transplanting them) are modest, $US 7.14/500 trees. Beginning in the second year, harvesting and feeding 2 kg dry *calliandra* per day as a supplement throughout the lactation period increases milk production by about 372 kg[3]/yr., an increase of about 12% over base milk yields. Incremental benefits per year after the first year are over 9 times higher than incremental costs. The net present value (NPV) assuming a 20% discount rate is $US 260. Net benefits per year after year 1 are $US 79.

In the partial budget assessing *calliandra* as a substitute for dairy meal, establishment, cutting, and feeding costs are the same as in the preceding analysis. By feeding *calliandra*, the farmer saves the money he would have spent buying and transporting 730 kg dairy meal during the year. Incremental benefits per year after the first year are over 13 times higher than incremental costs. Milk production does not increase but net benefits are slightly higher than in the supplementation case. The NPV assuming a 20% discount rate is $US 413. The net benefits per cow per year after year 1 are $US 125. Therefore, using *calliandra* increases farmers' annual income by about $US 79 to $US 125 per cow per year after the first year, depending

Table 1. Partial budget: Extra costs and benefits of using calliandra as a supplement for increasing milk production, central Kenya ($US/yr, 2001).

	Extra cost		Extra benefit		Net benefit
Year	*Item*	*$US*	*Item*	*$US*	*$US*
1	Tree seedlings	3.85			
	Planting labor	3.3			
	Subtotal	7.14		0	-7.14
2	Cutting/feeding labor	10.03	Extra milk produced (372 kg)	89.18	79.16

Net Benefit = extra benefits minus extra costs. Years 3-10 same as year 2. Net present value at 20% discount rate = $US 259.95 per year; Net benefit per year after year 1 = $US 79.16; Annualized net benefit treating establishment costs as depreciation = $US 76.77. Note: Base farm model: The farm has 500 calliandra trees and one dairy cow. The cow consumes a basal diet of 80 kg Napier grass per day and produces 10 kg milk/day. Coefficients are from Appendix A.

Table 2. Partial budget: Extra costs and benefits of using calliandra as a substitute for dairy meal in milk production, central Kenya ($US/yr, 2001)

	Extra cost		Extra benefits		Net benefit
Year	*Item*	*$US*	*Item*	*$US*	*$US*
1	Tree seedlings	3.85		0	
	Planting labor	3.3			
	Subtotal	7.14			-7.14
2	Cutting; feeding labor	10.03	Saved dairy meal cost	129.72	
			Saved dairy meal transport	4.02	
			Interest on capital	1.11	
	Subtotal	10.03		134.85	124.82

Years 3-10 same as year 2. Net present value at 20% discount rate = $US 413.36. Net benefit per year after year 1 = $US 124.82. Annualized net benefit treating establishment costs as depreciation = $US 122.44. Note: Base farm model: Same as in Table 1. Coefficients are from Appendix A.

on whether the farmer is supplementing or substituting. As the average farmer owns 1.7 cows, *calliandra* has the potential to increase a farmers' income by about $US 134 to $US 212 per year representing an increase of roughly 10% in total household income (Murithi, 1998).

The net benefits per cow per year after the first year are somewhat lower than those calculated for the years 1996-1998, as reported in Franzel et al. (2002a). Net benefits for 1996-1998 (expressed in 2001 dollars after adjusting for inflation) ranged from $US 114 to 183 per cow per year after the first year, depending on whether *calliandra* was used for supplementation or substitution. The 2001 figures, $US 79 to $US 125, represent a reduction of about 30% as compared to the 1996-98 figures. The main causes of the decline were an adjustment in the input-output coefficient (the amount of milk produced from *calliandra*) and a reduction in milk prices, associated with a decline in processing facilities following the collapse of Kenya's dairy marketing parastatal in the late 1990s.

The analyses confirm that the costs of establishing, maintaining, and feeding *calliandra* are low. In both the substitute and supplement scenarios, farmers recover their costs very quickly, in the second year after planting. In order to break even, a farmer using *calliandra* as a supplement needs to obtain only 0.08 kg of milk from 1.0 kg of *calliandra* (dry), rather than the 0.62 kg milk per kg (dry) of *calliandra* obtained in on-farm trials and assumed in the analysis (Paterson et al., 1996).

Several intangible or otherwise difficult to measure benefits and costs have been omitted from this analysis. *Calliandra* provides benefits to some farmers as firewood, in erosion control, as a boundary marker, a fence, and as an ornamental. It also increases the butterfat content of milk, giving it a richer taste and creamier texture. When used as a supplement, *calliandra* may improve animal health and fertility and reduce the calving interval. Finally, several farmers noted that *calliandra* had important benefits relative to dairy meal: it was available on the farm, cash was not needed to obtain it, and its nutritional content was more reliable than that of dairy meal. These views support the thesis that farmers prefer enterprises and practices that do not rely on uncertain governmental or market mechanisms (Haugerud, 1984). The main cost not assessed was the opportunity cost of the land occupied by the shrubs. However, this cost is likely to be low or none, especially when *calliandra* replaces or is added to an existing hedge or bund, is planted on contour bunds to conserve soil, or when *calliandra* hedges border on homesteads, roads, paths, or field boundaries. Another possible cost is the effect on nearby crops. But, because the shrubs are nitrogen fixing and are usually maintained at heights of only 1 m, they have little or no negative effects on adjacent crops. In a survey of *calliandra* growers, only 7% felt that the shrubs reduced the yields of nearby crops (Franzel et al., 2002a).

Sensitivity analysis was conducted to determine how changes in key parameters would affect the results (Table 3). A 30% reduction in the milk price would reduce the NPV by 35%. However, using *calliandra* would still be profitable. In the substitute scenario, changing the milk price would not affect the profitability of *calliandra* relative to dairy meal. A change in the price of dairy meal does not affect the use of *calliandra* as a supplement. However, in the substitution scenario, a 30% increase in dairy meal price raises the NPV by 32%. A reduction of price by 30% reduces the NPV by 32%. Overall, the sensitivity analysis shows that the net benefits of using *calliandra* as a supplement or as a substitute are very stable. Despite the range of negative situations tested, net present values and net benefits remain positive.

Table 3. Sensitivity analysis showing the effect of changes in key parameters on the profitability of using calliandra, central Kenya ($US per cow per year).

	Dairy meal supplement		*Dairy meal substitute*	
	Net present	*Annualized*	*Net present*	*Annualized*
Base Analysis	260	77	413	122
Milk price + 30%	350	103	413	122
Milk price –30%	170	50	413	122
Dairy meal + 30%	260	77	545	162
Dairy meal – 30%	260	77	281	83
Discount rate = 10%	408	77	644	122
Discount rate = 30%	178	76	286	122
Using potted seedlings	250	73	404	119
1 kg shrubs give 30% more milk	350	103	413	122
1 kg shrubs give 30% less milk	170	50	413	122
Labor cost + 30%	249	73	402	119
Labor cost – 30%	271	80	425	126

Note: Base analyses are shown in tables 1 and 2.

Fodder trees appear to be appropriate for smallholder dairy farmers throughout the highlands of eastern Africa – *calliandra*, for example, can grow at altitudes between 0 and 2200 m, requires only 1,000 mm rainfall, can withstand dry seasons up to four months long, and is suitable for cut-and-carry feeding systems or for grazing systems (Roothaert, Karanja, Kariuki, Paterson, Tuwei, Kiruiro et al., 1998). It is also suitable for dairy goat production, which is growing rapidly in Kenya. The potential impact of fodder trees thus appears to be very large. If all 625,000 smallholder dairy farmers were to adopt *calliandra* or similar fodder shrub species, the benefits would amount to about US $ 84 million per year. Moreover, fodder trees are being planted by dairy farmers at numerous other sites in east and southern Africa. Over 10,000 farmers have adopted fodder trees in Uganda and Tanzania; farmers are also planting them in Rwanda, Ethiopia, Malawi, and Zambia.

4.2. Rotational woodlots

Additional costs involved in rotational woodlots, relative to the maize-fallow system, included costs associated with producing tree seedlings, reduced maize yields, and labor for transplanting, gapping, pruning, and wood harvesting (Table 4). In the woodlot treatments, maize costs and yields are lower in the second year than in the first year because maize is planted at a lower density, less fertilizer is used, and because the trees interfere with the maize. In the maize fallow system, maize costs and yields were only measured during the first year of the cultivation; values in the second year are assumed to be the same as in the first year. Labor use in the woodlots system over the 5-year period is over 2.5 times that of the maize fallow system, primarily because of the labor required for wood harvesting in year 5, which accounts for over half of total labor. The total discounted input costs of rotational woodlots were 52% higher than for the maize-fallow system, mainly because of the costs of producing the potted seedlings and of harvesting the trees.

In the first year, the rotational woodlot incurred losses of $US 37 while the maize-fallow system's net benefits were $US 40. Additional benefits of the woodlots included the value of pruned wood in year 2 and wood yields in year 5. The payoff period for the woodlot, that is, the period required to earn positive net benefits, is 5 years as compared to less than 1 year for the maize-fallow system.

Table 4. Financial analysis of rotational woodlot as compared to a maize allow system, Tabora District, Tanzania ($US/ha). [a]

	Rational woodlots[b]			*Maize fallow system*[c]	
Benefits and costs	*Year 1*	*Year 2*	*Year 5*	*Year 1*	*Year 2*
Benefits					
Maize grain yield	142.54	88.85		158.39	158.39
Wood yield			806.62		
Pruning yield		23.53			
Total benefits	142.54	112.38	806.62	158.39	158.39
Labor costs					
Land preparation	8.59	8.59		8.59	8.59
Planting	2.53	1.9		2.53	2.53
Weeding	9.41	9.41		9.41	9.41
Fertilizer application	1.18	1.18		1.18	1.18
Harvesting	7.12	6.05		7.12	7.12
Threshing	3.71	2.33		4.12	4.12
Transplanting, watering, and digging microcatchments	4.18				
Gapping	1.42				
Pruning		5.18			
Wood harvesting			93.14		
Total	38.13	34.64	93.14	32.94	32.94
Other costs					
Tree seedlings	56.3				
Maize seed	4.62	3.7		4.62	4.62
Fertilizer	80.67	64.54		80.67	80.67
Total	141.6	68.24		85.29	85.29
Summary data					
Grand total cost	179.72	102.87	93.14	118.24	118.24
Discounted costs	275.11			180.64	

(Table 4, cont.)

	Rational woodlots[b]			*Maize fallow system*[c]	
Benefits and costs	*Year 1*	*Year 2*	*Year 5*	*Year 1*	*Year 2*
Net benefit	-37.17	9.51	713.48	40.16	40.16
Workdays	0.11	0.1	0.27	0.09	0.09
Net benefit to labor	0.96	44.14	806.62	73.1	73.1
Net ben. to labor/workday	0.02	0.75	5.09	1.31	1.31
Net present value	388.52			61.36	
Discounted workdays	0.31			0.14	
Discounted net benefit to labor	498.25			111.68	
Discounted net benefit and workday	2.67			1.31	

[a]*Prices and quantities of inputs and outputs are from Appendix B.*
[b]*Maize is intercropped with the trees during the first two years. There are no benefits or costs during years 3 and 4. All costs and benefits are discounted over a 5 year period.*
[c] *Maize is cultivated for two years followed by three years of fallow. There are no benefits or costs during years 3 through 5. All costs and benefits are discounted over a 5 year period.*

In spite of its higher costs and longer payoff period, the rotational woodlot's net present value is $US 388/ha, over 6 times higher than that of the maize fallow system. Returns to labor are more relevant to Tabora farmers than returns to land, because labor is much scarcer than land. The woodlot's returns to labor, expressed in discounted net benefits per discounted workday, were $US 2.67, over double that of the maize-fallow system.

An important advantage of the woodlots is that they allow farmers to substitute land and labor for cash, which they have great difficulty obtaining. Tobacco farmers can obtain firewood for curing only by purchasing it, whereas with the rotational woodlots, they can use their land and labor to produce it, using little if any cash in the process. The labor required for harvesting the wood is considerable but it can be spread over a long period during the farmers' slack season. The extra labor required for planting and maintaining the trees is relatively little.

Sensitivity analysis showed that the performance of rotational woodlots relative to the maize-fallow system is fairly stable across a wide range of changes in important parameters (Table 5). Increases or decreases of 50% in the price of maize, wood, or labor, or in the yields of maize or wood do not affect the superiority of rotational woodlots. Increasing the discount rate from 20% to 30% or reducing it to 10% also does not affect the rankings. Among the variables examined, the profitability of the woodlots is most sensitive to changes in the wood price and yield. The profitability of the maize-fallow system is sensitive to changes in maize price and yield.

The farm model (Table 6) shows that a household with 1.33 ha under woodlot, planting and harvesting 0.265 ha each year, would be able to provide enough wood to meet its tobacco curing and domestic needs each year. Such a household would use 75 workdays and earn $US 182, over triple the net returns that a family would earn using the same amount of labor to produce maize.

Table 5. Sensitivity analysis of the results of the financial analysis of rotational woodlots to changes in key parameters, Tabora District, Tanzania.

Parameter	*Rotational woodlots*		*Maize without trees*	
	Returns to land (Net present value, $US/ha)	*Returns to labor ($US/ workday)*	*Returns to land (Net present value, $US/ha)*	*Returns to labor ($US/ workday)*
Base analysis (from Table 4 data)	389	2.67	61	1.31
50% decrease maize yield	272	2.1	-56	-0.12
50% increase maize yield	476	3.49	179	2.56
50% decrease maize price	298	2.19	-60	-0.11
50% increase maize price	479	3.15	182	2.72
50% decrease wood yield	155	1.42	61	1.31
50% increase wood yield	622	3.92	61	1.31
50% decrease wood price	155	1.42	61	1.31
50% increase wood price	622	3.92	61	1.31
50% decrease wage rate	443	2.67	86	1.31
50% increase wage rate	334	2.67	36	1.31
30% discount rate	302	2.51	55	1.31
10% discount rate	510	2.84	70	1.31

Table 6. Farm models comparing net returns to labor of a farmer practicing rotational woodlots (planting a portion of the farm to rotational woodlots each year) to those of a farmer allocating the same amount of labor to cultivating maize without trees, Tabora District, Tanzania.

Farmer with rotational woodlots[a]				*Farmer using same amount of labor to cultivate maize*			
Crop	*Area (ha)*	*Labor (workdays)*	*Net returns to labor/ year ($US)*	*Crop*	*Area (ha)*	*Labor (workdays)*	*Net returns to labor/ year ($US)*
Woodlot, 1st year; intercropped with maize	0.265	17	-9.86				
Woodlot 2nd year; intercropped with maize	0.265	16	2.52				
Woodlot 3rd year	0.265	0	0				
Woodlot 4th year	0.265	0	0				
Woodlot 5th year	0.265	42	189.21				
Total: Woodlots	1.33	75	181.87	Maize	1.34	75	53.64

[a]A household with one hectare of tobacco produces about 610 kg tobacco leaves, requiring about 37.2 t yr^{-1} of firewood for curing and 3.3 t yr^{-1} for domestic use (Ramadhani et al., 2002). The woodlot produces 152.7 t wood per five years. Therefore, by planting 0.265 ha yr^{-1} of woodlot each year, a household meets its firewood needs.

4.3. Improved fallows

The benefits of improved fallows, relative to continuously cropped maize, were labor saved in years 1 and 2 because maize was not planted, firewood production in year 2, increases in maize yields in years 3 through 5, and reduced land preparation and weeding costs in the first post-fallow maize crop. Added costs included *sesbania* seed, labor for establishing the nursery, transplanting, and maintaining the fallow, and labor for harvesting and threshing the increased maize produced.

Maize yields in the year following the improved fallows averaged 3.6 t/ha, as compared to yields of 1.0 t/ha for continuous, unfertilized maize and 4.4 t/ha for continuous, fertilized maize. The post-fallow plot out-yielded the unfertilized plot on all 12 farms and the fertilized plot on 4 of the 12 farms. Results of the economic analysis of the 12 farms, using average values across farms, are summarized in Table 7; the detailed budgets for improved *sesbania* fallows and fertilized and unfertilized maize are shown in Appendix C. Over the 5-year period, a hectare under the improved fallow treatment required 13% less labor than a hectare of unfertilized maize and 33% less labor than fertilized maize (Table 7). Relative to unfertilized maize, the improved fallow increases total maize production per hectare over the 5-year period by 52%, even though it does not produce maize during the first 2 years of the fallow. But fertilized maize gives the highest yield over the 5-year period, triple that of improved fallows (Table 7). The value of firewood produced in the fallow was low, only about 3% of the value of maize following the improved fallow.[4]

Table 7. Labor requirements, maize production, and returns to land and labor of Sesbania sesban improved fallows and continuously cropped maize over a 5-year period, using an average farm budget, eastern Zambia.

Option	*Work-days per ha*	*Tons Maize per ha*	*Returns to land: net present value ($US/ha)*	*Returns to labor: discounted net returns ($US/ workday)*
	Over a 5-year period[b]		*1996*	*1996*
Continuous fcrtilizcd maizc	499	4.8	5	0.42
Improved 2-year Sesbania fallow	433	7.3	115	0.85
Continuous fertilized maize	649	21.9	203	0.93

[a] *The means of values from individual budgets of the twelve trial farmers were used. Details on budgets and coefficients are provided in Appendix C.*
[b] *A 5-year period is used because that is the period needed to complete a cycle of the improved fallow practice; two years of fallow and three years of cropping.*

Net present values (NPVs) per hectare for fertilized maize were 76% higher than those of improved fallows; both were much higher than for unfertilized maize. Five of twelve farmers obtained higher NPVs for improved fallows than for fertilized maize; 11 obtained higher NPVs for improved fallows than for unfertilized maize. NPVs were low relative to other years because 1996 was a year of high maize yields and thus low maize prices.

A main disadvantage of improved fallows relative to continuous maize is that farmers have to wait until after the fallow to recoup their investment; in continuous maize, farmers earn positive net benefits in the first year. The payback period, that is, the period required for improved fallows to yield higher cumulative net present values than unfertilized maize, was 3 years for 10 of the 12 farmers. This indicates that even if a farmer does not get higher yields than unfertilized maize during the second and third post-fallow maize harvests, improved fallows were still more profitable than unfertilized maize for these farmers.

Assessing returns to labor is more relevant to most Zambian farmers than returns to land, because labor tends to be scarcer than land. On returns to labor, improved fallows outperformed unfertilized maize by a wide margin and performed almost as well as fertilized maize, using average values across the 12 farms and 1996 prices (Table 7). Improved fallows gave higher net returns to labor than for unfertilized maize on 11of the 12 farms and higher net returns to labor than for fertilized maize on 7 of the 12 farms. Even assuming no maize yield response to improved fallows in year 4 and year 5, returns to labor on improved fallows were higher than those for unfertilized maize on 10 of 12 farms. In summary, improved fallows had much higher returns to land and labor than unfertilized maize but lower returns to land than fertilized maize. On returns to land, the improved fallows performed almost as well as fertilized maize.

One important farmer innovation in improved fallows is the intercropping of the trees with maize during the first year of fallow establishment. The maize and the trees compete in the plot and reduce maize yields by about 20% compared to unfertilized maize in pure stands (Franzel et al., 2002b). But farmers benefit from harvesting a maize crop from the tree plot. In fact the practice has significant financial benefits: the farmer reduces her first year losses from $US 52 to $US 35 and the NPV increases from $US 115 to $US 129/ha.

The performance of improved fallows relative to continuous, unfertilized maize is fairly stable under a wide range of possible changes in parameters (Table 8). For example, improved fallows have returns to land and labor at least double those of unfertilized maize under most tested changes, including a 50% increase or decrease in the discount rate, and the prices of fertilizer and labor. An increase in post-fallow maize yield of only 1.1 t/ha is needed in the third year to cover the costs of establishing and maintaining the fallow, relative to unfertilized maize, in terms of returns to land or labor. In contrast, the performance of improved fallows relative to continuous, fertilized maize is sensitive to changes in some key parameters. Increases in maize prices (such as occurred between 1996 and 1998) raise the returns to fertilized maize at a much faster rate than they raise the returns to

Table 8. Sensitivity analysis showing the effects of changes in parameters on the profitability of improved fallows, eastern Zambia ($US).

	Continuous unfertilized maize		*Improved fallows*		*Continuous fertilized maize*	
	Returns to land	*Returns to labor*	*Returns to Land*	*Returns to labor*	*Returns to land*	*Returns to labor*
Base analysis	5	0.42	115	0.85	204	0.93
Maize price + 50%	101	0.74	239	1.34	639	2.06
Maize price – 50%	-90	0.1	-9	0.37	-231	-0.2
Labor price + 50%	-54	0.42	68	0.88	126	0.93
Labor price – 50%	65	0.42	162	0.8	280	0.93
Discount rate 30% instead of 20%	4	0.34	80	0.64	166	0.76
Discount rate 10% instead of 20%	7	0.53	167	1.172	258	1.18
Seedling cost +50%	5	0.42	110	0.81	203	0.93
Seedling cost -50%	5	0.42	120	0.9	203	0.93
Fertilizer price + 50%	5	0.42	115	0.85	-10	0.37
Fertilizer price – 50%	5	0.42	115	0.85	417	1.48

improved fallows. Similarly, the relative profitability of the two practices is highly sensitive to the price of fertilizer; a 50% increase in price would make improved fallows much more profitable than fertilizer on returns to both land and labor. Changes in the discount rate and in the cost of labor and seedlings have little effect on the performance of improved fallows relative to fertilized maize.

The risk of drought is critical for farmers in Zambia; unfortunately the effects of drought in the season following an improved fallow cannot be assessed using the data collected for this study. But there are four reasons why improved fallows are likely to be much less risky than fertilized maize. First, in the event of a complete crop failure, a farmer using the recommended fertilizer rate would lose his investment in fertilizer, US$ 149/ha whereas a farmer with improved fallow would lose his investment in planting and maintaining the trees, only about US$ 52/ha (The actual savings would be less since farmers apply fertilizer at less than the recommended rate). In addition, both farmers would lose their investment in growing maize that year. Second, whereas nearly all of a farmer's investment in fertilizer is in cash terms, improved fallows require little or no cash input. The opportunity cost of cash is extremely high and if the farmer buys fertilizer on credit, loss of the maize crop may result in substantial losses in productive capacity in order to repay the loan. Third, the benefits of improved fallow are likely to be spread over a 3-year period whereas those of nitrogen fertilizer take place in a single year. Thus in the above case where a farmer's crop fails in the first post-fallow season, there is likely to be a substantial response the following year. Fourth, improved fallows improve the soil structure and organic matter content of the soil, thus enhancing the soil's ability to retain moisture during drought years (Kwesiga et al., in press). Finally, it is important to note another important risk of relying on fertilizer. In some years, fertilizer may be delivered too late in the season to have an effect on yields.

The analyses of profitability presented thus far assess returns per hectare and per workday; but how will adoption of improved fallows affect farm income once they have been incorporated into the farming system? A farm household cultivating manually and having 1.4 ha and 120 workdays available for cultivating maize would fully adopt improved fallows by planting 0.28 ha per year to improved fallows, she would thus have an equal portion of the area under a different phase of improved fallow each year. The farmer could earn US$ 189 per year using fertilized maize, US$ 118 per year growing improved fallows, or only US$ 50 cultivating continuous maize without fertilizer (Table 9). Even if there is no residual effect on maize yields in the third year following improved fallows, earnings are still almost twice as high as on unfertilized maize.

Table 9. Farm models comparing net returns to labor per year of a 1.4 ha farm practicing Sesbania sesban *improved fallows with farms cultivating continuous maize, with and without fertilizer, eastern Zambia*[a]*.*

Crop	*Area (ha)*	*Workdays/yr*	*Kg maize produced/yr*	*Net returns/yr $US*
Farming practicing improved fallows (farm adds 0.28 ha of improved fallow/yr)				
Fallow1st yr	0.28	35	0	-1
Fallow 2nd yr	0.28	1	0	2
Maize 1st post fallow	0.28	27	1,019	61
Maize 2nd post fallow	0.28	28	570	32
Maize 3rd post fallow	0.28	29	448	23
Total	1.4	120	2,037	118
Farm with unfertilized maize				
Maize	1.2	120	1,159	50
Farm with fertilized maize				
Maize	0.92	120	4,077	262

[a] *Household is assumed to have only 119 workdays available during the cropping season for maize production; the amount needed to manually cultivate 1.2 ha maize without using fertilizer. Costs and returns are from Appendix 1. Improved fallows are two years in length and are followed by three years of maize crops.*

Whereas the above analyses assess the profitability of alternative soil fertility practices, farmers do not necessarily view them as alternatives. For example, a farmer may use fertilizer, manure and improved fallows, allocating each to a different part of her farm. Researchers have found that there are important synergies between organic and inorganic inputs for improving soil fertility (Palm, Myers, & Nandwa, 1997). However few Zambian farmers apply mineral fertilizer following an improved fallow, probably because they lack sufficient soil fertility inputs for covering their entire cultivated area (Keil, 2001).

5. CONCLUSIONS

The three agroforestry practices assessed in this chapter have different objectives: feeding livestock, providing firewood, and improving soil fertility. They were adopted in very different environments ranging from semi-arid, low population density areas of Tanzania to the sub humid, high-density highlands of Kenya. Nevertheless, each provided important financial returns and is being adopted on a large scale. Full adopters of fodder shrubs in Kenya, rotational woodlots in Tanzania, and improved fallows in Zambia earn $US 68 – $US 212 per year more from these practices than from alternative, available practices. Actual benefits are lower, because most farmers do not, or have not yet, fully adopted. In addition to the financial returns, there are several other intangible types of benefits. First, in all three cases, the practices provide by-products and services which are difficult to value. For example, fodder shrubs serve as border markings, improve animal health and calving rates, provide firewood and curb soil erosion. Improved fallows improve soil structure and moisture retention and provide firewood. Rotational woodlots reduce deforestation, as home-produced firewood is substituted for firewood cleared from the forest and trucked to the farm.

Second, all three practices involve relatively low investments of land and labor in exchange for substantial cash savings. As most farmers have difficulty earning cash and have multiple demands on the small amounts of cash they earn, they greatly appreciate being able to invest home-sourced land and labor as substitutes for purchasing cash inputs. Farmers in Zambia mentioned that the profitability of mineral fertilizer is almost irrelevant to them; they simply did not have cash to purchase it. Women noted that even if credit were available, they would not purchase fertilizer because they would then risk losing their productive assets if there was a drought and they were unable to pay back their loan (Peterson, 1999). This highlights a third benefit of the agroforestry practices: they help reduce risk from uncertain rainfall. The benefits of improved fallows are spread over a 2-3 year period (or longer in the case of newly introduced species such as *Gliricidia sepium* which may be coppiced), whereas nitrogen fertilizer provides benefits for only a single year. A farmer experiencing a crop failure would lose her investment in fertilizer whereas a farmer planting maize following an improved fallow would lose only her investment in planting the trees (about one-third of the fertilizer cost). Finally, agroforestry practices in the three case studies help farmers minimize risk in input markets. Fertilizer and dairy meal prices fluctuate considerably and farmers appreciate being able to produce substitutes for them on their farms. Farmers also complain about timely availability of purchased inputs and, in the case of dairy meal, the quality of the purchased product.

The case studies also highlight several methodological issues concerning assessment of financial benefits. On-farm trials are useful for measuring benefits, because agroforestry practices can be readily compared with alternative ones. Researcher-designed, farmer-managed trials appear most appropriate for financial analysis. Because these trials are designed by researchers (in consultation with farmers), non-experimental practices (such as weeding) are relatively uniform across treatments. This uniformity ensures that differences among treatments are caused by the practices being tested and not by extraneous variables. The standardization of plot size and purchased inputs in such trials also helps facilitate the collection of data on the use of labor, the most complex input to measure. In contrast, farmer-designed trials vary greatly among farms in size, types of inputs, and management and are thus less conducive to assessing profitability. Farmer-managed trials are preferred to research-managed ones, because measurement of inputs and outputs more realistically reflects farmers' experiences with the practices (Franzel et al., 2001).

Calculating returns to labor is another critical feature of the case studies; these are especially important where land is relatively abundant, as in Tabora district. In Zambia, fertilizer offers much greater returns to land but improved fallows' performance in terms of returns to labor helps explain its attractiveness.

Finally, NPVs are useful for comparing the results of practices that have costs and benefits over a series of years. But NPVs do not provide information on how farmers' annual incomes are affected by a practice, and their interpretation is not intuitively obvious to policy makers. Calculating the effect of the practice on annual incomes is done in two ways. In the case of fodder shrubs, where establishment costs are relatively low and costs and benefits vary little following the first year, two measures are calculated: the annual benefit after year 1 and the annualized net benefit treating establishment costs as depreciation. In the case of rotational woodlots and improved fallows, benefits are not generated during the first 2-4 years and costs and benefits vary among years. Therefore an alternative method is used to assess annual income: a farm is assumed to adopt the practice in phases, allocating equal-sized plots of land to each phase of the practice each year. Thus a farmer with rotational woodlots would plant a portion of woodlot to his farm each year, thus ensuring that he harvests what he needs each year. This permits an assessment of the effect of the practice on annual income.

The results from the case studies also have important implications for researchers, extensionists and policy makers. Reducing labor costs is an important avenue for increasing profitability in all of three systems. For example, using bare-rooted seedlings has important benefits over potted seedlings in fodder shrubs and improved fallows; intercropping with maize reduces tree performance in improved fallows and rotational woodlots but has very positive benefits to farmers in increasing returns to labor and land. Researchers and extensionists need to emphasize to reducing labor costs in all three practices and farmers' own innovations are often the greatest source of such modifications in technology. In Zambia, for example, farmers were the first to use bare-root seedlings and to intercrop their trees with maize; researchers followed with experiments to confirm the effectiveness of these practices, and they are now widely used by farmers (Kwesiga, Akinnifesi, Mafongoya, McDermott, & Agumya, 1999).

Several features of the financial analyses suggest that credit for establishing agroforestry is not required. Establishment costs are low, $US 7 for planting sufficient numbers of fodder shrubs to feed a cow and $US 6 and $16 for 0.25 ha of improved fallows and rotational woodlots, respectively. In all three cases, all, or nearly all, of the establishment costs are for labor; no, or almost no, cash is required. Moreover, farmers can and do adopt in increments, beginning on a small scale and gradually increasing the areas they allocate to the practices (Franzel et al., 2002a; Ramadhani et al., 2002; Kwesiga et al., in press). These findings suggest that there is little justification for providing credit to smallholders for agroforestry, because they can adopt easily without access to finance. Payback periods were also relatively low in these case studies, 2 years for fodder shrubs, 3 years for improved fallows, and 5 years for rotational woodlots.

Finally, the assessments presented have two important limitations. First, they emphasize enterprise-specific budgets and thus may miss important interactions among enterprises within the farming systems. Whole-farm analyses, while more costly, can help avoid this pitfall. Second, analyses of profitability should not be considered as the sole criterion for assessing the feasibility, acceptability, and adoption potential of an agroforestry practice to farmers. Profitability is certainly an important criterion but other factors such as cultural taboos, farmer preferences, resource bottlenecks, policy constraints, and market failures also play important roles. Assessments of profitability need to be complemented by other types of studies to identify and assess these and other issues that farmers face in using agroforestry practices.

6. NOTES

[1] While financial analysis generally refers to analysis of profitability from the farmers' perspective, economic analysis refers to profitability analysis from society's perspective (Gittinger, 1982).

[2] Bare-rooted seedlings are grown in raised seedbeds instead of in polythene pots and are thus much cheaper to produce. Following transplanting, they may have lower survival rates than potted seedlings, depending on moisture availability and other factors.

[3] 1 kg of milk is about equal to 1 liter of milk.

[4] The value of sesbania wood varies: in some areas, farmers burn the wood in the field to get rid of it whereas in other areas, they carry it to the homestead to use as firewood.

7. APPENDIX A: COEFFICIENTS AND PRICES USED IN THE FINANCIAL ANALYSIS OF CALLIANDRA FOR INCREASING MILK PRODUCTION, CENTRAL KENYA

Items	*Values*	*Data sources*
Coefficients		
Period of analysis	10 years	Assumption
Lactation period	300 days	Paterson et al., 1996
Days fed *calliandra*	365 days	Assumption
Days fed dairy meal	365 days	Assumption
Calliandra quantity fed per cow per day	6 kg fresh (equiv. to 2 kg dry)	Paterson et al., 1996
Dairy meal quantity fed per cow per day(substitution scenario, equivalent to 6 kg fresh *calliandra*)	2 kg	Paterson et al., 1996
Milk output per day from 1 kg dry *calliandra*	0.62 kg	Paterson et al., 1996
Calliandra leafy biomass yield per tree in year 1	0 kg	Farmers' experience
Calliandra leafy biomass yield per tree per year, year 2-5	1.5 kg (dry)	Paterson et al., 1998
Trees required to feed 1 cow per year	500	Computed from above.
Tree survival rate	80%	Survey data
Calliandra planting labor	20 trees per hour	Farmers
Calliandra cutting and feeding labor	15 minutes per day	Farmers
Discount rate	20%	Rough estimate of value of capital in alternative uses
Interest on capital freed up by using *calliandra* instead of purchasing dairy meal	Capital tied up for an average of 2 weeks, 20% annual interest rate.	
Prices		
Milk	$US 0.240/kg	Farmers in 2001
Dairy meal	$US 0.178/kg	Farmers in 2001
Transport of dairy meal	$US 0.005/kg	Farmers in 2001
Seedling cost (bare-rooted)	$US 0.005/seedling	S. Koech (personal communication, 2003)

APPENDIX A, (cont.)

Items	*Values*	*Data sources*
Labor cost	$US 0.110/hour	Farmers in 2001 (3/4 of daily wage)
Milk price (farm gate)	0.24/kg	Farmers in 2001
	US 2.39	Use of capital recovery formula* (Spencer et al., 1979)
1 $US = 78 Kenya Shillings		Average exchange rate, 2001

* *$K=(rv)/(1-(1=r)^{-n})$ where K is the annual service user cost, V is the original (acquisition) cost of the fixed capital asset, r is the discount rate, and n is the expected life of the asset. This procedure allows both the depreciation on capital and the opportunity cost of capital to be costed out.*

8. APPENDIX B: COEFFICIENTS AND PRICES USED IN THE FOR ROTATIONAL WOODLOTS, TABORA DISTRICT, TANZANIA

Variable	*Amount ($US)*[a]	*Source of information*
Maize		
Maize seed price	$US 0.18/ha	Average of 1995/96 and 1996–/97 market prices
Maize seed rate year 1	25 kg/ha	Farmers' estimates
Maize seed rate year 2	20 kg/ha	Farmers' estimates
Fertilizer rate	4 bags urea/ha	Research recommendation
Fertilizer cost	$US 20.17/bag	Market price 1996/1997
Threshing	$US3.70/100 kg^{-}	Farmers' estimates
Maize yield, pure stand	1943 kg/ha	On-station data adjusted
Maize yield with trees, yr. 1	1749 kg/ha	On-station data adjusted
Maize yield with trees, yr. 2	1090 kg/ha	On-station data adjusted
Maize price	$US 0.081/kg	Average market price 1995/96 and 1996/97
Trees		
Transplanting, watering, and digging micro-catchments	88 trees/day	Farmers' estimates
Transplanting cost	$US 4.18/ha	On-farm trial data
Mortality rate	34 percent	On-farm trial data
Gapping rate	34 percent	On-farm trial data
Tree population	625 trees/ha	On-farm trial data
Wood price	$US 5.28/Mg	Avg. cost of wood cut and transp. from forest, 1995/96 and 1996/97
Wood yield	152.7 t/ha	On-farm trial data, fresh weight
Wood harvesting	$US 93.14/ha	On-farm trial data
Tree seedling price	$US0.067/ seedling	Market price 1995/96 and 1996/97

APPENDIX B, (cont.)

Variable	*Amount ($US)*[a]	*Source of information*
Other		
Wage rate	$US 0.59/day	Farmers' estimates
Discount rate	20%	Researchers' estimate
Labor requirements (work days/ha)		
Land preparation	14.6	Labor survey data 1997
Maize sowing	4.3	Labor survey data 1997
Weeding	16	Labor survey data 1997
Fertilizer application	2	Labor survey data 1997
Maize harvesting	12.1	Labor survey data 1997
Maize threshing	6.3	Labor survey data 1997
Tree seedling transplanting	7.1	Labor survey data 1997
Tree seedlings gapping	2.4	Labor survey data 1997
Tree pruning	8.8	Labor survey data 1997
Wood cutting	36.5	On-farm trial data
Wood chopping	121.9	On-farm trial data

[a] *$US 1 = Tshs 595 (1997).*

9. APPENDIX C. COST BENEFIT ANALYSIS OF IMPROVED FALLOW AND CROPPING OPTIONS, EASTERN ZAMBIA ($US/ha, 1996)

	Maize cropping without fertilizer					*Two-year sesbania fallow*					*Maize cropping with fertilizer*				
	Year 1	*Year 2*	*Year 3*	*Year 4*	*Year 5*	*Year 1*	*Year 2*	*Year 3*	*Year 4*	*Year 5*	*Year 1*	*Year 2*	*Year 3*	*Year 4*	*Year 5*
Maize (mz) yields (kg/ha)	964	964	964	964	964	0	0	3638	2037	1601	4384	4384	4384	4384	4384
COSTS															
Cash costs															
Maize seed	22.24	22.24	22.24	22.24	22.24			22.24	22.24	22.24	22.24	22.24	22.24	22.24	22.24
Nursery costs						2.93									
Fertilizer								0	0	0	142.82	142.82	142.82	142.82	142.82
Fert. transport								0	0	0	6.4	6.4	6.4	6.4	6.4
Total	22.24	22.24	22.24	22.24	22.24	2.93		22.24	22.24	22.24	171.46	171.46	171.46	171.46	171.46
Labor costs															
Tree nursery						10.52									
Land preparation	12	12	12	12	12	12		9	12	12	12	12	12	12	12
Ridging	4	4	4	4	4	4		3	4	4	4	4	4	4	4
Planting mz	2	2	2	2	2	0		2	2	2	2.8	2.8	2.8	2.8	2.8
Planting trees	0	0	0	0	0	11.24		0	0	0	0	0	0	0	0
1st weeding	8	8	8	8	8	8		6	8	8	10	10	10	10	10
2nd weeding	4	4	4	4	4	4		3	4	4	6	6	6	6	6
Tree cutting	0	0	0	0	0	0	2.08	0	0	0	0	0	0	0	0
Harvesting mz	5.96	5.96	5.96	5.96	5.96	0	0	8.63	7.03	6.6	9.58	9.58	9.58	9.58	9.58
Mz shelling	3.96	3.96	3.96	3.96	3.96	0	0	6.63	5.03	4.6	7.58	7.58	7.58	7.58	7.58
Total labor	39.92	39.92	39.92	39.92	39.92	49.76	2.08	38.28	42.07	41.2	51.96	51.96	51.96	51.96	51.96

9. APPENDIX C. COST BENEFIT ANALYSIS OF IMPROVED FALLOW AND CROPPING OPTIONS, EASTERN ZAMBIA ($US/ha, 1996)

	Maize cropping without fertilizer					*Two-year sesbania fallow*					*Maize cropping with fertilizer*				
Total costs	62.16	62.16	62.16	62.16	62.16	52.68	2.08	60.51	64.31	63.44	223.42	223.42	223.42	223.42	223.42
Labor days	99.8	99.8	99.8	99.8	99.8	124.4	5.2	95.7	105.2	103	129.9	129.9	129.9	129.9	129.9
BENEFITS															
Maize	63.74	63.74	63.74	63.74	63.74	0	0	241.56	135.27	106.28	291.11	291.11	291.11	291.11	304.11
Firewood	0	0	0	0	0	0	6.4	0	0	0	0	0	0	0	0
Total benefits	63.74	63.74	63.74	63.74	63.74	0	6.4	241.56	135.27	106.28	291.11	291.11	291.11	291.11	304.11
Net benefit (nb) to labor	41.5	41.5	41.5	41.5	41.5	-2.93	6.4	219.32	112.98	84.04	119.64	119.64	119.64	119.64	132.66
Net ret to lab/day	0.42	0.42	0.42	0.42	0.42	-0.02	1.23	2.29	1.07	0.82	0.93	0.93	0.93	0.93	1.02
Net benefits	1.58	1.58	1.58	1.58	1.58	-52.68	4.32	181.04	70.96	42.84	68.07	68.07	68.07	68.07	80.7
Workdays	499					433					649.5				
NPV	4.74					115					203.58				
Discounted days	298					255					386				
Discounted nb to lab	124.12					217					357.8				
Discounted nb/disc days	0.42					0.85					0.93				
Quantity mz	4.8	t/5 yr				7.3	t/5 yr				21.9	t/5 yr			

9.1. Notes to Appendix C

Annual Maize yields for fertilized and unfertilized maize are average yields across the years of the trial. Fourth year and fifth year maize yields in the improved fallow treatment are 56% and 44% of third year yields, respectively, as reported in an on-farm trial involving 48 farmers (Kwesiga, Franzel, Place, Phiri, & Simwanza, 2003).
Prices are from local markets for the 1996 cropping season. Exchange rate: US$1.00=1250 Zambian Kwacha (ZK) in 1996 and 1683 ZK in 1998.

Cash costs
Maize seed: Seed rate of 20 kg/ha. Cost : 1340 ZK/kg
Nursery cash costs: Total costs per seedling, including cash and labor costs, are 1.4 ZK, median from cost analysis of 8 farmer nurseries. Mean cost was 1.9 ZK, standard deviation (sd), 1.2. It is assumed that 12000 seedlings are raised in order to achieve a density of 10,000 seedlings/ha in the field. Nursery cash costs accounted for 22% of the total cost of the nursery and included rent of land in valley bottom and purchase of a watering can.
Fertilizer: the recommended rate is 112-40-20 kg of N-P_2O_5-K_2O per ha. In 1996, it required 200 kg of D compound purchased at 459 ZK/kg and 200 kg of urea purchased at 433 ZK/kg. 1998 prices were 580 ZK/kg and 520 Zk/kg, respectively.
Fertilizer transport: estimated at 1,000 ZK/50 kg bag, from Chipata to farm in 1996 and 1,350 ZK/bag in 1998.
Labor: Labor data for maize cultivation are assembled from several sources cited in Franzel et al. (2002b) and from survey farmers. Labor data concerning trees are from surveyed farmers.
Labor cost: Costed at 500 ZK/workday in 1996. A workday is assumed to involve 7 hours of work. Hiring labor is not common; reported wage rates were highly variable. 500 ZK per day represents the approximate average returns per labor in maize production for 1996, that is, the value of labor at which a farmer growing maize without fertilizer breaks even. In 1998, this value was about 1300 Kw/workday.
Nursery: See 'nursery cash costs' above. Activities included collecting and threshing seeds, constructing beds, collecting sand, compost, and soil, planting, covering with grass, watering, weeding, digging out the seedlings, and transporting them to the field. Mean number of workdays required to produce 12,000 seedlings, sufficient to plant and gap up one hectare, was 26.8. (sd 22.7)
Land preparation and ridging: 30 and 10 workdays/ha, respectively. They are 25% less during the year after the improved fallow, according to estimates of trial farmers.
Planting maize: 5 workdays/ha. When applying fertilizer, 7 workdays/ha.
Planting trees: 420 trees per day, median of data from 12 farmers (mean=499, sd =424).
Weeding: Assumed to be the same for trees as for maize, as claimed by farmers. Weeding requirements decline by 25% during the year after the improved fallow, according to estimates of trial farmers. Weeding requirements are assumed to increase 33% with fertilizer use.
Harvesting and post-harvest: Labor varies with quantity. A yield of 1 t/ha requires 15 workdays for harvesting and 10 days for post-harvest activities (shelling and transportation). A yield of 4.6 t/ha is estimated to require 60% more harvest labor and 90% more post-harvest labor.

Benefits
Eleven of the twelve trial farmers had two year fallows; one had a three year fallow. For the purpose of comparison with the other sample farms for drawing up enterprise budgets, we assumed that Phiri had a two-year fallow. This assumption increased the net present values in Table 7 by 1% and the net benefit/day by 1%.
Maize: Yields are from the twelve trial farmers for the season following the improved fallow and are compared with yields on continuously cropped adjacent fields, with and without fertilizer (Table 7). For the continuously cropped maize fields, yields are assumed to be constant over the 5-year period (964 kg ha-1 without fertilizer and 4,384 kg ha-1 with fertilizer). Maize yields following the improved fallows are as measured in on-farm trials. The maize price is 83 ZK/kg, the estimated farm-gate price during the harvest period, 1996. The 1998 price was 167 Kw/kg. Firewood: Firewood is not normally sold; yield is estimated at 4 t/ha and price at 2000 ZK/t. Discount rate: 20%

10. REFERENCES

Coe, R. (1998). Participatory on-farm experimentation in agroforestry: experiences and the role of biometrics. In *Proceedings of the International Biometrics Conference*, Invited papers (pp.1-10). 13-18 December 1998, Cape Town, South Africa: International Biometrics Association.

Cooper, P. J., Leakey, R. R. B., Rao, M. R., & Reynolds, L. (1996). Agroforestry and the mitigation of land degradation in the humid and sub-humid tropics of Africa. *Experimental Agriculture, 32*, 235-290.

Franzel, S., Wambugu, C., & Tuwei, P. (2003). The adoption and dissemination of fodder shrubs in central Kenya, Agricultural Research and Network (AGREN) Series Paper No. 131. London: Overseas Development Institute.

Franzel, S., Arimi, H., & Murithi, F. (2002a). *calliandra calothyrsus*: Assessing the early stages of adoption of a fodder tree in the highlands of central Kenya. In S. Franzel and S. J. Scherr (Eds.), *Trees on the Farm: Assessing the Adoption Potential of Agroforestry Practices in Africa* (pp. 125-144) Wallingford, UK: CABI.

Franzel, S., Phiri, D., & Kwesiga F. (2002b). Assessing the adoption potential of improved fallows in eastern Zambia. In S. Franzel and S. J. Scherr (Eds.), *Trees on the Farm: Assessing the Adoption Potential of Agroforestry Practices in Africa* (pp. 37-64) Wallingford, UK: CAB International.

Franzel, S. Coe, R. Cooper, P., Place, F. & Scherr, S. J. (2001). Assessing the adoption potential of agroforestry practices in sub-Saharan Africa. *Agricultural Systems, 69*, 37-62

Gittinger, J. P. (1982). Economic analysis of agricultural projects. Baltimore: Johns Hopkins Press.

Haugerud, A. (1984). *Household dynamics and rural political economy among Embu farmers in the Kenya highlands.* Unpublished PhD dissertation, Northwestern University, Chicago.

Keil, A. (2001). Improved fallows using leguminous trees in eastern Zambia: Do initial testers adopt the technology? Unpublished masters thesis. Faculty of Agriculture, Institute of Rural Development, Georg-August University, Goettingen, Germany.

Kwesiga, F., Akinnifesi, F. K., Mafongoya, P. L., McDermott, M. H., & Agumya, A. (2003). Agroforestry research in southern Africa during the 1990s. Review and challenges ahead. *Agroforestry Systems, 59,* 173-186.

Kwesiga, F., Franzel, S., Mafongoya, P., Ajayi, O., Phiri, D., & Katanga, R. et al. (in press). *Improved fallows in Eastern Zambia: Evolution, adoption, and impact.* (Discussion Paper Series on Successes in African Agriculture). Washington, DC: International Food Policy Research Institute.

Kwesiga, F. R., Franzel, S., Place, F., Phiri, D., & Simwanza, C. P. (1999). *Sesbania sesban* improved fallows in eastern Zambia: their inception, development, and farmer enthusiasm. *Agroforestry Systems, 47,* 49-66.

Murithi, F. M. (1998). Economic evaluation of the role of livestock in mixed smallholder farms of the central highlands of Kenya. Un published Ph.D. thesis, Department of Agriculture, University of Reading, Reading, United Kingdom.

Otsyina, R., Minae, S., & Cooper, P. (1996a). A never ending story: rotational woodlots for soil conservation, wood, and fodder. *Agroforestry Today, 8*(2), 8-10.

Otsyina, R., Msangi, R., Gama, B., Ramadhani, T., Madulu, J., & Mapunda, H. (1996b). *SADC/ICRAF Agroforestry Research Project, Tumbi, Tabora, Tanzania: Annual Report 1996* (AFRENA Report No. 105). Nairobi, Kenya: International Centre for Research in Agroforestry.

Otsyina, R., Msangi, R., Gama, B., Ramadhani, T., Nyadzi, G., & Shirma, D. (1997). *SADC/ICRAF Agroforestry Research Project, Tumbi, Tabora, Tanzania Annual Report 1997.* (AFRENA Report No. 114). Nairobi, Kenya: International Centre for Research in Agroforestry.

Palm, C. A., Myers, R. J. K., & Nandwa, S. M. (1997). Combined use of organic and inorganic nutrient sources for soil fertility maintenance and replenishment. In R.J. Buresh, P.A. Sanchez, and F.G. Calhoun (Eds.), *Replenishing Soil Fertility in Africa* (pp.193-217) Madison Wisconsin, USA: SSSA Special Publication Number 51. Soil Science Society of America and American Society of Agronomy.

Paterson, R. T., Karanja, G. M., Roothaert, R., Nyaata, O.Z., & Kariuki, I.W. (1998). A review of tree fodder production and utilization within smallholder agroforestry systems in Kenya. *Agroforestry Systems, 41,* 181-199.

Paterson, R. T., Roothaert, R., Nyaata, O. Z., Akyeampong, E., & Hove, L. (1996). Experience with *calliandra calothyrsus* as a feed for livestock in Africa. In D. O. Evans (Ed.), *Proceedings Of An International Workshop On The Genus calliandra* (pp. 195-209) 23-27 January 1996, Bogor, Indonesia, Morrilton, Arkansas: Winrock International.

Peterson, J.S. (1999). Kubweletza Nthaka: Ethnographic decision trees and improved fallows in the Eastern Province of Zambia. Gainesville, Florida: University of Florida/Gender and Soil Fertility in Africa Collaborative Research Support Program.

Place, F. Franzel, S., DeWolf, J., Rommelse, R. Kwesiga, F., Niang, A. et al. (2002). Agroforestry for soil fertility replenishment: Evidence on adoption processes in Kenya and Zambia. In: C.B. Barrett, F. Place, & A. A. Aboud (Eds.), *Natural Resources Management in African Agriculture: Understanding and Improving Current Practices* (pp. 155-168). Wallingford, UK: CAB International.

Ramadhani, T., Otsyina, R., & Franzel, S. (2002). Improving household incomes and reducing deforestation; the example of rotational woodlots in Tabora District, Tanzania. *Agriculture, Ecosystems, and the Environment, 89*(3), 227-237.

Roothaert, R., Karanja, G. M., Kariuki, I. W., Paterson, R., Tuwei, P., & Kiruiro, E. (1998). *calliandra* for Livestock (*Technical Bulletin No. 1)* Embu, Kenya: Kenya Agricultural Research Institute, Regional Research Centre.

Sanchez, P. A. (1996). Science in agroforestry. *Agroforestry Systems, 9,* 259-274.

Spencer, D. S. C., Byerlee, D., & Franzel, S. (1979). *Annual costs, returns, and seasonal labor requirements for selected farm and non-farm enterprises in rural Sierra Leone* (Working Paper No. 27). East Lansing, Michigan: Michigan State University, Department of Agricultural Economics.

Upton, M. (1987). *African farm management.* Cambridge, UK: Cambridge University Press.

Wambugu, C., Franzel, S., Tuwei, P., & Karanja, G. (2001). Scaling up the use of fodder shrubs in central Kenya. *Development in Practice, 11,*(4) 487-494.

11. AUTHOR'S NOTE

This chapter was made possible in part by the data collection and analysis assistance of Festus Murithi, Stanislas Phiri, Donald Phiri, Tunu Ramadhani, and Samuel Koech.

STEPHEN C. GRADO AND AMANDA L. HUSAK

ECONOMIC ANALYSES OF A SUSTAINABLE AGROFORESTRY SYSTEM IN THE SOUTHEASTERN UNITED STATES

1. INTRODUCTION

Agroforestry systems have proven their financial viability and attractiveness as important land use alternatives in various settings throughout the world (Garrett, 1997). Silvopasture is a unique agroforestry system that combines forage crops, trees, and livestock production (United States Department of Agriculture [USDA], 1997). Its application is most prominent in the southern and western United States (US), the Pacific Northwest, and potentially, the Midwest (Garrett & Buck, 1997). In the southern US, silvopasture is the most common form of agroforestry (Zinkhan & Mercer, 1997).

Lundgren, Connor, and Pearson (1983) and Pearson (1991) reported that the potential for forest grazing is greater in the southern US than in any other region of comparable size in the country. Of the 278 million acres (112 million hectares) of land area in the southern US, approximately 38 million acres (15 million hectares) are used for crops (including harvested cropland, crop failure, and cultivated summer fallow), 9 million acres (4 million hectares) are idle, 15 million acres (6 million hectares) are used only for pasture, 19 million acres (8 million hectares) are in grassland pasture (including grasslands, non-forest pasture, and range), 43 million acres (17 million hectares) are in other uses (including marshes, open swamps, bare rock areas, urban areas, and special use areas), and 154 million acres (62 million hectares) are in forestland (excluding reserved, special use, or park land; includes forested grazing land) (USDA, 1999). Conversion of only a portion of the idle or marginal cropland available in the south to multiple-use systems such as silvopasture could lead to increased economic and social benefits to landowners and ecological benefits to wildlife (Pearson, 1991).

Silvopastoral research in the southern US during the last century has concentrated on cattle and pines (Williams, Gordon, Garrett, & Buck, 1997). Silvopastoral systems represent a form of multiple-use management in which

J. R. R. Alavalapati & D. E. Mercer, Valuing Agroforestry Systems, 39–57.

landowners, animals, and plants interact and produce an array of diverse benefits. Silvopasture promotes timber production while providing landowners with the annual cash flows necessary to sustain operations. Silvopastoral systems also have the potential to incorporate other revenue producing activities such as fee hunting and pine straw production. Although past studies by Clason (1999), Haney (1980), Harwell and Dangerfield (1991), and Lundgren et al. (1983) demonstrated that silvopastoral systems are economically and biologically feasible, few have discussed the benefits that wildlife or pine straw production add to these systems. Limited information is available on the wildlife benefits or the potential yields and monetary benefits from pine straw production inherent in various silvopastoral systems. Several studies divulged the potential wildlife-related benefits reaped by landowners and regional and national economies from recreational activities (e.g., hunting) that occur on traditional agricultural and forested ownerships (Grado, Hovermale, & St. Louis, 2001; Jones, Munn, Jones, & Grado, 1998; Jones et al., 2001; United States Department of the Interior [USDI], Fish and Wildlife Service [FWS], & United States Department of Commerce [USDC], 2002). Studies show that communities can benefit from hunter expenditures on food, lodging, hunting materials, and equipment (Grado, Hurst, & Goodwin, 1997; Guynn & Yarrow, 1990). Additionally, some studies have focused upon the increased income to landowners from hunting fees (Grado et al., 2001; Jones et al., 1998, 2001; Steinback, 1999; Thomas, 1996). Pearson (1991) first mentioned the possibility of increased land values through hunting leases, and Grado et al. (2001) demonstrated the monetary value of wildlife to silvopastoral systems by including hunting leases in the overall management plan. Pine straw production has primarily been investigated in the context of pine plantation management rather than for silvopasture (Dickens, Dangerfield, & Moorhead, 2003; Duryea & Edwards, 1989; Mills & Robertson, 1991).

There are a myriad of other benefits from silvopastoral systems. They can increase marketing options, reduce soil erosion, convert forage crops to protein, reduce fire hazards, influence the understory vegetation, and increase nutrient availability by, for example, adding nitrogen to the biomass production system (Mosher, 1984). To effectively demonstrate the benefits and financial returns from silvopastoral systems, each component of the system must be assigned a corresponding monetary value. A discussion of reliable and consistent valuation methods for forestry, agriculture, and agroforestry investments follows.

1.1. Forestry Investment Valuation

Forestland values are affected by many factors including site quality, timber markets, proximity to the mill, topography, restrictive logging regulations, site preparation costs, and land taxes (Binkley & Vincent, 1988; Samuelson, 1976). Some forestland value indicators affected by these factors are stumpage values, forest holding values, forest liquidation values, bid prices, market values, net present value (NPV), equivalent annual income (EAI), rate of return (ROR), land expectation value (LEV), and willingness to pay for land or other assets (Bullard & Straka, 1998; Klemperer, 1996). For the purposes of this discussion, the focus will be placed upon NPV, EAI, ROR, and LEV.

Evaluating potential capital investments in forestry-related practices begins with an estimation of the present value of all revenues and costs associated with the practice to yield a NPV for the investment. All costs and revenues are discounted to the present with costs then subtracted from revenues.

Once NPV has been calculated, it can be used to derive an EAI for a project. EAI is often expressed as the landowner's earnings per acre (or hectare), per year when fully considering the time value of money (Bullard & Straka, 1998). Simply put, it is NPV expressed as an annual amount and is often used to compare the economic returns from forestry investments with those obtained from pasture rents, agricultural crops, or other land uses that yield annual returns (Bullard & Straka, 1998).

The ROR of any investment represents the rate of compound interest that is earned by the project's investment capital. For a project with only one cost and revenue, ROR can be calculated directly using a simple formula. For projects with multiple costs and revenues, ROR is estimated by finding the compound interest rate at which the total present value of costs equals the total present value of revenues. ROR is often used in analyses of investment projects and is popular with non-industrial private forest landowners (Bullard & Straka, 1998).

LEV is used to estimate the value of forestland using the NPV of all revenues and costs involved in producing outputs from a forest. It considers all present and future revenues and costs, with the exception of land cost, to be obtained from a particular tract of land dedicated to a particular activity into perpetuity. LEV can assist in selecting management regimes for a particular tree species on a specific site because it represents the bare land value for the site when committed to a particular regime into perpetuity. A comparison of all LEVs obtained for various regimes allows one to rank them on the basis of their potential returns. NPV and ROR are used only for accepting or rejecting investment decisions. EAI and LEV are used for ranking investment decisions.

1.2. Agricultural Investment Valuation

Agricultural investments, like forestry investments, can be deemed acceptable using NPV and ROR, while EAI and LEV can be used to rank agricultural investment alternatives (Bullard & Straka, 1998). Agricultural land uses can also be compared to other land uses such as forestry and agroforestry investments. In reality, farmers make decisions based on more complex criteria than financial assessments. Agricultural methods are often judged, as well, on how specific systems meet the proprietor's basic needs of food, shelter, and cash income (Arnold & Dewees, 1999).

1.3. Agroforestry Investment Valuation

Agroforestry investments, due to their integration of forestry, agricultural, wildlife components and other activities (e.g., pine straw production) must incorporate valuation techniques from all components to estimate the economic attractiveness of an investment. This will also permit an analyses of the contribution of each component to an overall system. Typically, economic analysis of agroforestry land use alternatives is conducted using NPV, EAI, ROR, and NPV of perpetual rotations or LEV. The congruity of such values creates an easy mode of comparison for numerous agroforestry options.

Many computer models have been designed to make financial comparisons of agroforestry options. They include MULBUD (Etherington & Matthews, 1983), MIDAS (Kingwell & Pannell, 1987), TREE$PLAN (Bulman, 1991), FARMTREE (Loane, 1991), FARMULA (Kubicki, Denby, Stevens, Haagensen, & Chatfield, 1993), and USAEM (Pearson, Knowles, Middlemiss, Balwin, & Busby, 1995). Of all computer models used to analyze agroforestry systems, spreadsheet-based models are the most commonly used because they allow a sensitivity analysis of changes to inputs on any operation in question (Thomas, 1991b). Although wildlife benefits created by agroforestry systems are considerable, many computer models are limited in their capacity to incorporate those benefits. Such deficiencies necessitate the development or adaptation of a suitable computer-based model that would include a valuation of wildlife and other non-market output benefits (e.g., savings from soil erosion reductions).

1.4. Economic Studies in Agroforestry

Numerous economic studies of agroforestry systems have been conducted (Harou, 1983; Hoekstra, 1987; Husak, 2000; Husak & Grado, 2002; Thomas, 1991a). Generally, these studies examine the financial costs required to establish, manage, and produce various combinations of agricultural and timber crops, potential revenues from different agroforestry alternatives, and profitability of adopting agroforestry practices. Few, if any studies, have looked at the monetary aspects of wildlife or pine straw production in agroforestry systems.

Edward (1991) analyzed and compared the profitability of a wide variety of agroforestry practices in Senegal using NPV, Benefit/Cost ratio, and ROR. The analyses were conducted from the farmer's viewpoint in an effort to bridge the information gap between agroforestry and landowners. This study concluded that integration of agroforestry practices into traditional farming systems yields greater rates of return than monoculture practices alone.

Jorge, Ramirez, and Carlos (1991) investigated the economic viability and technical feasibility of modern agroforestry practices in the Amazon. Their study determined that selected modern agroforestry practices (e.g., agrosilvicultural and silvopastoral) have the potential to increase wood and coffee production, improve labor efficiency, and reduce cash requirements during market lulls.

Price (1995) scrutinized the application of valuation techniques in estimating the costs and benefits associated with agroforestry systems. Examples were given for eight approaches to valuation of non-market benefits that could potentially be elicited from agroforestry systems. The study suggested that a systematic and quantitative investigation of all benefits and costs associated with agroforestry production is necessary to convince economists and landowners that agroforestry offers positive monetary and non-monetary benefits.

2. OBJECTIVES

The chapter objectives are to illustrate the profits associated with southern silvopasture under various scenarios and compare them to those accrued from traditional single-use agricultural and forest management systems. The analyses will also demonstrate the monetary benefits derived by private landowners from including and utilizing fee hunting and pine straw production in silvopastoral systems.

This information should serve to increase the attractiveness of silvopasture systems to landowners and farmers by furthering the knowledge on this subject by compiling past research results, illustrating the monetary benefits of these systems, providing a means of valuing monetary returns from incorporating silvopasture systems into traditional agricultural systems, and recommending future research areas. Additionally, it will be demonstrated that agroforestry systems can potentially increase monetary benefits to landowners by attracting more wildlife than monoculture systems. Pine straw production will also serve as another periodic revenue stream that can enhance the attractiveness of these systems.

3. METHODOLOGY

3.1. System Selection

Four hypothetical land management systems in the southern US were chosen to make investment comparisons to silvopasture and included soybeans, rice, cattle, and a loblolly pine (*Pinus taeda*) plantation. These four systems, in addition to silvopasture, are commonly found in the southern US and have generated a substantial amount of readily obtainable information (Byrd & Lewis, 1983; Mississippi Agricultural and Forestry Experimental Station [MAFES], 2001, 2002a, 2002b; Pearson, 1991). An average farm size of 259 acres (105 hectares) was used for each system, corresponding to the average farm size for the southern region of the US (USDA, 2003). A site index of 65 and a base age of 25 years were assumed for loblolly pine for all sites, similar to studies by Clason (1999), Harwell and Dangerfield (1991), and Pearson (1991). Regional data was taken from Alabama, Arkansas, Florida, Georgia, Kentucky, Louisiana, Mississippi, North Carolina, South Carolina, and Tennessee following regional divisions suggested by Merwin (1997). Average annual cash rents for this region are $52.50/acre ($130.00/hectare) for non-irrigated cropland, $61.20/acre ($151.00/hectare) for irrigated cropland,

$18.00/acre ($44.00/hectare) for pastureland, and $18.00/acre ($44.00/hectare) for timberland (Grado et al., 2001; USDA, 2002a).

3.1.1. Silvopasture

The parameters for this system were selected in light of real-world examples by Clason (1999), Grado et al. (2001), Harwell and Dangerfield (1991), and Lundgren et al. (1983) where loblolly pine was planted at a density of 454 trees per acre (1,122 trees per hectare) on a 4 ft x 8 ft x 20 ft (1 m x 2 m x 6 m) spacing and maintained on a 30-year sawtimber rotation. Commercial thinnings to a residual basal area of 70 ft^2 (6.50 m^2) were conducted at ages 15, 20, and 25 to improve the growth and value of the stand (Nebeker, Hodges, Karr, & Moehring, 1985). Final harvest occurred in year 30. Due to difficulties in modeling the typical spacing (see Clason, 1999; Harwell and Dangerfield, 1991; Lundgren et al., 1983) for silvopastoral systems, an initial stocking of 908 trees per acre (2,244 trees per hectare) was used to simulate competition between trees, and thinning volumes were halved for ages 15 and 20 to reflect typical yields from silvopasture. Timber yields per acre (or hectare) were estimated using WINYIELD (Hepp, 1994). Yields were: 8.46 cords (75.77 m^3) of pulpwood at age 15; 3.69 cords (33.05 m^3) of pulpwood at age 20; 6.35 cords (56.87 m^3) of chip 'n' saw at age 25; and 5.27 cords (47.20 m^3) of chip 'n' saw and 4.78 mbf (Doyle) (27.87 m^3) of sawtimber at age 30. Prescribed burning was used annually from ages 4 to 30 to reduce fire hazards and plant competition, kill brush, improve access, and stimulate forage growth (Grado et al., 2001). Residual trees were pruned following thinnings to reduce tree taper and increase volume (Valenti, 1986). Timber prices used were: $405/mbf ($171/m^3) (Doyle) for sawtimber; $79/cord ($22/ m^3) for chip 'n' saw; and $18/cord ($5/m^3) for pulpwood (Daniels, 2003). For consistency, timber prices were assumed to remain constant throughout the rotation.

Byrd and Lewis (1983), Clason (1999), and Pearson (1991) have shown that the introduction of cattle to silvopastoral systems has no negative effect on timber growth if introduction occurs after trees reach a height of 18 inches (46 cm). Cattle were introduced to the system in year two to allow time for forage and tree establishment (Grado et al., 2001; Pearson, 1991). In a previous study, Lundgren et al. (1983) used a stocking rate (SR) of 0.74 animal units per acre (AU/ac [1.83 AU/ha]) for a continuously grazed cow-calf operation with annual calf sales.

For this analysis, the SR was 0.36 AU/ac (0.89 AU/ha) for the second and third years and 0.59 AU/ac (1.46 AU/ha) for the remaining years. Cattle were grazed continuously and removed to a supplemental winter-feeding area on a small portion of the total acreage during the hunting season to avoid cattle-hunter interactions. Calves were produced each year and held for sale in their second year. Assuming that one bull can service 25 to 30 cows, the 259-acre (105-hectare) site initially supported 83 cows and 3 bulls or 87.5 AU (0.34 AU/ac [0.84 AU/ha]) in years two and three (Pearson, 1991). [Animal units were calculated assuming 1.0 AU for a cow weighing 1,000 lb (454 kg), either dry or with calf, 0.75 AU for a weaned animal weighing less than 900 lb (408 kg), and 1.5 AU for a bull weighing less than

2,000 lb (907 kg) (Manske, 2001)]. The annual calving rate was assumed to be 92% or 76 calves, increasing the herd to 159 total cows and calves and three bulls (145 AU) or a stocking rate of 0.56 AU/ac (1.38 AU/ha) (Lundgren et al., 1983; Pearson, 1991).

Calves were sold in their second year since two-year old steers and heifers, weighing 1,000 lb (454 kg) yield higher prices ($800/head) than yearling calves weighing 500 lb (227 kg) ($468/head) (USDA, 2002c). Cows were purchased for $560/head and sold in 10 years at $420/head (USDA, 2002c). Bulls were purchased for $1,060/head and sold in five years at $615/head (USDA, 2002c). Generally, cows produce for 10 to 12 years, and bulls produce for five years (Pearson, 1991). For consistency, cattle prices remained fixed throughout the rotation.

A permanent summer grass mixture composed of bahiagrass (*Paspalum notatum*), bermudagrass (*Cynodon dactylon*), dallisgrass (*Paspalum dilatatum*), and other mixed grasses and Mount Barker clover (*Trifolium subterraneum*) were planted, fertilized, and maintained annually beginning in year one. The summer grass mixture was established by seeding at a planting rate of 35-40 lb/ac (39-45 kg/ha); clover was established at a planting rate of 15-20 lb/ac (17-22 kg/ha) (SCS 1994). Fertilizer was applied at rates of 500 lb/ac (560 kg/ha) of 13-13-13 NPK at establishment; 200 lb/ac (224 kg/ha) of ammonium nitrate, 100 lb/ac (112 kg/ha) of phosphate, and 100 lb/ac (112 kg/ha) of potash were applied annually for maintenance (Soil Conservation Service [SCS], 1994; MAFES, 2001).

Annual maintenance costs, which included land rent during years 0 to 30, forage establishment and maintenance costs (e.g., fertilizer, herbicide, seed, soil testing, labor, diesel fuel, repair and maintenance to fences, tractors, and implements) were incurred between years 1 to 30 (Grado et al., 2001; MAFES, 2001). Revenues from the sale of steers and heifers occurred during years 3 to 30. Prescribed burning costs and hunting lease revenues were incurred annually in years 4 to 30 (Dubois, McNabb, & Straka 2003; Jones et al., 1998, 2001). Delaying prescribed burning and hunting leases until year 4 reduces the chance of tree damage and results in more suitable habitat for game species (Byrd & Lewis, 1983; Hazel, 1990; Pearson, 1991). Annual costs and revenues involved in the silvopasture system are reported in Table 1.

Periodic costs and revenues incurred in a silvopasture operation include site preparation, planting, and fertilizer costs in year zero. Total site preparation/establishment costs include minimal site preparation (i.e., prescribed burning only), planting costs of $14/acre ($35/hectare) and $36/acre ($89/hectare), respectively (Dubois et al., 2003), and an adjusted seedling cost of $27/500 individuals (South Carolina Forestry Commission [SCFC], 2003). Cattle costs were converted to a per acre (or hectare) basis by dividing the purchase price for cows or bulls by the number of acres (hectares) and multiplying this by 83 for cows and 3 for bulls. Steer and heifer prices were converted to a per acre (hectare) basis by multiplying the number of calves produced each year (n = 76) by the sale price ($800/head) and dividing this by the total number of acres (hectares) in the tract (n = 259 acres/105 hectares). Cattle revenues were obtained by reducing the initial

Table 1. Costs and Revenues for a Silvopastoral System (US 2002 Dollars).

Year	Activity	Cost $/acre	Cost ($/hectare)	Revenue $/acre	Revenue ($/hectare)
0	Establishment	77.73	(192.08)		
0 to 30	Land Rent	52.50	(129.73)		
1 to 30	Management	159.19	(393.37)		
2, 12, 22	Cow Purchase	179.46	(443.46)		
12, 22	Cow Sales			134.59	(332.58)
2 to 30	Supplemental Feed	21.60	(53.37)		
2 to 30	Animal Maintenance	5.40	(13.34)		
3 to 30	Steer/Heifer Sales			234.75	(580.08)
4 to 30	Prescribed Burning	13.25	(32.74)		
4 to 30	Hunting Leases			4.89	(12.08)
Every 5 Years	Bull Purchase	12.29	(30.37)		
Every 5 Years	Bull Sales			7.12	(17.59)
10, 14, 18, 22, 26, 30	Pine Straw			1.50[a]	(3.71)
15	Thinning			152.28	(376.29)
15	Pruning	38.08	(94.10)		
20	Thinning			66.42	(164.13)
20	Pruning	23.63	(58.39)		
25	Thinning			501.65	(1,239.60)
25	Pruning	16.15	(39.91)		
30	Harvest			2,352.23	(5,812.49)

[a]Net revenue per acre (hectare) after raking, baling, and fertilizing.

purchase price by 75% for cows and 58% for bulls; this percentage reduction in price represented the value depreciation of cattle (Pearson, 1991). Thinning revenues were obtained by multiplying the estimated yields by the timber prices. Pruning costs were estimated at $0.17 per tree and reflected the actual cost to the landowner for pruning their own trees (Grado et al., 2001). Periodic pine straw raking was based on the fact that the best time to start raking a pine stand is around 10 years of age for loblolly pine and then every four years to protect the soil (Duryea & Edwards, 1989). Yields of 140 to 150 bales per acre (346 to 371 bales per hectare) can be realized for pine plantations with 500 trees per acre (1,236 trees per hectare) (Duryea & Edwards, 1989; Mills & Robertson, 1991); however, in the silvopasture setting used in this analysis it was assumed that 125 bales per acre (309 bales per hectare) could be accrued at each raking interval (Mills & Robertson, 1991). Per acre (hectare) periodic costs and revenues for the silvopasture treatment were itemized in Table 1.

3.1.2. Soybeans

Soybeans were planted annually at a rate of 40 pounds per acre (lb/ac; 45 kilograms per hectare [kg/ha]) and yielded 35 bushels per acre (bu/ac; 3.0 m^3/ha) (MAFES, 2002b). Phosphorus (46% P205) and Potash (60% K20) fertilizers, Apron®XL fungicide, Roundup®Ultra 4SL herbicide, and Larvin®3.2 insecticide were applied to stimulate growth and protect against fungus, weeds, and insects (MAFES, 2002b). Soybean prices remained constant at $5.36/bu ($152.00/m^3) (MAFES, 2002b; USDA, 2002c). Total annual costs for soybean production were $165/ac ($408/ha); total annual revenues were $197/ac ($487/ha). Annual land rent was $53/ac ($131/ha).

3.1.3. Rice

Rice was planted annually at a rate of 90 lb/ac (101 kg/ha) and yielded 150 bu/ac (13 m^3/ha) (MAFES, 2002a). Solid urea (46% N) fertilizer, and Propanil®4E, Ordram®15-G, and Grandstand® herbicides were applied to stimulate growth and protect against weeds (MAFES, 2002a). Rice prices were assumed to remain constant at $2.94/bu ($83.43/m^3) (MAFES, 2002a; USDA, 2002c). Total annual costs were $381/ac ($941/ha); total annual revenues were $441/ac ($1,090/ha). Annual land rent was $61/ac ($151/ha).

3.1.4. Cattle

Cattle production was simulated using the combined information for the cattle and forage components for the silvopastoral system. A permanent summer grass mixture composed of bahiagrass, bermudagrass, dallisgrass, and other mixed grasses and Mount Barker clover, was planted, fertilized, and maintained annually beginning in year one. Grass was established at a rate of 35-40 lb/ac (39-45 kg/ha); clover was established at a rate of 15-20 lb/ac (17-22 kg/ha) (SCS, 1994). Fertilizer was applied at rates of 500 lb/ac (560 kg/ha) of 13-13-13 NPK at establishment, and 200 lb/ac (224 kg/ha) of ammonium nitrate, 100 lb/ac (112 kg/ha) of phosphate, and 100 lb/ac (112 kg/ha) of potash for annual maintenance (SCS, 1994; MAFES, 2001).

Cattle were introduced to the site in year 1 to allow time for forage growth. The site supported 156 AU or 147 cows and 6 bulls (0.6 AU/ac [1.48 AU/ha]) in years 1 and 2 (Grado et al., 2001; D.G. St. Louis, personal communication, 2001). Calves were produced each year and held for sale in their second year. The annual calving rate was assumed to be 92% or 135 calves, so that in years 3 to 30, the number of cattle would increase to 147 cows, 135 calves, 135 yearling calves, and 6 bulls (257 AU) or a stocking rate of 1.0 AU/ac (2.47 AU/ha) (Grado et al., 2001; St. Louis, personal communication, 2001). Cattle sale and purchase prices remained the same as those used for silvopasture.

Annual costs and revenues for cattle production include land rent of $18/ac ($44/ha) in years 0 to 30; forage establishment and fence maintenance costs of $183/ac ($452/ha) and forage maintenance costs of $84/ac ($208/ha) in years 1 to 30; animal supplement costs of $43/ac ($106/ha) and animal maintenance costs of

$43/ac ($106/ha) in years 2 to 30; and steer and heifer sales of $417/ac ($1,030/ha) in years 3 to 30. Periodic costs and revenues include sale prices of $238/ac ($588/ha) for cows and $14/ac ($35/ha) for bulls and purchase prices of $318/ac ($786/ha) for cows and $25/ac ($62/ha) for bulls at 10 and 5-yr increments, respectively.

3.1.5. Pine Plantation

Using a plantation design described by Harwell and Dangerfield (1991) for use in the southern US, loblolly pine was planted at a density of 605 trees per acre (1,495 trees per hectare) in year 0 and maintained on a 35 yr sawtimber rotation. A pre-commercial thinning was conducted in year 10. Commercial thinnings to a residual basal area of 70 ft^2 (6.5 m^2) were conducted at ages 15 and 25 (Harwell & Dangerfield, 1991). Final harvest occurred in year 35. Timber yields were estimated using WINYIELD (Hepp, 1994). Yields per acre (hectare) were 13.71 cords (122.79 m^3) of pulpwood at age 15; 14.77 cords (132.29 m^3) of chip 'n' saw at age 25; and 9.34 mbf (Doyle) (56.46 m^3) of sawtimber at age 35. Timber prices remained the same as those used for silvopasture.

Total establishment costs for the pine plantation included minimal site preparation (i.e., prescribed burning) and planting costs of $13/ac ($32/ha) and $26/ac ($64/ac), respectively (Dubois et al., 2003), and seedling costs of $27/500 individuals (SCFC, 2003). Other costs included land rent at $18/ac ($44/ha), herbicide spraying at $50/ac ($124/ha), pre-commercial thinning at $65/ac ($160/ha), and prescribed burning at $13/ac ($32/ha) (Dubois et al., 2003). Timber revenues were $247/ac, $1,167/ac ($2,884/ha), and $3,783/ac ($9,348/ha) in years 15, 25, and 35, respectively.

Using the costs and revenues for silvopasture, soybeans, rice, cattle, and pine plantation systems and real interest rates of 5%, 7%, and 9%, LEVs, EAIs, and RORs were calculated for each system. The analysis was conducted without regard to risk or inflation.

3.2. Supplemental Income

Silvopasture systems, due to their inherent diversity, provide nesting and breeding areas, food, and cover for numerous wildlife species throughout the rotation (e.g., White-tailed Deer, *Odocoileus virginiana*; Northern Bobwhite, *Colinus virginianus*; and Eastern Wild Turkey, *Meleagris gallopavo*) (Turcotte & Watts, 1999; Yarrow & Yarrow, 1999). Succession of the habitat from pasture to tree cover attracts different species at different times in the rotation and provides the landowner with a chance to view wildlife and incorporate hunting leases for some of the species attracted. In a silvopastoral setting, hunting lease rates and bare land value will increase as habitat quality increases.

Studies by Joyce, Flather, Flebbe, Hoekstra, and Ursic (1990) and Morrison (1992) have investigated the abundance and diversity of wildlife species in forests managed for timber production. Even-aged management, such as that used in pine plantations, will attract both breeding and non-breeding birds throughout the rotation (Morrison, 1992). As in the silvopasture system, Eastern Wild Turkey and White-tailed Deer will use pine plantations at various rotation ages.

The landscaping and horticultural industry prefers pine straw for mulch because it is attractive, stays in place well, and does not readily float away during rainstorms (North Carolina Division of Forest Resources [NCDFR], 1997). This popularity has also led pine straw to become an important and profitable forest product (NCDFR 1997). In fact, pine straw production, even under low yield and price conditions (50 bales/acre/year or 124 bales/hectare/year at $0.70/bale), can add as much as $35/acre/yr ($87/hectare/yr) to a forestland owner's income (NCDFR, 1997). Additionally, studies in Arkansas on large pine plantations, suggest that landowners could lease the land for pine straw harvesting and receive between $10 and $25/acre ($25 and $62/hectare), representing a new source of income for rural areas (Hays, 1997).

Hunting leases and pine straw production values were used in the cash flow model as an additional, supplemental income opportunity and demonstration of wildlife diversity for both the silvopastoral system and pine plantation. A hunting lease value of $4.89/ac ($12.08/ha) and a pine straw production value of $1.50/ac ($3.71/ha) were considered (Duryea & Edwards, 1989; Yarrow & Yarrow, 1999).

4. RESULTS AND DISCUSSION

4.1. Monetary benefits

LEVs, EAIs, and RORs for all five systems are reported in Table 2. At 5% interest, the LEV of $1,284 ($3,174) and EAI of $64 ($159) were greatest for the pine plantation, followed closely by silvopasture ($1,253 [$3,097]; $63 [$155]), soybean production ($1,158 [$2,861]; $53 [$130]), and cattle production ($1,127 [$2,785]; $56 [$139]). Rice production had the lowest LEV of $1,050 ($2,594) and EAI of $48 ($118) at 5%. At 7% and 9%, soybean production yields the highest LEVs of $843 ($2,082) and $668 ($1,650) and EAIs of $52 ($127) and $51 ($125), while the pine plantation yields the lowest LEVs of $615 ($1,519) and $299 ($738) and EAIs of $43 ($106) and $27 ($66). These values indicate that, at low interest rates like 5%, the preferred uses for the land are pine plantations, silvopasture, or soybean production. At higher interest rates, like 7% and 9%, soybean or cattle production is the preferred land use.

Although LEVs and EAIs consistently give the same ranking for potential investments, EAI is often included to compare forestry and agricultural investments (Bullard & Straka, 1998). Equivalent annual incomes represent the net present value (i.e., all revenues minus all costs discounted to the present) of an investment expressed as an annual amount (Bullard & Straka, 1998). EAIs for these systems follow the same trend as the LEVs.

Although RORs should not be used for ranking purposes, they provide some idea of the average rate of interest earned on capital over the life of the investment. For

Table 2. Land Expectation Values (LEV), Equivalent Annual Income (EAI), and Rates of Return (ROR) for All Production Systems (2002 US dollars).

System	Interest Rate	LEV		EAI		ROR*
		$/acre	($/hectare)	$/acre	($/hectare)	
Silvopasture	5%	1,253.11	(3,096.50)	62.66	(154.84)	5.9%
	7%	693.83	(1,714.49)	48.57	(120.02)	5.9%
	9%	411.83	(1,017.65)	37.06	(91.58)	5.9%
Soybeans	5%	1,157.73	(2,860.81)	52.50	(129.73)	1.4%
	7%	842.70	(2,082.36)	51.52	(127.31)	1.4%
	9%	667.69	(1,649.90)	50.58	(124.99)	1.4%
Rice	5%	1,049.58	(2,593.57)	47.60	(117.62)	-2.5%
	7%	763.98	(1,887.84)	46.71	(115.42)	-2.5%
	9%	605.31	(1,495.75)	45.85	(113.30)	-2.5%
Cattle	5%	1,126.95	(2,784.75)	56.35	(139.24)	4.0%
	7%	806.02	(1,991.72)	56.42	(139.42)	4.0%
	9%	619.57	(1,530.99)	55.76	(137.79)	4.0%
Pine Plantation	5%	1,284.43	(3,173.90)	64.22	(158.69)	10.1%
	7%	614.55	(1,518.59)	43.02	(106.30)	10.1%
	9%	298.84	(738.45)	26.90	(66.47)	10.1%

*Annual land rent was included in NPVs and RORs and excluded in LEVs and EAIs.

this analysis, RORs were 5.9% for silvopasture, 1.4% for soybeans, -2.5% for rice, 4.0% for cattle, and 10.1% for the pine plantation. The negative ROR for rice can be attributed to the high annual rent required for rice production, which creates negative annual returns on the investment. Annual land rent for each system was included only in NPV and ROR because LEV and EAI provide values of the land itself, thus never incorporating land rent.

4.2. Supplemental income benefits

As expected, when hunting leases and pine straw production values were included, LEVs and EAIs were greater at all interest rates for all values for both silvopasture in Table 3 and the pine plantation in Table 4. Without the lease or straw, silvopasture LEVs range from $1,162 ($2,871) at 5% to $359 ($888) at 9%, and pine plantation LEVs range from $1,174 ($2,901) at 5% to $229 ($567) at 9%. With the lease and straw included, silvopasture LEVs range from $1,258 ($3,108) at 5% to $414 ($1,023) at 9%, and pine plantation LEVs range from $1,284 ($3,174) at 5% to $299 ($738) at 9%. With the lease alone, silvopasture LEVs range from $1,253 ($3,097) at 5% to $412 ($1,018) at 9%, and pine plantation LEVs range from $1,260 ($3,113) at 5% to $281 ($693) at 9%. With the straw alone, silvopasture LEVs range

Table 3. Benefits with Supplemental Income for a Silvopastoral System for a 30-Year Production Period (2002 US dollars).

	Interest Rate	Land Expectation Values (LEV)		Equivalent Annual Income (EAI)		Difference between leasing options
		$/acre	($/hectare)	$/acre	($/hectare)	
Silvopasture	5%	1,161.66	(2,870.52)	58.08	(143.52)	
without lease;	7%	627.25	(1,549.97)	43.91	(108.50)	
without straw	9%	359.32	(887.90)	32.34	(79.91)	
Silvopasture	5%	1,257.77	(3,108.02)	62.89	(155.40)	8.3%
with lease;	7%	696.80	(1,721.83)	48.78	(120.54)	11.1%
with straw	9%	413.88	(1,022.72)	37.25	(92.05)	15.2%
Silvopasture	5%	1,253.11	(3,096.50)	62.66	(154.84)	7.9%
with lease;	7%	693.83	(1,714.49)	48.57	(120.02)	10.6%
without straw	9%	411.83	(1,017.65)	37.06	(91.58)	14.6%
Silvopasture	5%	1,166.32	(2,882.04)	58.32	(144.11)	0.4%
without lease;	7%	630.22	(1,557.31)	44.12	(109.02)	0.5%
with straw	9%	361.37	(892.96)	32.52	(80.36)	0.6%

Table 4. Benefits with Supplemental Income for a Pine Plantation for a 35-Year Production Period (2002 US dollars).

	Interest Rate	Land Expectation Values (LEV)		Equivalent Annual Income (EAI)		Difference between leasing options
		$/acre	($/hectare)	$/acre	($/hectare)	
Pine Plantation	5%	1,174.10	(2,901.26)	58.71	(145.08)	
without lease;	7%	529.60	(1,308.67)	37.07	(91.60)	
without straw	9%	229.48	(567.06)	20.65	(51.03)	
Pine Plantation	5%	1,284.43	(3,173.90)	64.22	(158.69)	9.4%
with lease;	7%	614.55	(1,518.59)	43.02	(106.30)	16.0%
with straw	9%	298.84	(738.45)	26.90	(66.47)	30.2%
Pine Plantation	5%	1,259.96	(3,113.43)	63.00	(155.68)	7.3%
with lease;	7%	593.40	(1,466.32)	41.54	(102.65)	12.0%
without straw	9%	280.53	(693.20)	25.25	(62.39)	22.2%
Pine Plantation	5%	1,198.57	(2,961.73)	59.93	(148.09)	2.1%
without lease;	7%	550.74	(1,360.91)	38.55	(95.36)	4.0%
with straw	9%	247.79	(612.30)	22.30	(55.10)	8.0%

from $1,166 ($2,882) at 5% to $361 ($893) at 9%, and pine plantation LEVs range from $1,199 ($2,962) at 5% to $248 ($612) at 9%. The LEVs and EAIs reflect monetary differences between incorporating or not incorporating hunting leases and/or pine straw production into a silvopastoral system and a pine plantation.

5. CONCLUSIONS

Agroforestry is gaining acceptance by landowners across the United States. Studies have shown that the adoption of these systems is both economically and biologically feasible. In fact, agroforestry systems provide habitat heterogeneity that promotes floral and faunal diversity. Results from this study illustrate the monetary benefits that can be gained from silvopasture systems.

At low interest rates, pine plantations yield slightly higher LEVs and EAIs than silvopasture or rice production. These higher values can be attributed to the undeniable profitability of growing timber and the ever-present demand for wood products. However, the profitability of silvopastoral systems should not be ignored. By providing annual and periodic revenues from the production of multiple outputs throughout the rotation, the output diversity of silvopasture increases the potential for profit and reduces the risk on investment. A multi-component investment like silvopasture will almost certainly be affected less by market and price fluctuations for cattle and timber than more traditional monoculture systems. Additionally, landowners will receive further satisfaction in efficiently and effectively increasing the quantity and quality of the products from their land. Like pine plantations, silvopastoral systems create the opportunity for additional income through hunting leases. On average, silvopastoral systems incorporating hunting leases yield 7.9% to 14.6% more value per acre (hectare) than conventional silvopasture, which is comparable to the 7.3% to 22.2% increase in value gained through hunting leases on pine plantations. In addition, the literature suggests that silvopasture provides both quantity and quality of wildlife habitat that cannot be found in the other systems used in this analysis.

Another added benefit to both the silvopastoral and pine plantation systems is the production and harvest of pine straw. On average, the incorporation of periodic pine straw harvest yields 0.4% to 0.6% more value per acre (hectare) than conventional silvopasture, which is comparable to the 2.1% to 8.0% increase for pine plantations. Although percentage increases in per acre (hectare) value from inclusion of pine straw is small relative to that of hunting leases, the inclusion of both increases per acre (hectare) values by as much as 15.2% for silvopasture and 30.2% for pine plantations. In down markets, the addition of this supplemental income may help to defer expenses, ultimately allowing the landowner the luxury of waiting for a market up-turn.

The ultimate decision to invest in any land-use system lies in a landowner's vision for the land, experience with land-use options, knowledge of potential economic returns from the land, awareness of available markets, and demand for products from this land. Additionally, changes in interest rates, time to maturity of the investment, as well as desired rate of return on the investment, will make land-use systems more or less attractive to a particular landowner. Of the factors surrounding the decision to invest, the two that may most heavily influence a landowner are their experience with land-use options and their flexibility regarding the time to maturity for the investment. A landowner who possesses greater experience with multiple land-use options and who can afford to delay the ultimate maturity of the investment for several decades may be more likely to choose agroforestry options such as silvopasture. For this reason, several lease and interest rates were used in this study to illustrate the variability and potential for returns from the described systems. Depending on a landowner's knowledge, desires, and needs, the rates of return calculated for the systems in this analysis may or may not represent attractive alternatives to current land uses.

In conclusion, silvopasture is an environmentally and economically feasible multiple-use system with great potential for application in the southern US. Silvopasture compares favorably to other land use systems and provides additional benefits made possible by the diversity and productivity of the system. Although excluded from this study, other opportunities for supplementing income, such as floriculture, exist and may be incorporated into silvopastoral systems. In addition, future price increases for cattle, timber, and agricultural crops would further increase income opportunities from silvopastoral systems.

As technological advances in the implementation and maintenance of agroforestry systems are introduced, widespread adoption of silvopasture may become as common as other land use systems today. This study serves as a methodology for observing these systems and providing landowners with information on the potential monetary benefits produced.

6. FUTURE RESEARCH

There are a number of issues related to silvopastoral practices that require an ambitious level of investment to make these systems more attractive to landowners and society (Association for Temperate Agroforestry [AFTA], 2000).

One area of interest relates to silvicultural practices. This would include research on spacing, pruning, fertilization, and species choices for forest trees and grasses. Any improvement in the mix of these items, tailored for specific sites, would most certainly improve the financial viability of silvopasture in general, and relative to other common land use practices. Investigations into various configurations of multi-row and multi-species versus single-row and single-species systems would have the same effect (AFTA, 2000).

The increasing popularity of pine straw also needs to be investigated in further detail. There are issues related to the promotion and harvesting of pine straw that could impact site quality for trees and also reduce site attractiveness for wildlife (Schafale & Weekly, 1990; NCDFR, 1997).

There are also a number of other basic research areas that need to be explored. Typical examples include investigating forage preferences of animals in silvopastoral practices. While this has been previously examined to some extent, there are potential areas where improvement could be realized. In addition, fundamental cultural practices such as mowing, herbicide use, and cultivation affect establishment and early growth of tree seedlings in a pasture setting should be investigated further (AFTA, 2000).

In summary, these are just a few of the research-based endeavors that need to be undertaken to establish silvopastoral systems in the southern US and other areas where it may be a compatible land use. Any improvements to these systems will lead to both financial and public acceptability, thus furthering the likelihood that farmers and forest landowners will embrace silvopasture in the future.

7. REFERENCES

Arnold, J.E.M., & Dewees, P.A. (1999). Trees in managed landscapes: Factors in farmer decision making. In L.E. Buck, Lassoie, J.P., & Fernandes, E.C.M. (Eds.), *Agroforestry in Sustainable Agricultural Systems* (pp. 277-294). Boca Raton, FL: CRC Press LLC.

Association for Temperate Agroforestry. (2000). *Agroforestry in the United States: Research and technology transfer needs for the next millennium*. Columbia, MO: Author.

Binkley, C.S., & Vincent, J.R. (1988). Timber prices in the U.S. South: Past trends and outlook for the future. *Southern Journal of Applied Forestry, 12*, 15-18.

Bullard, S.H., & Straka, T.J. (1998). *Basic concepts in forest valuation and investment analysis* (Ed. 2.21). Auburn, AL: Preceda Education and Training.

Bulman, P.A. (1991). TREE$PLAN – A model to compare farm tree designs in southern Australia. *Agricultural System and Information Technology Newsletter*, *3*, 16-18.

Byrd, N.A., & Lewis, C.E. (1983). *Managing pine trees and bahiagrass for timber and cattle production*. (General Report R8-GR 2). Asheville, NC: USDA Forest Service, Southern Region.

Clason, T.R. (1999). Silvopastoral practices sustain timber and forage production in commercial loblolly pine plantations of northwest Louisiana. *Agroforestry Systems*, *44*, 293-303.

Daniels, R.A. (2003). *Mississippi timber price report*. Retrieved March 1, 2003, from the MSUcares (Coordinated Access to the Research and Extension System), Mississippi State University Extension Service and Mississippi Agriculture and Forestry Experiment Station Web site: http://msucares.com/forestry/prices/reports/2003/janfeb.pdf

Dickens, E.D., Dangerfield, Jr., C.W., & Moorhead, D.J. (2003). Short-rotation management options for slash and loblolly pine in southeast Georgia, USA. In G.S. Amacher & J. Sullivan (Eds.), *Proceedings of the 2003 Southern Forest Economics Workshop* (pp 61-65). Virginia Beach, VA.

Dubois, M.R., McNabb, K., & Straka, T.J. (2003). Costs and cost trends for forestry practices in the South. *Forest Landowner Manual Edition*, *62*, 3-13.

Duryea, M.L., & Edwards, J.C. (1989). *Pine-straw management in Florida's forests*. (IFAS Circular 831). Gainesville, FL: University of Florida, Cooperative Extension Service.

Edward, K.G. 1991. Comparison of agroforestry practices in Senegal using financial analysis. In G.M. Sullivan, S.M. Huke, & J.M. Fox (Eds.). *Financial and Economic Analyses of Agroforestry Systems: Proceedings of a Workshop* (pp. 109-124). Paia, HI: Nitrogen Fixing Tree Association.

Etherington, D.M., & Matthews, P.J. (1983). Approaches to the economic valuation of agroforestry farming systems. *Agroforestry Systems*, *1*, 347-360.

Garrett, H.E.G. (1997). *Agroforestry: An integrated land-use management system for production and farmland conservation*. (USDA-SCS 68-3A7S-3-134). Fort Collins, CO: United States Department of Agriculture – Soil Conservation Science.

Garrett, H.E.G., & Buck, L. (1997). Agroforestry practice and policy in the United States of America. *Forest Ecology and Management*, *91*, 5-15.

Grado, S.C., Hovermale, C.H., & St. Louis, D.G. (2001). A financial analysis of a silvopasture system in south Mississippi. *Agroforestry Systems*, *53*, 313-322.

Grado, S.C., Hurst, G.A., and Godwin, K.D. (1997). Economic impact and associated values of the wild turkey in Mississippi. *Proceedings of the. Annual Conference of the Southeastern Association of Fish and Wildlife Agencies, 51*, 438-448.

Guynn, Jr., D.C., & Yarrow, G.K. (Eds.). (1990). Fee hunting. *Proceedings of 1990 Conference on Fee Hunting on Private Lands in the South.* Clemson, SC: Clemson University.

Haney, H.L., Jr. (1980). Economics of integrated cattle-timber land use. In R.D. Child & E.K. Byington (Eds.), *Proceedings of the Southern Forest Range and Pasture Symposium* (pp. 165-184). Morrilton, AR: Winrock International.

Harou, P.A. (1983). Economic principles to appraise agroforestry projects. *Agricultural Administration, 12,* 127-139.

Harwell, R.L., & Dangerfield, Jr., C.W. (1991). Multiple-use on marginal land: A case for cattle and loblolly pine. *The Forestry Chronicle, 67*, 249-253.

Hays, S.R. (1997). Pine needles – A hot, new commodity. *Agricultural Research, May*: 16-17.

Hazel, R.B. (1990). *Timber and wildlife. Landowner's handbook for managing southern pines.* Asheville, NC: Cooperative Extension Service and United States Department of Agriculture Forest Service, Southern Region.

Hepp, T.E. (1994). WINYIELD 1.0, A Windows-based Forest Growth, Yield, and Financial Analysis Tool for Southern Forests [Computer Software]. Norris, TN: Tennessee Valley Authority.

Hoekstra, D.A. (1987). Economics of agroforestry. *Agroforestry Systems, 5*, 293-300.

Husak, A.L. (2000). Wildlife and economic benefits from agroforestry systems. Unpublished master's thesis, Mississippi State University, MS.

Husak, A.L., & Grado, S.C. (2002). Monetary benefits in a southern silvopastoral system. *Southern Journal of Applied Forestry, 26*, 159-164.

Jones, W.D., Munn, I.A., Grado S.C., & Jones, J.C. (2001). *Fee Hunting: An Income Source for Mississippi's Non-Industrial Private Landowners.* (FWRC Research Bulletin Series FO164). Mississippi State, MS: Forest and Wildlife Research Center.

Jones, W.D., Munn, I.A., Jones, J.C., & Grado, S.C. (1998). A survey to determine fee hunting and wildlife management activities by private, non-industrial landowners in Mississippi. *Proceedings of the Annual Conference of the Southeastern Association of Fish and Wildlife Agencies, 52*, 421-429.

Jorge, U., Ramirez A., & Carlos, S. (1991). Are modern agroforestry practices economically viable? A case study in the Ecuadorian Amazon. In G.M. Sullivan, Huke, S.M., and Fox, J.M. (Eds.), *Financial and economic analyses of agroforestry systems. Proceedings of a workshop*, (pp. 273-292). Paia, HI: Nitrogen Fixing Tree Association.

Joyce, L.A., Flather, C.H., Flebbe, P.A., Hoekstra, T.W., & Ursic, S.J. (1990). Integrating forage, wildlife, water, and fish projections with timber projections at the regional level: A case study in the southern United States. *Environmental Management, 14*, 489-500.

Kingwell, R.S., & Pannell, D.J. (1987). MIDAS, a bio-economic model of a dryland farming system [Computer software]. Pudoc: Wageningen.

Klemperer, W.D. (1996). *Forest resource economics and finance*. New York, NY: McGraw-Hill

Kubicki, A., Denby, C., Stevens, M., Haagensen, A., & Chatfield, J. (1993). Determining the long-term costs and benefits of alternative farm plans. In R.J. Hobbs, & Saunders, D.A. (Eds.), *Re-integrating Fragmented Landscapes: Towards Sustainable Production and Nature Conservation*, (pp. 245-258). New York, NY: Spring-Verlag.

Loane, B. (1991). Economic evaluation of farm trees – Methodology and data for FARMTREE model [Computer software]. Melbourne, Australia: Victorian Department of Agriculture and Department of Conservation and Environment.

Lundgren, G.K., Connor, J.R., & Pearson, H.A. (1983). An economic analysis of forest grazing on four timber management situations. *Southern Journal of Applied Forestry, 7*, 119-124.

Manske, L.L. 2001. Animal Unit Equivalent for Beef Cattle Based on Metabolic Weight. Dickinson, SD: North Dakota State University, Livestock Production Extension Service.

Merwin, M.L. (1997). *The status, opportunities, and needs for agroforestry in the United States.* Columbia, MO: Association for Temperate Agroforestry.

Mills, R., & Robertson, D.R. (1991). *Production and marketing of Louisiana pine straw.* (Publication 2430). Baton Rouge, LA: Louisiana State University, Cooperative Extension Service.

Mississippi Agricultural and Forestry Experiment Station (MAFES). (2001). *Forage 2001 planning budgets*. (Agricultural Economics Report 125). Mississippi State, MS: Author.

Mississippi Agricultural and Forestry Experiment Station (MAFES). (2002a). *Rice 2002 planning budgets*. (Budget Report 2001-003). Mississippi State, MS: Author.

Mississippi Agricultural and Forestry Experiment Station (MAFES). (2002b). *Soybeans 2002 planning budgets*. (Budget Report 2001-002). Mississippi State, MS: Author.

Morrison, M.L. (1992). Bird abundance in forests managed for timber and wildlife resources. *Biological Conservation, 60*, 127-134.

Mosher, W.D. (1984). What is agroforestry? In N. E. Linnartz & Johnson, M.K. (Eds.), *Agroforestry in the southern United States*, (pp. 2-10). Baton Rouge, LA: Louisiana State Agricultural Center.

Nebeker, T.E., Hodges, J.D., Karr, B.K., & Moehring, D.M. (1985). *Thinning practices in southern pines – With pest management recommendations*. (Technical Bulletin 1703). Asheville, NC: United States Department of Agriculture, Forest Service.

North Carolina Division of Forest Resources. (1997). *Pine straw production in North Carolina.* (Forestry Leaflet FM-19). Raleigh, NC: Author.

Pearson, H.A. (1991). Silvopasture: forest grazing and agroforestry in the southern coastal plain. In D.R. Henderson (Ed.), *Proceedings of the Mid-South Conference on Agroforestry Practices and Policies*, (pp. 25-42). Morrilton, AR: Winrock International.

Pearson, H.A., Knowles, R.L., Middlemiss, P.G., Baldwin, V.C., & Busby, R.L. (1995). United States agroforestry estate model. *The Compiler, 13*, 27-37.

Price, C. (1995). Economic evaluation of financial and non-financial costs and benefits in agroforestry development and the value of sustainability. *Agroforestry Systems, 30*, 75-86.

Samuelson, P.A. (1976). Economics of forestry in an evolving society. *Economic Inquiry, 14*, 466-492.

Schafale, M.P. & Weekly, A.S. (1990). *Ecological concerns about pine straw raking in a southeastern longleaf pine ecosystem*. Raleigh, NC: North Carolina Natural Heritage Program, North Carolina Division of Parks and Recreation.

Soil Conservation Service. (1994). *Mississippi planting guide for wildlife food plots, livestock, erosion control, and beautification*. Jackson, MS: Author.

South Carolina Forestry Commission. (2003). *South Carolina Forestry Commission price guide for tree seedlings, equipment, and services*. Retrieved March 1, 2003, from the South Carolina Forestry Commission Web site: http://www.state.sc.us/forest/refprice.htm#seedlings.

Steinback, S.R. (1999). Regional economic impact assessments of recreational fisheries: An application of the IMPLAN modeling system to marine party and charter boat fishing in Maine. *North American Journal of Fisheries Management, 19*, 724-736.

Thomas, M. (1996). Hunting management. *Forest Landowner, 55*, 12-14.

Thomas, T.H. (1991a). Agroforestry – does it pay? *Outlook on Agriculture, 19*, 161-170.

Thomas, T.H. (1991b). A spreadsheet approach to the evaluation of the economics of temperate agroforestry. *Forest Ecology and Management, 45*, 207-235.

Turcotte, W.H., & Watts, D.L. (1999). *Birds of Mississippi*. Jackson, MS: University Press of Mississippi.

United States Department of Agriculture. (1997). *Agroforestry – functions and values*. (Issue Brief 14). Washington, D.C.: United States Department of Agriculture Natural Resources Conservation Service/Soil and Water Resources Conservation Act.

United States Department of Agriculture, Economics Research Service. (1999). *Agricultural statistics*. Washington D.C.: USDA National Agricultural Statistics Service, Economic Research Division. U.S. Government Printing Office.

United States Department of Agriculture. (2002a). *Agricultural cash rents*. Washington D.C.: USDA National Agricultural Statistics Service, Economic Research Division. U.S. Government Printing Office.

United States Department of Agriculture. (2002b). *Agricultural land values*. Washington D.C.: USDA National Agricultural Statistics Service, Economic Research Division. U.S. Government Printing Office.

United States Department of Agriculture. (2002c). *Agricultural prices*. Washington D.C.: USDA National Agricultural Statistics Service, Economic Research Division. U.S. Government Printing Office.

United States Department of Agriculture. (2003). *Farms and land in farms*. Washington D.C.: USDA National Agricultural Statistics Service, Economic Research Division. U.S. Government Printing Office.

United States Department of the Interior, Fish and Wildlife Service, & United States Department of Commerce, Bureau of the Census. (2002). *2001 national survey of fishing, hunting, and wildlife-associated recreation*. Washington D.C.: Author.

Valenti, M.A. (1986). A comparison of the effects of one-step and two-step pruning on loblolly pine stem form. *Southern Journal of Applied Forestry*, *10*, 215-253.

Williams, P.A., Gordon, A.M., Garrett, H.E.G., & Buck, L.E. (1997). Agroforestry in North America and its role in farming systems. In A.M. Gordon & Newman, S.M. (Eds.), *Temperate Agroforestry Systems*, (pp. 9-84). Oxon, UK: CAB International.

Yarrow, G.K., & Yarrow, D.T. (1999). *Managing wildlife*. Birmingham, AL: Sweetwater Press.

Zinkhan, C.F., & Mercer, D.E. 1997. An assessment of agroforestry systems in the southern USA. *Agroforestry Systems*, *35*, 303-321.

W. M. DREW, J. R.R. ALAVALAPATI, AND P. K. R. NAIR

DETERMINING AGROFORESTRY PROFITABILITY USING THE POLICY ANALYSIS MATRIX

A Case Study from Pohnpei, Federated States of Micronesia

1. INTRODUCTION

Small-island developing states (SIDS) in the Pacific region have a number of valuable natural resources, both aquatic and terrestrial. Changing economic conditions, rapidly increasing populations, and the fragility of local ecosystems combined with market isolation contribute to increasing demands on natural resources for formal and informal economic activities. The challenge for many SIDS is to ensure that these resources are developed in a manner that meets the needs of current and future generations in the pursuit of economic development. Agriculture remains the single largest sector of many island states in the Pacific, often contributing substantially to local employment and being responsible for a significant portion of foreign exchange earnings. As a result, agriculture is recognized as an essential component of overall economic development strategies for many of these island states (Food and Agriculture Organization [FAO], 1994; United Nations Development Program [UNDP], 1994).

Furthering the development of agricultural sectors in the Pacific will entail either introducing new technologies and/or improving existing ones. The introduction of new agricultural technologies has often been based on results from "on-farm" or "on-station" research under somewhat controlled conditions. Much of this research has focused on biophysical or socioeconomic attributes at the micro- or farming system-level. In contrast, there has been much less effort toward investigating the role that policy intervention may have on the adoption, use, and allocation of scarce natural resources, and, in turn, farmer profits. Knowledge of current policy impacts

J. R. R. Alavalapati & D. E. Mercer, Valuing Agroforestry Systems, 59–78.

as well as alternative policies under consideration may provide an important linkage between the micro-level conditions and macro-level decision-making.

Agroforestry has been practiced in the Pacific for millennia contributing at one time to Pacific Islanders being among the most self-sufficient and well-nourished people in the world. Indigenous Pacific island agroforestry systems represent a tremendous diversity, and often have evolved in response to ecological, cultural, and socioeconomic changes (Clarke & Thaman, 1993; Nair, 1993; Manner, 1993; Raynor, 1989; Raynor & Fownes, 1991a, 1991b; Thamen, 1975). These systems are generally considered to be ecologically sustainable. As globalization and the drive to increase economic well-being continues, many Pacific Islanders in both the private and public sectors are looking to introduce new technologies that have yet to prove their economic superiority, let alone their ecological sustainability. In part, the lack of appreciation for the existing agroforestry practices in the Pacific (and elsewhere) can be attributed to incomplete understanding about the important role that agroforestry practices currently play in food production, local economies, and ecological balance, or the potential of these systems for future agricultural development.

In Pohnpei, Federated States of Micronesia (FSM), agroforestry has been practiced for centuries (Haun, 1984) just like elsewhere in the region. Today it remains the predominant land use activity (Asian Development Bank [ADB], 1996). Over time, numerous crops and technological introductions have been made to Pohnpei's agroforests through continued waves of migration, and more recently, through direct and indirect efforts of colonial administrations (Barrau, 1961). Pohnpei, like many other Pacific islands is at a cross roads whereby traditional norms are changing to more western oriented desires. Whereas only a decade ago, prestige was largely defined by who provided the largest traditional crops to the local chiefdom, today where possible, these crops are being sold to purchase cars, televisions, and CD players. Moreover, with more than 60% of the population under the age of 25 and very limited employment opportunities in the formal economy (Government of the Federated States of Micronesia, 1999), Pohnpeaens are searching for alternative ways to accommodate their changing lifestyles. It has yet to be seen how the dynamic situation will ultimately affect Pohnpei. For the present time these changes are leading to increasing pressure to intensify (or in some cases replace) current agricultural practices in ways that may not be suitable to Pohnpei's ecological, social, cultural, or economic setting (Kostka & Raynor, 2000). Given the long history of agriculture on Pohnpei and the uncertainties about its economic future, it has been suggested that future developments of Pohnpei's agricultural sector should be based on a more informed understanding of agriculture's current role and how policy may influence future farmer decisions (Pohnpei Office of Agriculture and Forestry, 1996).

A stated objective of the national and state governments is to promote import substitution policies to reduce dependence on imported food and increase household incomes (Government of FSM, 1999; Office of the Governor, 1996). At the state level, alternative strategies are desired to enhance both commercial and traditional agricultural activities. Yet specific strategies have not been identified, largely due to the lack of baseline data about the production and economic value of existing agricultural systems for which comparisons can be measured.

The primary objective of this chapter is to examine the profitability of three different agroforestry systems on Pohnpei, FSM. The Policy Analysis Matrix (PAM) methodology is used to assess two different measures of profits: private profitability (PP), or those profits that are actually realized by practicing farmers; and social profitability (SP), or those profits that in theory would be realized in the absence of policy distortion and/or market failure. This will provide a means to quantify the impacts of current public sector policies and market failures (or lack thereof) for dominant agroforestry systems on Pohnpei. In addition, essential information will also be provided to better understand the current motives of farmers and how potential government strategies may impact the agricultural sector. Moreover, the comparison of profits reveals the underlying comparative advantage of different commodities/systems in question. A secondary objective of this chapter is to provide a framework to expand the PAM to non-market benefits, such as carbon sequestration[1], and to internalize external costs and benefits within the PAM framework.

During the past few decades, the PAM has been used to quantify profitability and the impacts of policy interventions on dairy systems in Kenya (Staal, 1996), rice in Thailand (Yao, 1997), and rice and sugar in Indonesia (Nelson & Panggabean, 1991; Pearson, Falcon, Heytens, Monke, & Naylor, 1991). Whereas the PAM methodology has been used to measure a single crop produced within an agroforestry system (Adesina & Coulibaly, 1998) to the best of our knowledge, the PAM approach not been used to quantify profitability of multiple crops cultivated within complex agroforestry systems anywhere in the world, let alone within the Pacific region.

What follows is first, a brief description of Pohnpei and the farming systems being analyzed followed by a section describing the PAM methodology. Next, we present sources of data and the policy scenarios used within our analysis followed by our results and a discussion on the respective findings. The chapter concludes with a few final thoughts regarding the Policy Analysis Matrix as a tool for profitability analyses.

1.1. The setting

Pohnpei is located in the central region of the Caroline Island group at 6° 54' N latitude and 158° 14'E longitude. At 355 square kilometers (km^2), Pohnpei is the second largest island in the Carolines. It is a high volcanic island having a steep mountainous interior region, the highest point being 772 meters above sea level. Average annual rainfall is approximately 4,800 millimeters (mm) evenly distributed over 300 days of the year (National Oceanic and Atmospheric Administration [NOAA], 1987).

In 1995, intact upland forests covered 15% of the island, with an additional 15% consisting of disturbed upland forests (ADB, 1996). Agroforests, consisting primarily of homegardens resembling multistoried tree gardens, cover 37% of the island, with coastal forests (16%) and secondary forests (12%) comprising much of the remaining land area. Extensive mangrove forests and a fringing reef surround much of the island. In the lowlands, highly weathered Oxisols are the most common soils found, whereas Inceptisols are more commonly found in the interior and upland regions of the island (Laird, 1987).

In 1986, the Federated States of Micronesia entered into a 15-year *Compact of Free Association* with the United States, which has guaranteed payments of approximately $1.4 billion in exchange for exclusive military access to the area's waterways. To date, the Compact has provided most of the funds necessary to operate the country's economy (Osman, 1995) primarily through the development of a large civil servant workforce. As a result of the most recent Compact negotiations, the national and state level governments will be receiving fewer funds combined with more financial oversight from the United States. With the steady decline in Compact funding, a primary objective of the Pohnpei government has been to reduce the scale of public employment. These efforts have resulted in a 20% reduction in public sector employment between 1996 and 1999 (Office of the Governor, 2003). During the same period, private sector employment has only seen marginal growth.

Of all pre-Compact economic activities, agriculture has probably been the most adversely affected. Once the Compact was implemented, the public sector was not only enlarged, but offered higher wages relative to the private sector. As a result, income from non-agriculturally related activities became more dominant while simultaneously reducing the reliance on subsistence-based food production. With the changes in funding and without a marked increase in private sector expansion, the downsizing of the public sector will eliminate what has been for the last 15 years a primary source of income for much of the island's population. As this occurs, it is agriculture will likely take on an increasingly important role for both household food and income production. Indeed, the 2000 census indicates that of the total workforce on Pohnpei, 49.6% are engaged in either formal or informal agricultural and/or fishing activities (55% for subsistence and 45% for income generation), an increase of close to 25% in recent years (Office of the Governor, 2003).

1.2. Farming systems

Practices known as "homegardens" or multistoried tree gardens represent the most expansive and common form of agroforestry on Pohnpei. These systems incorporate multipurpose trees and shrubs in intimate association with annual and perennial crops, and often domestic animals, adjacent to homes (Fernandes & Nair, 1986; Raynor, 1989; Falanruw, 1993). Traditionally, almost all production, be it for subsistence, or commercial purposes, is by family units using localized agroforestry practices, providing employment, food security, and income while maintaining the cultural and ecological integrity of the island's resources (Raynor, 1991).

The economically dominant species common to these systems include an array of tree and agricultural crops. These include: coconut *(Cocos nucifera)* and breadfruit *(Artocarpus altilis)* in the upper canopy; Ylang-ylang *(Cananga odorata)*, betel nut *(Areca catechu)*, and yams *(Dioscorea* spp.*)* in the upper and sub-canopy strata; and bananas and plantains (*Musa* spp.), hibiscus (*Hibiscus tiliaceus*), Indian mulberry (*Morinda citrifolia*), yam vines (*Dioscorea* spp.), and soursop (*Annona muricata*) in the lower sub-canopy strata. Closer to ground level typically below 2.5 meters, sakau *(Piper methysticum)* is very common in addition to root crops such as wild taro (*Alocasia macrorrhiza*), sweet taro, (*Colocasia esculenta*), and swamp taro (*Cyrtosperma chamissonis*) (Raynor & Fownes, 1991a, 1991b).

For centuries in island settings the production of different taro species has played an important role towards food security. A tuber, farmers have often relied on these crops during food shortages, particularly in the aftermath of typhoons and floods that decimate many aboveground crops. For generations, the maintenance of wetland-food production systems, consisting largely of taros, has remained an important component of agricultural activities ensuring that at least some food will always be available (Drew, Ewel, Naylor, & Sigrah, manuscript submitted for publication).

In Pohnpei, like in many other island locales, people have established homes in areas where wetlands exist, combining wetland food production practices with those resembling other homegarden systems. Data collected for this research indicated that homegardens on Pohnpei could be further differentiated by the presence of wetland areas that have been converted into taro patches. Together, these two homegarden systems, homegardens (HG) and homegardens with wetlands (HGW), dominate the agricultural landscape on Pohnpei and therefore both are examined in this study.

A third system of interest is centered on the cultivation of sakau, primarily in the uplands of Pohnpei. The significance of sakau to Pohnpei's economy is growing in response to increasing demand, primarily from the domestic market. At one time used almost exclusively for traditional ceremonial activities, sakau's increasing importance has become a mixed blessing. On the one hand, increasing sakau production is contributing both directly and indirectly to economic activity on Pohnpei. On the other hand, recent research has shown that production of sakau has led to the conversion of a significant amount of Pohnpei's remaining intact-upland forests to the upland sakau system (UPS) (ADB, 1996; Kostka & Raynor, 2000), best characterized as having much less diversity relative to either of the other two homegarden systems, with the primary objective of producing a single crop of sakau.

In recent times, there has been a push to expand sakau production into the upland areas due to cooler temperatures and greater rainfall, both of which reduce cultivation time before harvests. However, the drive to maximize sakau production in the uplands has led to clearing large areas that in turn are exposed to heavy rains and significant soil erosion. Indeed, based on recent evidence (Kostka & Raynor,

2000), commercial production of sakau in Pohnpei's upland forests is not sustainable, with estimates of production declining 5% – 10% annually. In contrast, sakau production within Pohnpei's homegardens seems to be ecologically sustainable, having been produced within these systems for decades (Raynor & Fownes, 1991a, 1991b).

Non-governmental organizations on Pohnpei are attempting to promote lowland sakau production within homegarden agroforests but continued upland clearing remains a problem. The government of Pohnpei now recognizes the negative impacts associated with upland sakau farming while also recognizing the need on the farmers' part to generate income. Comparing private and social profits for these three systems will provide a building block to develop appropriate strategies for the future use of Pohnpei's resources.

2. METHODOLOGY

Monke and Pearson (1989) developed the Policy Analysis Matrix methodology in the late 1970s primarily for use in the profitability analysis of single crops within specific cropping systems. The PAM has proven to be an effective tool for analyzing the influence of policy distortions on farmer profits and the willingness of farmers to modify existing practices. For example, trade, domestic factor market, as well as exchange rate policies can be analyzed in relation to their impacts on farm profits (Pearson et al., 1991). Additionally, the PAM allows for in-depth analysis of the domestic or international "production chain", those activities involved in the movement of products between the farmgate and markets. As a result, the macro-micro linkages can be clearly illustrated, helping to inform policy makers about the potential tradeoffs and impacts of alternative technologies, as well as revealing the underlying comparative advantage of different commodity systems.

According to Monke and Pearson (1989), the mechanics of the PAM consists of two accounting identities (Table 1). The first measures profits based on total revenues less the costs of tradable inputs and domestic factors. The top row of the PAM represents private costs and returns, where profitability (D) is defined as

Table 1. The Policy Analysis Matrix (PAM) Framework.

	Revenues	*Costs*		*Profits*
	Output	*Tradable inputs*	*Domestic factors*	
Private prices	A	B	C	D
Social prices	E	F	G	H
Effects of policy distortions and market failures	I	J	K	L

Private profits "D" = (A-B-C) *Social profits "H" = (E-F-G)*
Output transfers "I" = (A-E) *Tradable input transfers "J" = (B-F)*
Factor transfers "K" = (C-G) *Net transfers "L" = (I-J-K) or (D-H)*

revenue (A) less total costs (B+C). The prices used for calculating private profitability are based on observations, inclusive of any current policy interventions and/or market failures. The second row of the PAM measures social profitability based on the difference between social revenue (E) less social costs (F+G).

In contrast to private prices, social prices are "efficiency prices" or those that would be observed without any policy distortions or market failures, as illustrated in Figure 1. It follows that social revenues and costs reflect the underlying scarcity values and if used in actual decision-making would reflect the optimal allocation of resources from an economic efficiency standpoint. While relevant values for the calculation of private profitability are derived directly from actual observed prices at the farm and markets, determining social values is more complex.

Social prices for those commodities traded in the international market, both inputs and outputs, are derived using world market prices; CIF (cost, insurance, freight) price for imports and/or the FOB (freight on board) for exports. Both CIF and FOB prices provide the basis for calculating the import and export parity prices needed for determining farm level revenues and costs. For example, the "social" farmgate cost of imported fertilizer used within a farming system is its CIF price plus the marketing and distribution costs of moving the goods from the dock to the farmgate. In contrast, farmers exporting their goods would expect to realize the FOB price less the costs of processing, marketing, and transporting their goods to the

Figure 1. The effect of internalizing environmental services on prices and profitability of agroforestry practices.

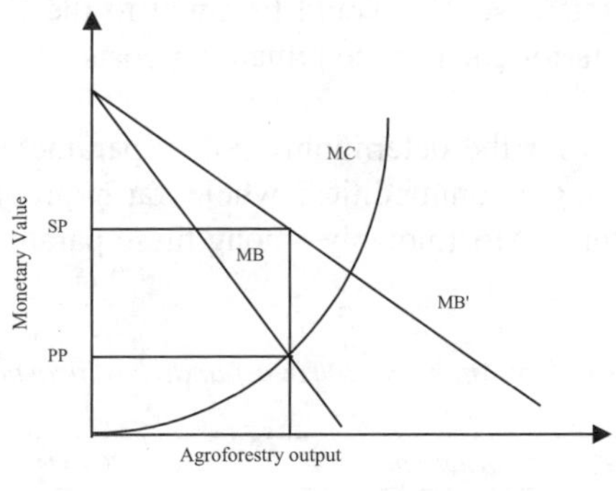

The horizontal axis measures the supply of agroforestry outputs, while the vertical axis measures the costs/benefits of agroforestry outputs. The curve, MC, shows the marginal cost of producing agroforestry output. If the environmental benefits of are not considered, the marginal benefit of agroforestry outputs is MB and the price is PP (private price). However, if environmental benefits (carbon sequestration for example) are considered the marginal benefit would be MB' and the price will be SP (social price). In other words, internalizing carbon sequestration benefits would increase the price of agroforestry output and thus the profitability. If there are external costs of agroforestry practices and if they are not internalized, the SP would be lower than PP.

dock in a distortion free setting. Calculating parity prices allows a policy analyst to identify the impacts of policy intervention(s) at all stages of production.

Social prices for domestic factors are based on social opportunity costs reflecting underlying domestic factor market conditions. For example, in the case of labor, assuming a functioning labor market freedom of movement among laborers, the social price would be equivalent to the going market wage rate.

The second accounting identity (third row of the PAM) allows for the calculation of divergences between private and social revenues, costs, and profits resulting from policy impact(s) and/or market failures. The third row of the PAM also allows for determination of the direction of transfers between producers, government budgets, and consumers. The signs of each of the variables in the third row facilitate determining whether producers are being "taxed" or "subsidized."

To quantify costs and benefits resulting specifically from externalities, these costs and benefits can be disaggregated from policy distortions and labor, land, and credit market failures by incorporating an additional private and social prices row (Table 2). For example, in the case of an environmental tax on outputs being assessed to mitigate the costs of soil erosion, the farmer would realize these costs, thereby internalizing them and, as such, the outcome would be reflected in the private prices$_2$ row within "A_2". On the other hand, where externalities are generated from, for example, carbon sequestration, yet the farmer is not internalizing benefits derived from this additional output of the farming system, the values of these externalities would be included in the social prices$_2$ row within "E_2". The net effects of both policy distortions and market failures are still found within the second accounting row, but in this case, "I" would be equal to the $\Sigma(A_1+A_2) - \Sigma(E_1+E_2)$. Likewise, the net transfers for each of the remaining costs and in turn profits would be calculated accordingly.

The PAM also allows for the determining policy parameters for measuring the impact of policies on single-commodities, whole farm production systems, and comparing multiple systems. Most notably among these parameters are the nominal

Table 2. The Policy Analysis Matrix expanded to incorporate externalities.

	Revenues	*Costs*		*Profits*
	Output	*Tradable inputs*	*Domestic factors*	
Private prices$_1$	A_1	B_1	C_1	D_1
Private prices$_2$	A_2	B_2	C_2	D_2
Social prices$_1$	E_1	F_1	G_1	H_1
Social prices$_2$	E_2	F_2	G_2	H_2
Effects of policy distortions and market failures	I	J	K	L

Table 3. Coefficients derived from the Policy Analysis Matrix used to measure the relative impact(s) of policies or market failures.

NPC =	A/E: a ratio that contrasts private and social output values. A NPC > 1 is indicative of private prices of output being greater than social prices reflecting that producers are positively protected.
EPC =	(A-B)/(E-F): the ratio of value added measured at private vs. social prices. Unlike the NPC and the NPI that measure the effect of divergences for output and tradable input prices respectively, the EPC measures the total effects of intervention in both markets. The implication of EPC > 1 is that there is an overall artificial incentive to produce a commodity due to the presence or absence of policies.
PCR =	C/(A-B): the ratio of domestic factor prices to value added at private prices. It identifies the cost of domestic resources in private prices necessary to produce a unit of value added. A PCR between 0 and 1 indicates that in private terms, domestic resources generate more than their value in value added.
DRC =	G/(E-F): the ratio of domestic factor prices to value added at social prices. It shows the cost of domestic resources in social prices needed to produce a unit of value added. If the DRC > 1 the commodity system is not desirable from an economic efficiency standpoint.
PC =	D/H: measures the incentive effect of all policies and provides a ratio to determine the relative net policy transfers. A PC = 1 indicates no net transfers

protection coefficient (NPC), effective protection coefficient (EPC), the private cost ratio (PCR), the domestic resource cost (DRC) and the profitability coefficient (PC). Each of these parameters is briefly described in Table 3.

3. DATA SPECIFICATIONS AND POLICY SCENARIOS

Data used for this analysis were collected from randomly selected households over a one-year period. Weight scales, clipboards, and record forms (prepared in the local language) were provided to participating household members to record the number of hours spent farming, along with either the crop weight or number harvested, per crop, depending on how it is sold within local markets (e.g. values for coconuts were recorded based on the number harvested whereas values for bananas were recorded by weight). Specific data were collected for the following crops: bananas, betel nut, breadfruit, coconuts, swamp taro, sweet taro, sakau, and yams. In each of the homegarden systems all of these crops are being cultivated albeit in different quantities as a function of the presence/absence of the wetlands. In contrast, in most areas where the UPS are located, sakau is the primary, if not only crop being cultivated. Nonetheless, data were collected for each of the above-mentioned crops in all three systems where present. All households were visited every week to two weeks to pick up the record forms, provide new ones, and ensure data were being recorded appropriately. Land area per household was based on either owner documents or using local landmarks as proxies to estimate land area used/owned by each household. Data for all three systems were derived from "mature" agroforests in which the perennial components had attained fruit-bearing age.

Averages for labor inputs and crop outputs per hectare were determined for each of the three respective systems providing the basis for cost and revenue estimates. Prices used to determine values for the crops harvested were based on those observed by farmers throughout the same one-year time period. The price of labor was based on Pohnpei's minimum wage rate[2]. All farm operations were manual; no form of mechanized farming was used. Only 5% of households indicated that they used any production inputs other than labor (the purchase of planting materials is almost nonexistent on Pohnpei where farmers obtain needed planting materials from their own farm plots). Farm profits therefore were derived using the total value of outputs less costs of labor.

There are no current trade or pricing policies contributing to divergences between private and social prices involving the HG and the HGW systems. Yet, these systems are providing a positive externality in the form of carbon sequestration. Farmers practicing either the HG or the HGW systems are not currently receiving payments for carbon being sequestered within these systems, and therefore, payments for carbon are not considered within the private profitability calculation. Instead, assuming a functioning carbon market in Pohnpei (correcting for a market failure), farmers would receive payment and thus, values for sequestration are to be included in the social profit calculation. More specifically, given carbon sequestration is produced within a given system, in the PAM it is accounted for in "E_2" or in the social "output" calculation. For complex tropical agroforests, average values for carbon sequestration rates have recently been estimated to range between 1.5 to 4 Mg ha^{-1} yr^{-1} (Palm, Woomes, Alegre, Arevaldo, Castilla, Cordeiro et al., 2000; Ruark, Schoeneberger, & Nair, 2003). For carbon sequestration value determination in the HG system, an average of these values (2.75 Mg ha^{-1} yr^{-1}) was used. Research conducted on another island in Federated States of Micronesia (Kosrae) has shown that wetland-based agroforests similar to those in Pohnpei have carbon sequestration rates approximating 2.3 Mg ha^{-1} yr^{-1} (Chimmer & Ewel, 2002), which is used for the HGW system in this analysis. Prices for carbon sequestration are generally between US$10 and US$30 per Mg (Niles, Brown, Pretty, Ball, & Fay, 2001).

In contrast to the two homegarden systems, carbon values were not included for the PAMs derived from the UPS system. The process of clearing much of the existing vegetation from the land, as well as removing sakau roots and concomitant soil disturbance during harvest, may actually lead to a net carbon loss in these systems. Moreover, cultivating sakau in the uplands has increased downstream erosion, negatively affecting the coral and fishery resources in local lagoons. Thus, carbon is not sequestered and negative externalities stemming from erosion generate costs not currently internalized by those farming in the uplands.

The "external" cost(s) of erosion resulting from the UPS are incurred by Pohnpeaens yet are not being paid for by the farmers responsible for them. As such, profits realized by the farmers actually reflect greater values than if they were paying the full cost of their cultivation activities (i.e., private profits are greater than social profits by an amount equivalent to the costs of the erosion impacts). While exact costs of erosion stemming from upland sakau production has not been determined, for our analysis, we have assumed a difference of 10% between the private and social output price of sakau for the UPS to reflect these costs. This assumption is based on the premise that if farmers in the uplands were to be assessed a 10% "environmental" tax (reflected in the "A_2" calculation) on the production of sakau the receipts from such a tax would be sufficient to pay for the "downstream" impacts associated with upland farming. Profits for these three systems are therefore not only affected by the composition and quantity of the different products being produced, but also the effects of the presence of two externalities, one positive (carbon sequestration) and one negative (erosion).

Based on numerous discussions with local farmers and extension agents on Pohnpei, evidence to date suggests that each of the two homegarden systems (HG and HMW) are producing a consistent quantity of outputs from year to year using the same level of labor inputs. As such, we assume that the annual data collected for this study represent inputs and outputs that remain constant from one year to another for both of the homegarden systems. On the other hand, Pohnpei farmers reported an unsustainable 10% annual decline in sakau cultivated in the UPS.

It is valuable to examine how changes in crop productivity as well as any proposed policies may influence farmer profits, and in turn, how these influences are likely to factor into farming decisions over time. Given the consequences of upland sakau production, policy makers on Pohnpei are interested in determining what strategies may keep farmers from further expanding into the uplands. From an economic policy standpoint, pricing policies in the form of taxes and/or subsidies are often used as incentives (or disincentives) to achieve a desired outcome. Here, in addition to examining how a 10% per annum decline in sakau yields in the UPS influences farmer profits, we also explore how two such policies, a subsidy on outputs from the two homegarden systems and an "environmental" tax on sakau from the UPS are likely to influence future farming activities on Pohnpei.

Within the context of the PAM methodology, it is necessary to calculate net present values (whereby future values are discounted into today's currency terms) for revenues, tradable and non-tradable inputs, and domestic factors to measure divergences that may exist over time and how they influence profitability outcomes. In theory, if elasticities of supply and demand were known for Pohnpei's economy, one could examine the dynamic effects associated with price and production changes over time. However, such data are nonexistent in Pohnpei. As a result, we assume constant output prices for all three systems, constant input and output coefficients for the HG and HGW systems, and annual sakau yields declining by 10% for the UPS. In the Federated States of Micronesia, the currency used is the US dollar; therefore all financial and economic values are in US dollar denominations.

4. RESULTS

Table 4 provides the profitability results for a single year's outcome under the current and potential policy scenarios Pohnpei. For all three systems, under existing policies (first column of results), both private and social profits are positive, indicating that farmers on Pohnpei have financial incentives to continue farming all three systems. Profits are greatest for the UPS (\$1,412 ha^{-1} yr^{-1}) followed by the HG (\$954 ha^{-1} yr^{-1}) and HGW (\$906 ha^{-1} yr^{-1}), explaining why farmers continue to move into the upland areas to farm. Comparing the second column of results, even if a carbon market was established and farmers in the uplands were taxed 10% for their erosion impacts, the social profit values generated from the UPS are still greater than the HG and HGW indicating that even with a 10% erosion tax, further expansion into the uplands is likely to continue.

Since policy makers are interested in minimizing impacts of sakau planting in the uplands and profits are currently far greater for the UPS than for either the HG or HGW, policy intervention in the form of economic incentive/disincentives may be necessary. The last three columns in Table 3 present how a range of subsidies on outputs from the two homegarden systems and erosion taxes on sakau in the upland system, are likely to influence both potential profits and farmer motivation for land use decisions.

Table 4. Estimates of private and social profitability calculated using the Policy Analysis Matrix (PAM) methodology, for a one-year period over the three dominant agroforestry land-use systems in Pohnpei, Federated States of Micronesia.

	Policy Scenarios			
Farming System	*Existing: no carbon market (US \$ ha^{-1} yr^{-1})*	*10.0% output price subsidy (US \$ ha^{-1} yr^{-1})*	*15.0% output price subsidy (US \$ ha^{-1} yr^{-1})*	*31.0% output price subsidy (US \$ ha^{-1} yr^{-1})*
Home Garden				
Private	\$954	\$1,100	\$1,173	\$1,407
Social	\$1,009	\$1,009	\$1,009	\$1,009
Divergence	-\$55	\$91	\$164	\$398
Home Garden Wetland				
Private	\$906	\$1,071	\$1,153	\$1,417
Social	\$953	\$953	\$953	\$953
Divergence	-\$47	\$118	\$200	\$464
Upland System				
Private	\$1,412	\$1,229	\$1,138	\$906
Social	\$1,229	\$1,229	\$1,229	\$1,229
Divergence	\$183	\$0.0	-\$91	-\$324

First, consider a subsidy of HG and HGW in the absence of sakau erosion taxes. The subsidy rate necessary to persuade farmers to abandon upland activities would need to be at least 31%. At a rate of 31%, profits from all three systems would be essentially the same (\$1,407 ha^{-1} yr^{-1}, \$1,417 ha^{-1} yr^{-1}, and \$1,412 ha^{-1} yr^{-1} for the HG, HGW, and UPS respectively). Under this policy, the government would also incur a budgetary cost of between \$398 ha^{-1} yr^{-1} for the HG farmers to \$464 ha^{-1} yr^{-1} for the HGW farmers. Given the existing limitations on the government budget on Pohnpei, combined with increasing demands to minimize government spending as a result of the new Compact agreement, this scenario is highly unlikely to be feasible without the provision of overseas grants.

A more practical alternative from a government budgetary perspective would be to assess a tax on sakau produced in the uplands. Under this scenario, the tax rate required to persuade farmers to abandon their upland activities is 27.5%, a value equivalent to a profit reduction of \$507 ha^{-1} yr^{-1} for upland farmers (Table 4). Any tax rates less than 27.5% would result in greater profits for the UPS relative to the HG and HGW while a rate higher than 27.5% would, in theory, provide enough of an incentive to persuade farmers to adopt either of the two homegarden systems. If farmers still wanted to maintain activities at or near the margin in the uplands, the government could generate revenues from this policy that are above and beyond what is likely to be required to mitigate erosion impacts from these activities.

The results presented above are based on profits for a 1-year time period. We now turn to how these profits are influenced for two longer time periods, (1) a 5-year time horizon and (2) a 10-year time horizon, testing the sustainability of these three production systems. We noted earlier that anecdotal evidence suggests that while both of the homegarden systems produce a sustainable flow of outputs, sakau yields in the uplands decline by about 10 % annually.

For the HG and HGW, a sustainable stream of private profits of \$954 and \$906 for 5 years respectively would produce a net present value of \$3,616 ha^{-1} and \$3,436 ha^{-1} if discounted at 10% (Table 5). The net present value of profits for the UPS is \$4,213 ha^{-1} using the same discount rate and time period but with production declining 10% annually[3]. Even taking into account the declining productivity of upland sakau over a five-year period, under the existing setting on Pohnpei, farmers in the uplands would still generate greater profits relative to the other systems and therefore will continue to farm in these areas.

Although during the one-year period, establishing a carbon market and accounting for erosion costs does not change the order of profits for each of the three systems, these policy interventions combined with the declining sakau production in the UPS results in the HG generating the greatest profits with the HGW and UPS profits differing by only \$20 ha^{-1} (Table 5). Thus, in trying to achieve the government's objective of persuading upland farmers to move down to the lowlands, correcting for the existing market failures considered here would result in profits for all three systems being almost equal if farmers operate under a 5 year time horizon and a 10% discount rate.

Table 5. Estimates of five-year net present values based on a 10% discount rate for private and social profitability calculated using the Policy Analysis Matrix (PAM) methodology, for the three dominant agroforestry land-use systems in Pohnpei, FSM.

	Policy Scenarios			
Farming System	*Existing: no carbon market (US \$ ha^{-1})*	*10.0% output price subsidy (US \$ ha^{-1})*	*12.5% output price subsidy (US \$ ha^{-1})*	*15.0% output price subsidy (US \$ ha^{-1})*
Home Garden				
Private	\$3,616	\$4,170	\$4,309	\$4,448
Social	\$3,824	\$3,824	\$3,824	\$3,824
Divergence	-\$208	\$346	\$485	\$623
Home Garden Wetland				
Private	\$3,436	\$4,060	\$4,216	\$4,372
Social	\$3,614	\$3,614	\$3,614	\$3,614
Divergence	-\$178	\$446	\$602	\$759
Upland System				
Private	\$4,213	\$3,634	\$3,490	\$3,345
Social	\$3,634	\$3,634	\$3,634	\$3,634
Divergence	\$578	\$0.0	-\$145	-\$289

If, alternatively, the government preferred a pricing policy in the form of a subsidy on homegarden outputs, the sustained output of crops in the two homegarden systems combined with a decline of sakau production would result in a subsidy rate far below that of the one-year period. Table 4. shows that a price subsidy on outputs from the HG and HGW systems of 10% would not provide a sufficient financial incentive for upland farmers to halt their activities. However a subsidy rate of 12.5% or more would result in greater profits for the two homegarden systems relative to the upland system and in doing so, provide a financial justification for upland farmers to adopt either of the two homegarden systems.

Under the subsidy policy discussed above, and in order to attract upland farmers to adopt either of the two homegarden systems, the government would incur costs in the amount of between \$485 ha^{-1} and \$602 ha^{-1} over five years. While for a one-year period, the subsidy approach appears unlikely, payments based on a 12.5% subsidy over a five-year period may be more feasible. Indeed, on average, annual payments per hectare would range between \$97 for the HG and \$120 for the HGW.

Considering a tax on sakau produced in the UPS, a rate of 10% would equate to private and social profits being equal at \$3,634 ha^{-1} (Table 4). Moreover, with a tax of only10%, the UPS still generates greater profits than either the HG or HGW (in the absence of lowland subsidies). The impact of a 12.5% tax on sakau would result in the HG being more profitable than the UPS by a margin of \$126 ha^{-1}. At the same time, the UPS would still generate greater profits relative to the HGW by a margin of \$54 ha^{-1}. Thus in order to generate profits for both the HG and the HGW greater than the UPS, and thereby inducing upland farmers to adopt either the HG or HGW a tax of 15% would be necessary equating to a reduction in UPS profits by \$289 ha^{-1}.

As more time is taken into consideration, the ecological sustainability of these systems becomes much more pronounced. Table 6 shows that as a result of declining sakau yields in the UPS under the current policy setting, the net present values for both the HG and the HGW systems becomes greater than for the UPS (\$5,861 ha^{-1}, \$5,569 ha^{-1}, and \$5,358 ha^{-1} respectively)[4]. For those farmers and policy makers interested in generating greater profits over the long-term, the fact that sakau yields decline at such a rate in the UPS would, in theory, provide enough of an impetus to convince more farmers to forgo planting in the upland areas. However, given the limited opportunities to generate income, both in the short and long-term time period, farmers may still be inclined to continue farming the uplands as a means to generate greater income in the immediate time frame. In order to further persuade farmers to leave the upland areas, it may be necessary for the government to provide additional incentives.

Thus far we have examined profitability of three agroforestry systems based on actual dollar values. When comparing systems having identical commodities being produced, both the private and social profits are suitable indicators of relative competitiveness. When considering systems producing different outputs, or similar commodities using differing technologies or in different ecological settings, it is

Table 6. Estimates of ten-year net present values based on a 10% discount rate for private and social profitability calculated using the Policy Analysis Matrix methodology, for the three dominant agroforestry land-use systems in Pohnpei, FSM.

	Policy Scenario	
Farming System	*Existing: no carbon market (US \$ ha^{-1})*	*Existing: with erosion costs not internalized (US \$ ha^{-1})*
Home Garden		
Private	\$5,861	
Social	\$6,199	
Divergence	-\$338	
Home Garden Wetland		
Private	\$5,569	
Social	\$5,852	
Divergence	-\$283	
Upland System		
Private		\$5,358
Social		\$4,567
Divergence		\$790

Table 7. Ratio indicators derived from Policy Analyses Matrices for three dominant agroforestry land-use systems for a ten-year period discounted at 10%, in Pohnpei, FSM.

Ratio Indicators	*Homegarden*	*Homegarden Wetland*	*Upland Sakau*
Profitability coefficient (PC)	0.96	0.97	1.11
Private cost ratio (PCR)	0.35	0.45	.35
Domestic resource cost ratio (DRC)	0.34	0.44	.38

useful to examine the ratios (defined in Table 3) to ascertain the *relative* competitiveness of systems and the extent of policy divergences. Three such coefficients, the profitability coefficient (PC), the private cost ratio (PCR), and the domestic resource cost ratio (DCR), derived from the PAMs under the current policy setting (ten-year time period) are shown in Table 7. For the profitability coefficient, an indicator of the net transfer effects, it is evident that the UPS system, having a value > 1 indicates that farmers are receiving an implicit subsidy whereas farmers engaged in the two homegarden systems are implicitly being taxed due to the non-existent carbon market. Moreover, the relative value of the impacts is greatest for the upland system, inferring that correcting for the lack of a carbon market would have less of an impact on the two homegarden systems than correcting for the external costs associated with erosion. There is less of a relative difference when considering the private cost ratio and domestic resource cost ratio for these three systems.

Indeed, even though the UPS system has the lowest PCR and DRC ratios implying that for each unit of domestic resources used to produce sakau in the uplands greater profits are realized, both of these ratios infer that domestic resources are being used to generate positive profits both in private and social terms[5].

5. DISCUSSION

The Policy Analysis Matrix methodology has been applied here to determine the relative profits and use of domestic resources associated with three agroforestry systems on Pohnpei. In Pohnpei, agroforestry systems have traditionally played an important role, primarily as the main source of food security, and to a lesser degree for commercial purposes. As the local economy continues to become more integrated into the global economy, so have preferences for western goods and subsequently the desire to increase household incomes for their purchase. To develop Pohnpei's economy, one such strategy that historically has been promoted by government officials involves taking advantage of Compact funds from the United States to inflate the public sector work force, thereby providing employment to local citizens. The results have led to a declining interest in agriculture on the part of the private sector, particularly for commercial purposes. The United States mandated a smaller public sector as a condition of the newly negotiated Compact, which directly impacts the government's role as the largest employer on the island.

With already very limited private sector opportunities, local families are turning to the upland areas of Pohnpei establishing unsustainable agricultural systems to increase income levels and/or replace prior income sources due to the loss of public sector jobs. To date these actions are generating income for local families. Yet, they are also negatively impacting the local environment, and in turn other natural resources heavily depended upon by much of Pohnpei's population.

In such situations where limited economic opportunities exist, private individuals often place a greater value on achieving short-term needs, even if it entails forgoing personal potential future benefits, or those that may be accrued by other members of society. In contrast, policy makers are often more concerned with potential benefits that may be accrued to all of society. In this sense, a policy maker is often more inclined to promote agricultural systems that use available resources in economically efficient manner.

The results presented here illustrate a dilemma often faced by policy makers, and now facing those on Pohnpei. On the one hand, a stated goal of the government is to promote increasing income by developing the agricultural sector. Based on our results, the upland system is most apt to do so in the short-term. On the other hand, the government is interested in economically efficient and ecologically sustainable use of limited local resources. If this last objective is to be realized, it is critical to consider the profitability over a longer time, particularly if a known unsustainable system is currently expanding.

When considering the results of a one-year and a five-year period under the current situation on Pohnpei, profits remain greater for the UPS than either of the two homegarden systems. Assuming the government intends to promote agricultural development, farmers practicing the UPS interested in maximizing short-term profits will likely require policy intervention to adopt either of the two homegarden systems. In this chapter we have presented how different policy strategies may contribute to achieving this objective, namely (1) correcting for the current market failures, (2) establishing a subsidy for outputs derived from both homegarden systems, and (3) levying a tax on sakau produced in the upland system. On the other hand, for those farmers interested in maintaining a steady income over a longer time period, maintaining, or adopting either of the homegarden systems is likely to insure such an outcome. In either case, policy makers can now be more informed about why farmers are engaged in their respective systems and what might be done to influence the future of Pohnpei's agricultural sector.

6. COMMENTS ON THE POLICY ANALYSIS MATRIX METHODOLOGY

The Policy Analysis Matrix provides a means by which the impacts of policy and market failures can be quantified. In doing so, it allows not only for the quantification of individual policies, but also the quantification of the aggregate impacts of all policies affecting farmers currently engaged in different agricultural practices. Having the ability to analyze individual as well as the total impacts of policies can provide an important function in settings where different public institutions have established policies in an uncoordinated fashion, all of which have a common objective, but that ultimately work against one another. In this regard, the PAM can aid in untangling policies unsuccessful in their symbiotic goal.

Another useful aspect of the PAM is its utility in analyzing the potential of agricultural technologies under consideration. Once developed, sensitivity analysis can easily be carried out allowing for the impacts of new or modified policies to be explored. In addition, the PAM, while somewhat complicated "behind the scenes" provides a fairly simple means to examine dynamics taking place at all levels of the production chain. This can be particularly important to policy planners when attempting to identify where and how policy intervention is most likely to result in the desired outcome.

The case study presented here provided a unique opportunity to apply the PAM methodology within the context of agroforests over a multi-year time frame. This chapter also illustrated how the traditional PAM model could be expanded upon to include benefits derived from an ecological service even when a market for them is absent (i.e., correcting for the absence of a carbon market). In our case, carbon sequestration represents a benefit realized by society at large but results from an individual's actions. Other externalities, be they positive or negative, could also be incorporated into the PAM framework. The difficulty does not necessarily lie in determining how or where to include the cost or benefit of an externality within the PAM framework, but instead ascertaining relevant and accurate values to be used.

One limitation of the PAM methodology stems from the frequent use of fixed input-output coefficients making it more difficult to determine the dynamic effects attributable to policy shifts. As noted above, one means to address this dilemma is to determine price elasticities of supply and demand, as well as cross price elasticities of demand allowing for estimates of how farmers may respond to various policy interventions. Although this may be the case, in settings such as Pohnpei where the cash-economy is still developing, commodity choices are limited and motives of farmers are influenced by traditional norms, reliable estimates of elasticities are often difficult to obtain and thereby may limit the utility of PAMs under dynamic scenarios. This is not to say that certain assumptions pertaining to various elasticities could not be made and the robustness of the results tested through sensitivity analysis. Yet the PAM methodology was not developed as an ultimate means to derive perfect results under all circumstances. Instead, it is a tool to empirically determine how public intervention(s), or lack thereof, may influence the decisions of those participating in the agricultural sector and in turn, how those decisions may affect government resources.

7. NOTES

[1] While it is acknowledged that other external costs and benefits beyond carbon sequestration may result from practicing the three systems in question (for example biodiversity maintenance or loss, aesthetics etc), in our case we were limited in our analysis to only include carbon and assumed erosion values.

[2] Pohnpei's minimum wage in 2001 was US$1.35 hr^{-1}. Given the significant unemployment rate, the actual opportunity cost of labor is likely to be closer to US$0.00. Regardless, there is essentially no hiring

of labor for farming purposes on Pohnpei; instead, the immediate household members, friends, or extended family provide the labor.

[3] For a comparison, the private real present value of \$954 ha^{-1} yr^{-1}, \$906 ha^{-1} yr^{-1}, and \$1,412 ha^{-1} yr^{-1} incomes for the HG, HGW and UPS systems respectively discounted at 15% would be \$3,197 ha^{-1}, \$3,038 ha^{-1}, and \$3,771 ha^{-1}: discounted at 5.0% private real present values would be \$4,130 ha^{-1}, \$3,924 ha^{-1}, and \$4,748 ha^{-1} for the HG, HGW, and UPS systems respectively.

[4] Once again for comparison, the private real present value of \$954 ha^{-1} yr^{-1}, \$906 ha^{-1} yr^{-1}, and \$1,412 ha^{-1} yr^{-1} incomes for the HG, HGW and UPS systems respectively discounted at 15% would be \$4,787 ha^{-1}, \$4,549 ha^{-1}, and \$4,596 ha^{-1}: discounted at 5.0% private real present values would be \$7,365 ha^{-1}, \$6,998 ha^{-1}, and \$6,370 ha^{-1} for the HG, HGW, and UPS systems respectively.

[5] A PCR or DRC value of 1 indicates that a marginal increase in domestic resource use generate the exact same amount in value-added. Similarly, a PCR/DRC < 1 reflects a relative efficiency value for the use of domestic resources. A smaller value indicates greater value-added per unit of domestic resources.

8. REFERENCES

Adesina, A. A., & Coulibaly, O. N. (1998). Policy and competitiveness of agroforestry-based technologies for maize production in Cameroon: An application of policy analysis matrix. *Agricultural Economics 19*, 1-13.

Asian Development Bank (ADB). (1996). Pohnpei's watershed spatial plan and management guidelines. (T. A. No. FSM-1925). Watershed Management and Environment.

Barrau, J. (1961). Subsistence agriculture in Polynesia and Micronesia. *Bishop Museum Bulletin 223*.

Chimmer, R. A., & Ewel, K. C. (2002). Mechanisms of carbon accumulation in Micronesian tropical forested peatlands. Oral presentation, Society of Wetland Scientists 23rd Annual Meeting, June 3-7, Lake Placid, New York.

Clarke, W. C., & Thaman, R. R. (1993). Agroforestry in the Pacific Islands; systems for sustainability. New York: United Nations University Press.

Drew, W. M., Ewel, K. C., Naylor, R., & Sigrah, A. *The importance of a freshwater tropical wetland: assessing the direct-use values and understanding of ecological goods and services*. Manuscript submitted for publication.

Food and Agriculture Organization (FAO). (1994). The role of agriculture, forestry and fisheries in the sustainable development of small island developing states. Rome: FAO.

Falanruw, M. V. C. (1993). Micronesian agroforestry: Evidence from the past, implications for the future. In *Proceedings of the workshop on research methodologies and applications for Pacific Islands agroforestry; July 16-20, 1990: Kolonia, Pohnpei, Federated States of Micronesia*. (Gen. Tech. Rep. PSA-GTR-140). Pacific Southwest Research Station: Forest Service, USDA.

Fernandes, E. C. M., & Nair, P. K. (1986). An evaluation of the structure and function of tropical homegardens. *Agroforestry Systems 21*, 279-301.

Government of Federated States of Micronesia. (1999). *Proceedings of the 2nd FSM Economic Summit, Federated States of Micronesia*. September 13-17. Palikir, Pohnpei, Federated States of Micronesia.

Government of Federated States of Micronesia. 1998. *Gross Domestic Product Federated States of Micronesia 1996: Conceptual framework, methodology and results*. Palikir, Pohnpei, FSM.

Haun, A. (1984). *Prehistoric subsistence, population, and socio-political evolution on Ponape, Micronesia*. Unpublished doctoral dissertation. Eugene, Oregon: University of Oregon.

Kostka, M., & Raynor, W. C. (2000). *Pohnpei Forest Monitoring Report*. Pohnpei, FSM: The Nature Conservancy.

Laird, W.E. (1987). Soil survey of Island of Ponape, FSM. Pohnpei , FSM: USDA-SCS.

Manner, H. I. (1993). A review of traditional agroforestry in Micronesia. In *Proceedings of the workshop on research methodologies and applications for Pacific Islands agroforestry; July 16-20, 1990: Kolonia, Pohnpei, Federated States of Micronesia*. (Gen. Tech. Rep. PSA-GTR-140). Pacific Southwest Research Station, Berkeley, California: Forest Service, U.S. Department of Agriculture.

Monke, E. A., & Pearson, S. R. (1989). The policy analysis matrix for agricultural development. Ithaca, New York: Cornell University Press.

Nair, P. K. R. (1993). *An introduction to agroforestry*. London: Kluwer Academic Publishers.

National Oceanic and Atmospheric Administration (NOAA). (1987). *Local climatological data: annual summary with comparative data: Pohnpei, Eastern Caroline Islands, Pacific*. Asheville, North Carolina: NOAA, National Climatic Data Center.

Nelson, G. C., & Panggabean, M. (1991). The costs of Indonesia sugar policy: A policy analysis matrix approach. *American Journal of Agricultural Economics, Aug*, 703-712.

Niles, J. O., Brown, S., Pretty, J., Ball, A., & Fay, J. (2001). *Potential carbon mitigation and income in developing countries from changes in use and management of agricultural and forest lands*. (Center for Environment and Society Occasional Paper 2001-04). Colchester, UK: University of Essex.

Office of the Governor. (2003). Preliminary Census Report, 2000 FSM Census of Population and Housing. National Government, Federated States of Micronesia, Palikir, Pohnpei, FSM.

Office of the Governor. (1996). National Census Report, 1994 FSM Census of Population and Housing. *Palikir, Pohnpei, FSM: National Government, Federated States of Micronesia.*

Osman, W. M. (1995). *Federated States of Micronesia Economic Report*. Honolulu: Bank of Hawaii.

Palm, C. A., Woomes, P. L., Alegre, J., Arevaldo, L., Castilla, C., Cordeiro, D. G. et al. (2000). *Carbon sequestering and trace gas emissions in slash and burn alternative land uses in the humid tropics*. (ASB Climate Change Working Group Report. Final Report, Phase II). Nairobi, Kenya.

Pearson, S. R., Falcon, W. P., Heytens, P., Monke, E., & Naylor, R. (1991). *Rice policy in Indonesia*. Ithaca, New York: Cornell University Press.

Pohnpei Office of Agriculture and Forestry (1996). *Five-Year Strategic Action Plan (1997-2001)*. Pohnpei, FSM: University of Oregon, Micronesia and South Pacific Program, and the Pohnpei Office of Agriculture and Forestry.

Ruark, G. A., Schoeneberger, M. M., & Nair, P. K. R. (2003). *Roles for agroforestry in helping to achieve sustainable forest management*. Oral presentation, Intercessional Experts Meeting, Role of Planted Forests in Sustainable Forest Management, 24-30, March. Wellington: New Zealand.

Raynor, W. C. (1989). Structure, production, and seasonality in an indigenous Pacific island agroforestry system: A case example on Pohnpei island, FSM. *Unpublished master's thesis, Department of Agronomy and Soil Science, University of Hawaii, Honolulu, HI.*

Raynor, W. C. (1991). Agroforestry systems in Pohnpei: Practices and strategies for development. (RAS/86/036, Field Document 4). United Nations Development Programme: South Pacific Forestry Development Program.

Raynor, W. C., & Fownes, J. H. (1991a). Indigenous agroforestry of Pohnpei: 1. Plant species and cultivars. *Agroforestry Systems 16*, 139-157.

Raynor, W. C., & Fownes, J. H. (1991b). Indigenous agroforestry of Pohnpei: 2. Spatial and successional vegetation patterns. *Agroforestry Systems 16*, 159-165.

Staal, S. J. (1996). Periurban dairying and public policy in Ethiopia and Kenya: A comparative economic and institutional analysis. Unpublished doctoral dissertation, University of Florida, Gainesville, Florida.

Thaman, R. (1975). The Tongan agricultural system: With special emphasis on plant assemblages. Unpublished doctoral dissertation, University of California, Los Angeles, California.

United Nations Development Programme (UNDP). (1994). Report of the global conference on the sustainable development of small island developing states. April 25-May 6, Bridgetown, Barbados.

Yao, S. (1997). Rice production in Thailand seen through a policy analysis matrix. *Food Policy 22*, 6, 547-560.

9. AUTHOR'S NOTE

This research was made possible by support from Katherine Ewel, funding from the Institute for Pacific Islands Forestry, and the cooperation of the people of Pohnpei.

GEOFF BRIGHT

EXPLORING THE ECONOMICS OF AGROFORESTRY SYSTEMS USING A PRODUCTION FUNCTION APPROACH

1. INTRODUCTION

Economists have shown how economic theory may be applied to the study of agroforestry systems (Gregory, 1987; Mercer & Hyde, 1992; Nautiyal, 1998), and economic analysis tends to be incorporated in agroforestry models (World Agroforestry Program [ICRAF], 2004; Thomas, 1991; Institute of Water and Environment [IWE], 2003), yet further insights might be gained by drawing together the various economic concepts as applied to agroforestry, and, along with a few extensions, applying the complete framework to an agroforestry example. This chapter, therefore, seeks to combine the relevant elements of economic theory to construct a framework of the economics of an agroforestry system and then use this to construct a simulation model of a wheat-maize-unpruned leucaena (*Leucaena leucocephala)* system in the Himalayan foothills of India (Narain, Singh, Sindhwal, & Joshie, 1998). From this a number of research questions can be addressed, namely: do the results support the conclusions reached by the researchers conducting the original experiment, does this analysis provide any additional insights for modelers or policy makers which the original experiment was unable to provide, and would there be additional data implications if this approach were to be adopted from the outset?

The first part of the chapter builds up the theoretical picture by taking the basic economic model for analyzing the product-product relationship, the Production Possibility Frontier (PPF), and then extending this to take account of the special characteristics of agroforestry systems. The second part applies the economic framework to the case study mentioned above, and the final section reviews the results and considers their implications.

J. R. R. Alavalapati & D. E. Mercer, Valuing Agroforestry Systems, 79–93.

2. ECONOMIC THEORY

2.1. The Production Possibility Frontier

Determination of the optimum combination of forestry and agriculture enterprises has typically been modeled by economists using the Production Possibility Frontier (PPF) (Thomas, 1991). The PPF takes its normal convex shape from the underlying production functions (response curves) of the enterprises competing for land and other resources (Upton, 1987). This underlying shape then only changes in an agroforestry context as a result of the introduction of complementarities and intensified competition between those enterprises in intimate mixes. Economists have found difficulty in dealing with some of the complexities of agroforestry systems within the PPF approach both from a theoretical and modeling perspective. Theoretical suggestions have included trying to use a multi-time period series of PPFs (Etherington & Matthews, 1983), and modelers have tended to use a straight line PPF which only shows curvilinearity with respect to the introduction of light and water competition and nutrient and, possibly, shade complementarity (Thomas & Bright, 1991; Dyack et al., 1999; ICRAF, 2004).

The output from a tree or agriculture enterprise arises as a consequence of the combining of inputs. Output rises as additional units of input are added, but if one or more of the inputs are fixed in supply (i.e., the short run), then additions of variable inputs will eventually give rise to smaller and smaller additions to output: a phenomenon known as diminishing returns which give rise to the familiar declining sloped response curve, as shown for the wheat enterprise in Figure 1 below.

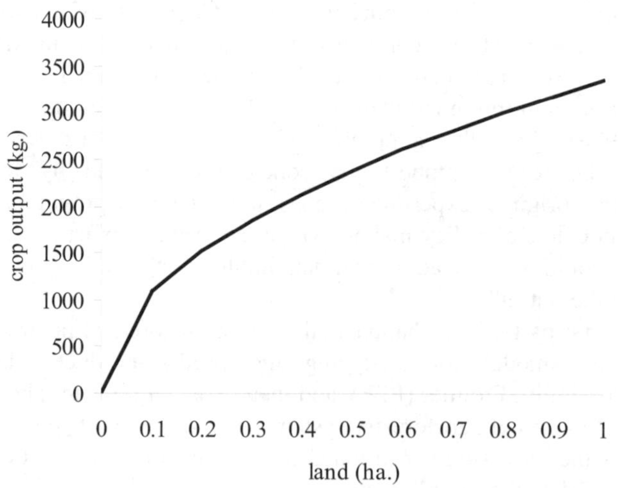

Figure 1. Wheat production function displaying diminishing returns.

A commonly used equation form to express this relationship in economics is known as the Cobb-Douglas production function, exemplified in equation 1

$$O = aL^b F^c B^d \tag{1}$$

where O = output, L, F and B are inputs such as land, fertilizer and labor, and for the input coefficients the condition $0<b,c,d<1$ applies, this constraint implying diminishing returns.

Now, if two enterprises, such as leucaena and wheat in the example, compete for a fixed amount of one of the inputs, such as land, then as extra amounts are allocated to leucaena, then the same amount of that input will be removed from the wheat-maize crop, and the resulting output will be determined by moving up and down the respective production functions. In other words, the PPF is determined from the two production functions for leucaena and wheat along with the link equation:

$$L_f = A - L_a \tag{2}$$

where A is the total area of land available[1].

If the production functions were straight lines for both enterprises, then the resulting PPF would also be a straight line, but since the production functions exhibit diminishing returns, the PPF will also be curvilinear, concave to the origin[2]. And so, as the limited resource is moved from one enterprise into the other, larger and larger amounts of one output will have to be given up for smaller and smaller increases in the other: economists call this the diminishing marginal rate of product transformation (Ritson, 1997; Penson, Capps, & Rosson, 1995).

Once the PPF has been constructed from the two production functions, it is then possible to determine the optimum combination of enterprises by introducing the relative prices of the products[3], assuming that costs are either fixed for the enterprise or for the plot as a whole. The isorevenue line, having a negative slope given by the ratio of the prices of the two products (p_x/p_y) is moved outwards away from the origin until it is tangential to the PPF. This point (A in Figure 2) denotes the combination of enterprise outputs giving the greatest revenue.

Because of the curvature of the PPF, the optimum point may denote a combination of enterprises or it may suggest that all resources be devoted to only one enterprise (a 'corner' solution). But note that, so far, there is no suggestion of an intimate mix of the two enterprises (polyculture); rather they are assumed to occupy separate sections of the area of land in monocultures. Only when there are complementarities might it be worth adopting polyculture: how this can be dealt with in the enterprise production functions will be shown later.

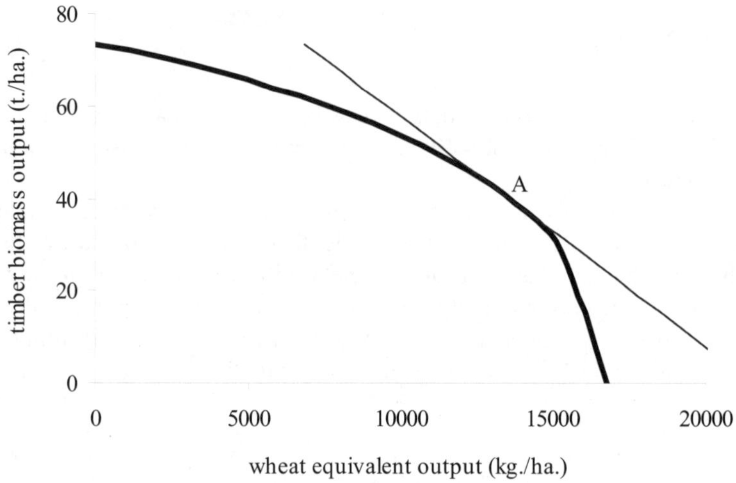

Figure 2. Production possibility frontier and isorevenue line for wheat and leucaena.

2.2. PPF and time

Unlike most agriculture systems with annual crops, in agroforestry the time dimension plays a particularly important role because the tree enterprise has a lifetime in excess of one year. This then means that there is not one, but a series of production periods, with the harvest of the tree crop occurring only once at the end of the period, or intermittently later on in the rotation, while the agriculture crop will be harvested within the first year, and for subsequent years thereafter, albeit with declining productivity as it faces increasing competition from the trees. Figure 3 shows an example of the pattern of revenues arising from a wheat-maize – unpruned leucaena system over a number of years, with crop revenues occurring, but at a declining level, in years 1 to 5 and tree harvest occurring in year 5.

Mercer & Hyde (1992), following Etherington & Matthew (1983), attempted to deal with the time dimension within the PPF approach by positing a series of PPF's over time, giving rise to a three dimensional PPF surface, rather than a curve. However, it is possible, instead, to present the information in the form of just one PPF at the start of the intervention. To do this one first needs to calculate the levels of production of the two enterprises for the points in time when they arise. Each

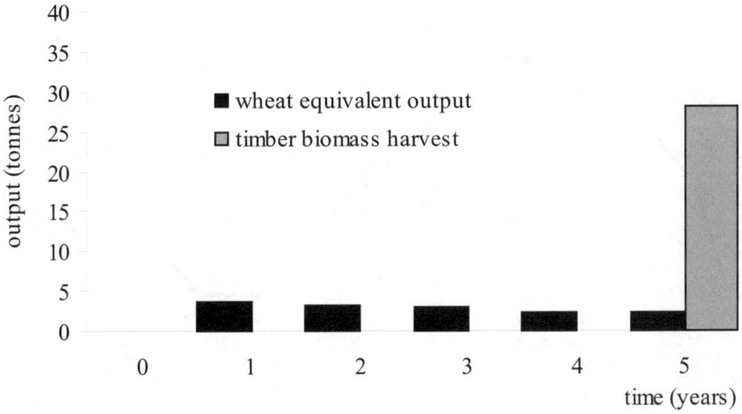

Figure 3. Pattern of revenues for tree and crop outputs over time.

period's output is then discounted to the present and summed to give the net discounted output (NDO) for each enterprise, using the formula

$$NDO = \sum_{t=0}^{t=T} \frac{O_t}{(1+r)^t} \tag{3}$$

where O_t is the output at time t, r is the real discount rate (real opportunity cost of capital) and T is the lifetime (optimum rotation length) of the tree enterprise.

This procedure is carried out for both enterprises and for each combination of land shares. The resulting figures can then be inserted into the normal PPF, but with net discounted outputs rather than annual outputs on the axes.

2.3. Combining the timber yield model with its production function

The production function may give a level of output for a particular input, but for tree crops we need to specify, not only the output at some optimal end point, but also the time-based growth function so that shade and other effects can be calculated for each time period. The two relationships may be linked together by first determining the tree production function and then deriving an associated set of growth functions for various points on the production function.

This can be achieved if one growth function has been derived for a given level of inputs, giving output O_{fT}, and assuming the shape of the growth function will be the same for different outputs at a particular point in time. Then, for different outputs from the production function, the growth curve is determined by adjusting the coefficients in the latter by the ratio of the desired output level and O_{fT}. Thus,

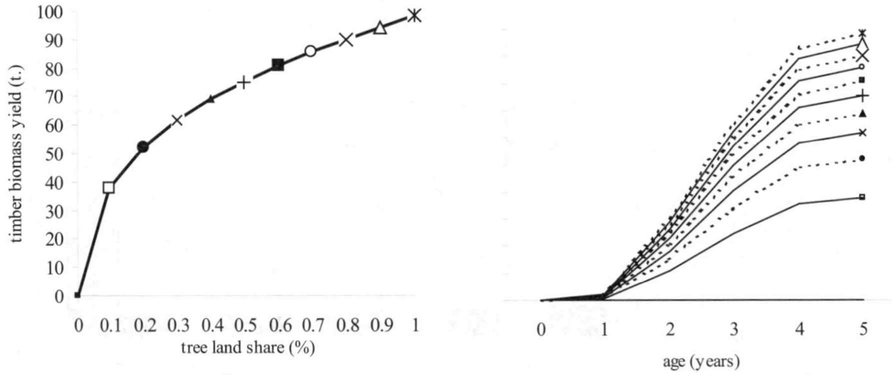

Figure 4. Tree production function points and associated growth curves for unpruned leucaena in India

Figure 4 shows how different output levels for the production function have counterpart growth curves in the example of unpruned leucaena in India: each of the points shown on the production function on the left correspond with the year 5 yield growth curve figures on the right.

2.4. Encompassing complementarities and other competition in the production function

Normally, production functions only show the effect on output of economic inputs, such as land, labor, seed, fertilizer and so on. In an agroforestry context, however, other resources necessary for growth (light, water, shade) need to be introduced into the equation as inputs (Rao, Nair, & Ong, 1998). On the other hand, some of these resources are also outputs of particular enterprises (e.g., nitrogen or shelter) and should, therefore, be shown as products of those enterprises. Both of these aspects can be represented by adjustments to the production function. Use of resources, such as light and water, can be included by adding another variable to the right hand side of the equation for each resource, and production of resources can be shown in a further equation for that particular resource, linked to the output of that enterprise (Stone, Kyle, & Conrad, 1993). If, for instance, nitrogen is produced by a tree enterprise, there might be an extra nitrogen 'production' function, thus:

$$N_{t+1} = jO_{f_t} \tag{4}$$

The amount of nitrogen made available to the crop is then added[4] to that applied directly to give the total nitrogen input used in the crop production function.

Nitrogen fixing is an example of complementarity between enterprises. On the other hand, for resources in limited supply, enterprises are more often in competition. Land is one example already discussed. Another is light: trees grow and spread their canopy which then shades agriculture crops. A shade function can be posited, thus:

$$H_{a_t} = F(O_{f_t}) \quad (5)$$

where H denotes shaded area and $O_{ft,}$ the cumulative tree output at time t. We can then insert an extra input in the agriculture production function, namely light, denoted by G to take account of the effect of the reduction in light available to the crop as tree shade increases. Now, if unrestricted access to light for the whole cropped area is given a value 1, then this value will be reduced as H rises:

$$G_{a_t} = 1 - \frac{H_{a_t}}{A} \quad (6)$$

and this then feeds into the crop production function:

$$O_{a_t} = aL^b_{a_t} F^c_{a_t} B^d_{a_t} G^f_{a_t} ...... \quad (7)$$

Figure 5a below illustrates the effect of complementarity (nitrogen fixing) and Figure 5b shows the effect of extra competition (tree shade) on the agriculture crops for different combinations of trees and crops. Moving from a high to a low crop:tree land share ratio the tree density increases and the trees fix increasing amounts of nitrogen which has the effect, *ceteris paribus*, of increasing crop yields with increasing age (Figure 5a).

However, the growing trees also reduce the light available to the agriculture crop, eventually leading to the stabilization of crop yields at a low level or even the complete shading–out of the crop, as shown in Figure 5b.

In addition, recognition of shade effects of trees on crops also necessitates consideration of self-shade effects of trees on trees. This can be taken into account in a similar way.

$$H_{f_t} = F(O_{f_t}) \quad (8)$$

As a simplified approach, the self-shade effect on tree growth can be related to the proportion of the crown overlapping with trees in the parallel strip. So the light variable is calculated as:

$$G_{f_t} = 1 - \frac{H_{f_{t-1}}}{C_{t-1}} \quad (9)$$

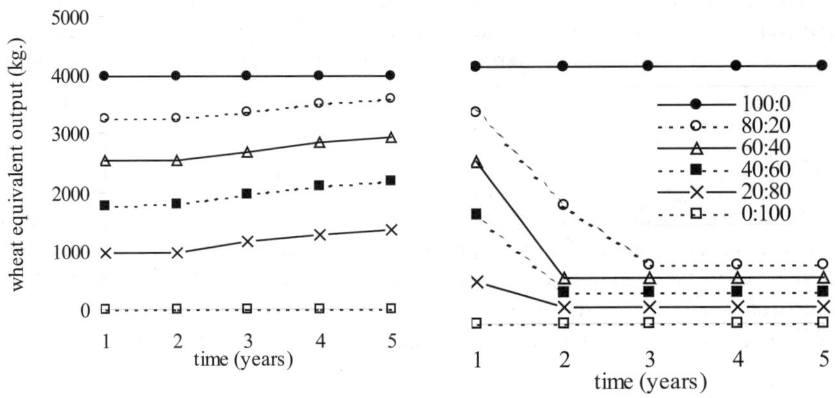

Figure 5. (a). Tree nitrogen effect on crop yields (b). Tree shade effects on crop yields.

Note that the shade effect is here lagged by one year; this is simply to avoid simultaneous determination of the interrelated shade and output (where C is the crown diameter).

G, the light 'input', is then inserted into the tree production function. However, because of tree growth occurring over a number of years and so being affected by differing amounts of shade in different time periods, the production function and its associated growth curve need to take account of this. This can be achieved by calculating the increment of growth for that particular year, based on the inputs (including the light variable, G) pertaining to the production function at the beginning of that year. Final output is then determined by summing these annual increments.

The effect of including these extra complementarity and competition effects on the PPF can be seen in Figure 6 below. The dotted gray curve is the PPF[5] excluding the effects of complementarities and extra competition (besides that for the normal economic inputs, land, labor, etc.) whereas the darker, continuous curve includes both of these effects. In this case, the extra competition (for light and water, for instance) serves to pull the PPF closer to the axes, while the complementarity effect of the trees works in the opposite direction, and so pulls the PPF further from the y-axis.

The two straight lines are the isorevenue lines showing maximum revenue achievable on each PPF, given crop and timber product prices, the furthest from the origin denoting the highest revenue. In this instance polyculture (point A) provides more revenue than a monoculture solution (point B) which would involve more land devoted to trees and less to crops, but in separate areas of the plot of land, rather than intimately mixed.

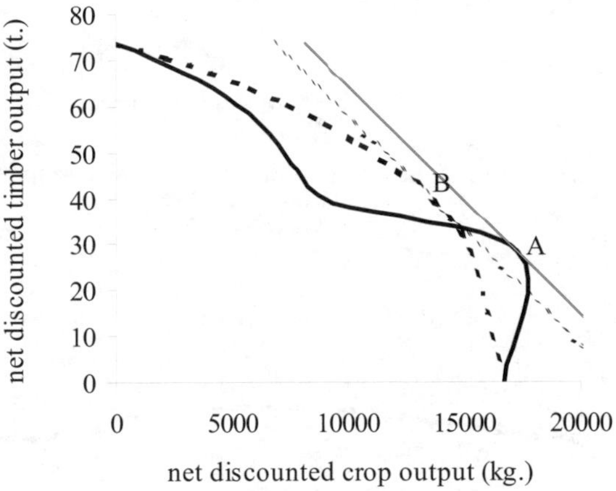

Figure 6. Revenue maximization for monoculture and polyculture PPFs.

If relative prices were to change so that crops became less valuable relative to trees (reduced isorevenue slope), then the monoculture might become more profitable than the polyculture. Further deterioration in the price of the crop relative to tree products (and hence a further lessening of the slope of the isorevenue lines) would give a corner solution, in which only trees would be grown on the land.

This form of diagram can be helpful in showing why, for a particular plot of land or farm, an agroforestry polyculture, mixed monocultures (2 or more separate monocultures) or a sole monoculture might be most profitable under a particular set of conditions. It also shows that even when relatively small changes in prices occur, there may be marginal (e.g., more trees within a polyculture) or radical (e.g. polyculture to monoculture[s]) change in the system. Furthermore, the distance apart of the isorevenue lines shows the extent to which total revenue differs for the different solutions.

However, the one drawback of the PPF-isorevenue diagram comes to light when not only are all costs not fixed within enterprises or within the plot, but if they also do not vary directly with output. In such a case, although the diagram indicates maximum revenue, it does not necessarily indicate the combination giving maximum profitability. To overcome this shortcoming, and as an alternative to this form of diagram, the discounted revenues or discounted profits (see Klemperer, 1996) of the different crop-tree combinations can be plotted for each combination, either in money terms or as a percentage of the highest value. Figure 7 illustrates such a discounted revenue or profit graph for the case study, showing the figures as percentages of the maximum discounted revenue for the monoculture option.

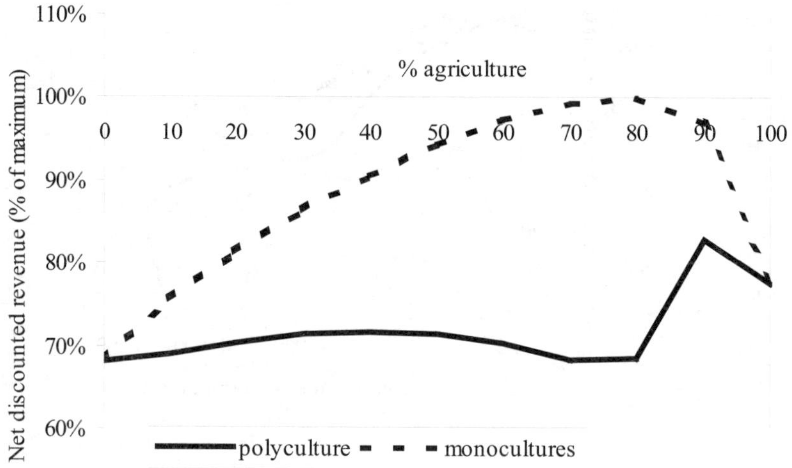

Figure 7 Discounted revenue graph (more details must be given)

3. CASE STUDY

Narain et al. (1998) reported the results of an agroforestry experiment conducted in the western Himalayan Valley Region of India between 1982 and 1992. Among the systems tested one consisted of unpruned leucaena, using 1.5 meter spacing and thinned to half the stocking rate after 6 years. This was compared with double leucaena rows and 4.5 meter alleys (widened during the experiment) cropped with wheat followed by maize each year. The case study here uses the main parameters from the experiment to construct a simulation model. Single leucaena rows spaced at 1.5 meters along 0.75 meter strips are left unpruned until harvest at 5 years. Alleys are cropped with wheat and maize, as above. A number of further assumptions have been made to tie the economic framework in with the key values obtained in the experiment.

Cobb-Douglas production functions were derived. For the leucaena the function is:

$$O_f = 61.9L^{0.47}G^{0.3}B^{0.15} \tag{10}$$

Timber output (total dried biomass) is expressed in tones, total land is 1 hectare, initial light is linked to the area of land devoted to the enterprise, but initially fixed at 1, and labor is fixed at 50 units for forestry. For simplicity, other inputs are assumed fixed, their influence being taken up by the constant term, while the effect of water competition is assumed to be picked up by the light variable.

The wheat-maize biomass production function is:

$$O_a = 2686.5L^{0.4}G^{0.28}\left(\frac{N}{100}\right)^{0.2}B^{0.1} \tag{11}$$

Crop output is expressed as kilograms of wheat grain equivalent, based upon the wheat grain production multiplied by 2.7 to take account of maize grain and other biomass from both crops, adjusted for their values relative to wheat grain deduced from the experiment results. Land for crops is determined from equation 2 with total area, A, taken as 1 hectare. Initial nitrogen is fixed at 100 kg. Labor is fixed at 50 units for agriculture. Other crop assumptions are the same as for the tree enterprise and all costs are assumed to be fixed.

The outputs from each of equations 10 and 11, subject to the overall land constraint (equation 2), for combinations ranging in 10% steps from 100% leucaena to 100% wheat-maize were generated for each of the five years of the rotation, discounted at 6%, and the enterprise discounted outputs then summed. These net discounted outputs (NDO's) are then plotted to produce the PPF before complementarity and extra competition effects: this has been shown earlier as Figure 2. The isorevenue line is based on the relative tree:crop produce prices per kilogram of 1:5 implied in Narain et al.

A growth curve was fitted in an attempt to match the tree biomass production figures reported by Narain et al., thus:

$$O_{f_t} = -18t + 23t^2 - 2.99t^3 \tag{12}$$

Where t is the age of the stand.

The leucaena production function (equation 10) was derived to match the total biomass output at 5 years given by equation 12 with all of the land devoted to leucaena. Equation 12 was then adjusted for other land share percentages by multiplying by the ratio of the production function output figures for that % share to the 100% share output figure. The production function and the linked growth curves were shown earlier in Figure 4.

Nitrogen was taken as a proxy for the complementarity factors. Nitrogen output from the trees was specified as:

$$N_{t+1} = 0.01O_{f_t} \tag{13}$$

The resulting amount was then added to that applied to the wheat-maize in the following year. Figure 5a shows the resulting wheat equivalent yields over the rotation and for different land shares, excluding any extra competition effects.

Water and light appeared to be the major extra competitive factors in the case study. For simplicity, light was taken as the proxy for all extra competitive factors. A shade function was constructed as a function of crown diameter. This, in turn was taken as a function of tree volume, which is based upon the production-growth function and the density of trees. Figure 5b shows the modeled shade effects on wheat yield, with the nitrogen-fixing effect on yield excluded. In the model these two effects were combined via the wheat-maize production function.

A spreadsheet simulation model was constructed using the above relationships and was run for each of the land shares, ranging from 100% leucaena to 100% wheat-maize for two scenarios: first, without the nitrogen and light complementarity and competition effects, and second, with those effects. The resulting data were then used to construct two PPFs and their associated maximum (tangential) isorevenue lines.

4. RESULTS

Figure 8 below shows the PPFs generated by the simulation exercise for the monoculture (gray dotted curve) and polyculture (smooth black curve) options with their associated isorevenue lines.

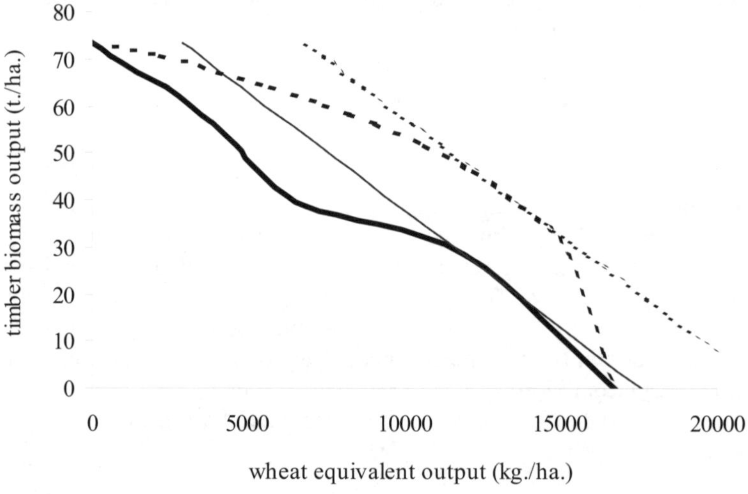

Figure 8. Case study PPF's and isorevenue lines for monoculture and polyculture.

In the agroforestry system, the optimum combination of crops and trees involves wide crop alleys, and a relatively low stocking rate for the trees, much in line with the system adopted by Narain et al. (1998)[6]. However, as the discounted revenue graph in Figure 7 shows, this system does not give a total discounted revenue much greater than that obtained by cropping alone. Furthermore, the higher isorevenue line reached by the monoculture PPF (and shown more clearly in Figure 7) suggests that farmers could obtain a much superior income by devoting about 80% of their plot to wheat-maize and 20% to leucaena, but as separate monocultures, although this result is partly dependent on the underlying production function assumptions. Nevertheless, this is in accord with Narain's conclusion that, "If grain production is the priority, crops and trees should be grown separately in blocks for wood, fuel, fodder, etc." (Narain et al., 1998).

With respect to sensitivity, if relative prices for leucaena products increased marginally this would drive only marginal changes to the system, but once a rise of 40% was reached a radical move to sole crop leucaena would be signaled. In the other direction, major change to cropping wheat-maize alone would only need a 20% fall in the leucaena price. Further, if a discount rate of 12% were used, more in tune with peasant householder time preference rates, a tree–crop mix even closer to pure cropping would be optimal, due to the heavy time penalty incurred by the tree output accruing after 5 years.

5. CONCLUSIONS

Returning to the three research questions posed at the start, the first one relates to whether this approach would give results consistent with those arrived at in the experiment. As has just been shown, the answer to this is yes, although the model used here would tend to give a more specific tree-crop mixture due to its use of continuous functions rather than three single data points for sole crop leucaena, sole crop wheat-maize, and one combination of leucaena and crops. However, this 'richer' result is reliant on a greater amount of detail, which has data implications, as will be seen later.

The second question asks whether further insights for modelers and policy makers can be gained by taking this approach. Because of the use of continuous functions and the isolation of the contributions of the individual economic factors as well as the specific complementarity and extra competition relationships, it is possible to gauge the effect of each of these and to predict the results of enhancing those which have favorable impacts and releasing the constraints imposed by those which are limiting, as well as changes in prices and costs. This also enables researchers to highlight those factors and enterprise mixtures, the study of which would be likely to bear most fruit.

The final question concerns the data implications of the economic, production possibility frontier approach. As one might expect, if greater insights are to be gained, it is at the cost of larger amounts of data. The production function approach has been used in agriculture (Dillon & Anderson, 1990), but does require the collection of data on specific inputs, products, and prices at the farm household level. Statistical analysis, such as multiple linear regression, on the transformed data is then required to derive the PPF. Furthermore, attention would need to be paid to gaining improved understanding of farm and farm household resource availability, constraints, and preferences[7]. Nevertheless, such research would have the added advantage of providing researchers with insights into the levels of inputs and constraints faced by farm households, which could then be used in the design of more realistic experiments. On the experimental side, multiple tree-crop combinations would need to be included, rather than the sole crop-one agroforestry mixture-sole tree pattern, as well as designs to isolate the part played by specific complementary and competitive factors. Experiments should also be designed with reference to the farm surveys so that conditions should mimic those faced by the farm households themselves.

This chapter has shown how an economist might approach the modeling of agroforestry systems using a production function approach. It has set out the ways in which the fundamental resource-competitive/complementary relationships between different enterprises can be presented. This is not new, but by bringing together and extending the various theoretical elements and applying the resulting framework to a practical example it has shown how this approach can be used and what insights might be gained for researchers and policy makers.

6. NOTES

[1] Note that there might be other inputs for which the two enterprises are competing. If so, there would need to be other link equations of a similar form to that for the land. If, however, the assumption was that such inputs vary directly with the amount of land, then, instead of such a link equation, the inputs would need to be linked together within the production function.

[2] If many of the other resources used in production are also fixed, then the resulting PPF could also be a straight line or even concave to the origin.

[3] Or relative gross margins if some of the costs are variable (Barnard & Nix, 1978), although this may not be strictly correct if those costs do not vary directly with output. See Bright (1994) for the effects of introducing variable costs into the analysis.

[4] Although it might be lagged to take account of delayed availability for uptake by the crop.

[5] This PPF is therefore that for monoculture combinations of crops and trees on separate areas of the same plot of land, although, if shading effects of the more dense stands of trees in monoculture systems were taken account of the tree output might be somewhat reduced.

[6] Their systems were changed twice during the course of the experiment, so it is difficult to make a direct comparison.

[7] In fact, to gain even greater understanding it would be wise to also introduce risk into the profitability calculations, although these then tend to become rather complicated (one possible approach is shown by Bright & Price, 2000).

7. REFERENCES

Bright, G. A. (1994). *The Short-run Production Possibility Curve: Convex and Static?* Unpublished. School of Agricultural and Forest Sciences, University of Wales, Bangor, United Kingdom.

Bright, G. A. & Price, C. (2000) Valuing forest land under hazards to crop survival. *Forestry,* 73(4), 361-370.

Dillon, J. L., & Anderson, J. L. (1990). *The Analysis of Response in Crop and Livestock Production.* Oxford, United Kingdom: Pergamon Press.

Dyack, B. J., Rollins, K., & Gordon, A. M. (1999). A model to calculate *ex ante* the threshold value of interaction effects necessary for proposed intercropping projects to be feasible to the landowner and desirable to society. *Agroforestry Systems, 44*, 197-214.

Etherington, D. & Matthews, P. J. (1983). Approaches to the economic evaluation of agroforestry farming systems. *Agroforestry Systems, 1*, 347-360.

Gregory, G. R. (1987). *Resource Economics for Foresters.* New York: John Wiley & Sons.

Institute of Water and Environment (IWE) (2003). *Silvoarable Agroforestry for Europe (SAFE).* Retrieved February, 2004 from the Institute of Water and Environment website, Cranfield University, Silsoe, United Kingdom: http://www.silsoe.cranfield.ac.uk/iwe/research/safe.htm.

Klemperer, W. D. (1996) *Forest Resource Economics and Finance.* New York: McGraw-Hill.

Mercer, D. E. & Hyde, W. F. (1992). *The Economics of Agroforestry.* In W. R. Burch & J. K. Parker (Eds.), *Social Science Applications in Asian Agroforestry*. New Delhi: Oxford & IBH Publishing.

Narain, P., Singh, R. K., Sindhwal, N. S. & Joshie, P. (1998). Agroforestry for Soil and Water Conservation in the Western Himalayan Valley Region of India. 2. Crop and tree production. *Agroforestry Systems, 39*, 191-203.

Nautiyal, J. C. (1998). Forest Economics: Principles and Applications. *Toronto, Canada: Canadian Scholars Press.*

Penson, J. B., Capps, O. & Rosson, C. P. (1995). *Introduction to Agricultural Economics*. Upper Saddle River, New Jersey: Prentice-Hall.

Rao, M. R., Nair, P. K. R., & Ong, C. K.(1998). Biophysical Interactions in Tropical Agroforestry Systems. *Agroforestry Systems, 39*, 3-50.

Ritson, C. (1977). *Agricultural Economics: Principles and Policy*. London: Collins.

Stone, S. W., Kyle, S. C., & Conrad, J. M. (1993). Application of the Faustmann principle to a short rotation tree species: an analytical tool for economists, with reference to Kenya and leucaena. *Agroforestry Systems, 21*, 79-90.

Thomas, T. H. (1991). A spreadsheet approach to the evaluation of the economics of temperate agroforestry [Special issue]. *Forest Ecology and Management, 45*,1-4.

Thomas, T. H., & Bright, G. A. (1991). *A spreadsheet approach to the economic modeling of agroforestry systems*. In *Agroforestry for sustainable production*. London: Commonwealth Science Council.

Upton, M. (1987) *African Farm Management*. Cambridge, United Kingdom: Cambridge University Press.

World Agroforestry Center (ICRAF) (2004). *WaNuLCAS, a model of water nutrient and light capture in agroforestry systems*. Retrieved February 13, 2004 from: http://www.worldagroforestrycentre.org/sea/products/afmodels/wanulcas/index.htm.

PETER BLANDON

ANALYZING RISK IN AGROFORESTRY SYSTEMS USING A PORTFOLIO APPROACH

A Case Study from the United Kingdom

1. INTRODUCTION

Sanchez (1995) chose to open his well-known paper in the journal *Agroforestry Systems* with a definition of agroforestry as "the traditional practice of growing trees on farms for the benefit of the farm family". Although not explicitly stating agroforestry is a discipline focused mainly on developing countries, it is easy to make such an inference. This poses a problem that is central to this chapter. The design, benefits and uptake of agroforestry systems depend, in part, on risk. However, models that deal with risk are intensely "data hungry". For example, in response to the 1989 question, "does agroforestry pay?" it was stated that agroforesters could produce "little concrete data by way of an answer" ("Does Agroforestry Pay", 1989). Similar comments have been made more recently (Scherr & Current, 1997) but, unfortunately, the data requirements of a full risk analysis are far in excess of those for straightforward profitability calculations and beyond any reasonable expectations of the data that might come to hand in the tropical/subtropical arena.

This is even true of the case study chosen here, a silvopastoral system in the United Kingdom. Even after extensive research, the data for a full risk analysis are not available. Furthermore, in the UK it can be assumed that crops are sold in markets and not used for subsistence or household needs thus allowing the analysis to proceed in terms of well-defined monetary values. A somewhat stronger assumption is that the farmer has access to perfect capital markets. This is, of course, unlikely to be the case in many applications of agroforestry when the absence of capital markets, or indeed markets in general, mean that farmers will put a premium on the immediate output of crops, often to the disadvantage of tree-based

J. R. R. Alavalapati & D. E. Mercer, Valuing Agroforestry Systems, 95–122.

elements of the system (Izac, 2003). Therefore, the main part of the chapter will deal with a temperate agroforestry system, thus permitting many simplifying assumptions. In the conclusions to the chapter a number of points will be made that deal with extending the method outlined here to what might be thought to be more "typical", tropical agroforestry situations.

The model of risk analysis to be applied here is portfolio theory. This is a theory that rigorously defines the concept of diversification in the context of reducing risk in financial portfolios. At the core of the theory lies the correlation of the returns of the various financial assets constituting the portfolio. In essence, if two stocks are negatively correlated, the risk of a portfolio of both of them is less than that of the individual stocks.

This theory is thought to be of relevance to agroforestry for two reasons. First, the risk of an agroforestry system is often alluded to in the literature, especially in the context of such systems being desirable from the farmer's point of view. Secondly, even if the economic aspects of diversification are not considered, the biological side of agroforestry systems is often described in terms of plant/animal interactions in a way that is reminiscent of portfolio theory.

In this chapter, the concepts of portfolio theory will be extended to cover the situation of agroforestry and it will be shown that the theory can offer important insights into the evaluation and design of agroforestry systems. The desirability of agroforestry as opposed to what Price (1995) referred to as "coarse-level" mixing, or simply growing crops on different parts of the farm, can be determined using a portfolio approach. It will also be seen that what is sometimes claimed to be a benefit of agroforestry – the crop interaction – can be a disadvantage.

However, as portfolio theory is likely to be an area with which those working in agroforestry are unfamiliar, a brief introduction follows.

2. PORTFOLIO THEORY AND RISK ANALYSIS

The hallmark of economic theory over the last fifty years or so has been the widespread introduction of risk analysis. Nowhere has this been more true than in the area of financial markets. A landmark work in this field is Markowitz's (1959) portfolio analysis in which the principles of reducing the risk of a portfolio of stocks were outlined (for a good overview of the area in the context of finance see Elton & Gruber, 1995).

2.1. The basic model

Markowitz argued that investors were interested in the overall return and risk of their portfolio and not, directly, the returns and risk of the individual stocks. In the simple two-asset case Markowitz's model takes the following form. Let w_1 be the proportion of the total portfolio invested in asset one and w_2 be the proportion invested in asset two. If the return on asset one is R_1 and that for asset two is R_2 the return on the portfolio R_p will be given by equation 1.

$$R_p = w_1 R_1 + w_2 R_2 \tag{1}$$

The contribution of Markowitz was in introducing what is a simple piece of statistics into finance; the definition of the risk of the portfolio.

$$\sigma_p^2 = w_1^2 \sigma_1^2 + w_2^2 \sigma_2^2 + 2 w_1 w_2 \sigma_{1,2} \tag{2}$$

Here, σ_1 represents the standard deviation of the returns of asset one and $\sigma_{1,2}$ is the covariance of returns of the two assets. This term can be replaced by $\sigma_1\sigma_2\rho_{1,2}$, where $\rho_{1,2}$ is the correlation of the returns R_1 and R_2.

This correlation of returns is the core of Markowitz's ideas. Unless the correlation of stock returns is perfect and positive (i.e., unless $\rho_{1,2} = +1$) there will always be a risk reducing effect to diversification. Indeed, if there is perfect negative correlation it is easy to show that a portfolio with asset one in the proportion $\sigma_2/\sigma_1+\sigma_2$ will have a risk of zero even though its individual constituents are, in isolation, risky. Indeed, this has long been known in agricultural markets where the use of futures to hedge against price fluctuations is merely an application of this principle.

An insight from portfolio theory that seems a little counterintuitive is the role of currency markets in international diversification. While there are many aspects to the production of crops for international markets, it is likely that the addition of the exchange rate into the equation will actually reduce risk for farmers. As long as the output of crops is not perfectly positively correlated with exchange rate fluctuations (which is likely to be the case) the revenues from the sale of crops on the international market expressed in the local currency will be somewhat more stable as the result of the less-than-perfect correlation of crop yields and currency movements. Indeed, in the analysis of diversification of stock portfolios it is usually found that international diversification reduces risk *because* of the riskiness of the currency markets (see Elton & Gruber, 1995 for references on this topic).

2.2. Larger portfolios

The main part of this chapter deals with what, on the face of it, is a two-asset case: a sheep/tree silvopastoral system, However, as a forty-year rotation is used, the system is, from a portfolio point of view, a forty-one asset system – forty lamb crops and one timber crop. So a generalization of the equations above into an *n*-asset system is needed. This is shown in equations 3 and 4.

$$R_p = \sum_{i=1}^{i=n} w_i R_i \tag{3}$$

Table 1. Variance/covariances, correlations, and expected net present value (ENPV) for a hypothetical three-crop system.

	Variance/covariance			*Correlation*		
Crop	*A*	*B*	*C*	*B*	*C*	*ENPV*
A	144	64.8	0	0.6	0	100
B		81	0		0	75
C			64			50

$$\sigma_p^2 = \sum_{i=1}^{i=n} \sum_{j=1}^{j=n} w_i w_j \sigma_{i,j} \tag{4}$$

In Blandon (1985) a hypothetical three-crop system with the characteristics described in Table 1 was examined.

The figures in Table 1 give rise to Figure 1. On the *y*-axis the expected net present value (ENPV) of the agroforestry system is given. On the *x*-axis the risk is measured as the standard deviation of the NPV. The points at the end of the line in the diagram, labeled A and C, represent the risk and return from crops A and C respectively if they are grown as monocultures (i.e., $w_a = 1$ or $w_c = 1$). The line joining these two points represents mixtures of the crops that minimize risk for any given ENPV.

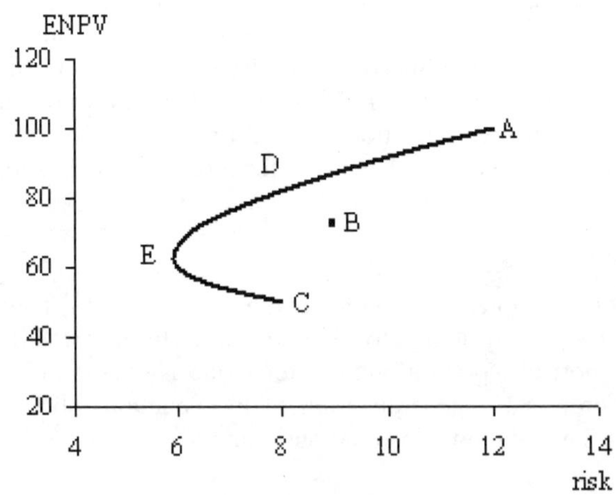

Figure 1. Risk analysis for a hypothetical three-crop system.

The mathematics and logic of the model ensure that the line will be weakly convex from below. In the extreme case, in which there is perfect positive correlation between A and C, the line joining points A and C would be straight. When correlation is less than plus one, there are some benefits from diversification and so the ENPV obtained from a crop combination will have less risk than that which would be implied by points on a straight line.

It is important to note that is point B, representing a pure crop B, is not a rational choice. It is dominated by points along the line CA, meaning that there are combinations of crops that give a higher return for less risk. For example, point D represents a combination of 59.3% of crop A, 21.5% of crop B and 19.3% of crop C (with rounding errors). This gives a return of 85 and a risk of 8.55 as a standard deviation or a variance of 73.2, better than B.

The line CA "bends back" on itself. Point E represents a minimum risk agroforestry system. The point actually represents 8% crop A, 37% crop B and 55% crop C. The segment of the line EA is known as the *efficient frontier* as it shows the combination of crops that give the maximum return for any given level of risk and would be the set from which any rational, risk-averse farmer would choose. The upshot of this is that the only rational monoculture is crop A. All other monocultures are bettered by some agroforestry system on the segment from E to A.

Note that this bending back does not always occur. It is caused here by the lack of correlation between two of the crops.

3. PORTFOLIO THEORY AND AGROFORESTRY

It is often stated that one of the benefits of agroforestry lies in the ability of different crops to exploit different aspects of the resources available. Early writings tended to concentrate on the competition for moisture and the combination of crops with different rooting characteristics. Similarly, silvopastoral or silvoarable systems emphasize the complementary nature of the crops, in the case of the sheep to be looked at here, the shading and shelter provided by the trees is said to increase the conversion of pasture into body mass and pasture growth is aided by trees in the early stages of the rotation. This is reminiscent of portfolio theory with the correlation of output substituted for the correlation of price. However, there are some significant differences between finance and agroforestry that make it impossible to transplant Markowitz's ideas into agroforestry without some modification.

3.1. Returns and profits (the expected values)

The model developed by Markowitz deals with the single variable of returns but, in agroforestry, it is likely that profits are the relevant decision variable. However, profit consists of prices multiplied by volumes, both of which might be considered stochastic in nature. Both Blandon (1985) and Lilieholm and Reeves (1991) chose net present value as the basic stochastic variable and so ignored the sources of correlation between the profits of two crops. Correlation can come from price interaction or volume interaction. Thus, two threads can be identified; price interrelationships, the basis of portfolio theory in finance, and volume interrelationships, which have been the main thrust in agroforestry. This chapter attempts explicitly to combine these two threads.

Similarly, there might be correlations between the price and volume of a single crop. For example, high prices might call forth more output; alternatively, high output might depress prices. Having said this, however, for simplicity it will be assumed in all of the following that the correlation between price and volume *for a given crop* is zero.

The explicit introduction of volume interactions means that the proportions of the assets, w_i, lose their meaning. As Price (1995) points out, the diversification possibilities offered by price variation alone are available by "coarse-scale mixing of *separate* single-product systems under one ownership", but most of the work in agroforestry has been directed at finding systems where the output is increased relative to that obtained at the coarse level. This has been quantified in competition studies by the concept of the *land equivalent ratio* (LER) (Mead & Willey, 1980; Moseley, 1994) shown in equation 5.

$$LER = \frac{Y_1}{M_1} + \frac{Y_2}{M_2} \tag{5}$$

Here, Y_1 shows the physical yield of crop 1 when grown in some form of mixture and M_1 shows the crop's yield on the same area when grown as a monoculture. The LER shows the relative land area that would be required to grow the same total volume of the two crops that are produced in an intimate mixture of some sort. So, a land equivalent ratio of 1.2 indicates that 20% more land would be required to produce the same volume of crops in a coarse-level mixture. In coarse-level mixing it is clear that the elements of the equation, Y_i/M_i would all be equal to the proportion of the total farm area devoted to the crops and would analogous to w_i.

But it is clear the land equivalent ratio describes only one part of the story. It is not true to state, as does Moseley (1994) that "a result of 0.91 … suggests that a farmer would not choose the agroforestry system over the monoculture". If the reduction in output is more than compensated for by a reduction in risk, agroforestry could be indicated.

Logically, of course, the cost elements should also be considered to be stochastic in nature and this would introduce a further set of correlations into the system. Here, for simplicity, costs will be considered to be non-stochastic.

3.2. Sequential Agroforestry Systems (the variances and covariances)

Using the terminology of Sanchez's (1995) paper, most agroforestry systems are sequential. Thus, for the derivation of the overall profit of the system, a common time frame is needed. In the system looked at below, a 40 year rotation period is chosen. The investment becomes, therefore, a 41-asset portfolio. There are forty outputs from the agricultural component of the system and one from the forestry part.

While a 41-asset portfolio can easily be handled by equations 3 and 4, the introduction of more than one time period also introduces the possibility of serial correlation in the volume and price figures. For reasons outlined below, these correlations can be taken to be zero.

4. COARSE-LEVEL MIXING IN A TWO-CROP SYSTEM

The following three sections look at a two-crop system, ash (*Fraxinus excelsior*) and rye-grass pasture supporting sheep. The outputs of the system are lambs and ash timber for joinery. Most of the figures relate to experiments in the system undertaken in the United Kingdom and specifically the experimental sites in North Wales. The basic example closely follows the figures in Thomas and Willis (2000).

The overall objective is the calculation of two efficient frontiers. The first, to be considered in this section, is the simple "coarse-level mixing" frontier where it is assumed that the crops are grown in spatially distinct areas. The only difficulty here lies in the derivation of the variance and covariance terms for net present values derived from two stochastic variables. The second efficient frontier, to be examined in the next section, is the agroforestry one in which the crops are assumed to be intimately mixed and interactions are taken into account.

In terms of the points on Figure 1, these two efficient frontiers will have the same end-points, referring to monocultures. It is useful, therefore, to look at the end-point solutions first after discussing the basic approach to the price and volume variables.

4.1. Price and cost variables

The prices of timber and sheep are treated in a specific way, based on the *efficient market hypothesis* from financial economics (see Elton & Gruber, 1995 for a description in finance). Under this hypothesis price changes follow a geometric Brownian motion that can be modeled as follows. Let the price in period t be P_t and let the logarithm of the proportional change in price or log price-relative, $ln(P_t/P_{t-1})$, be an independent drawing from a normally distributed random variable with a mean of μ and a variance of σ^2. The implication of such a structure is that the expected price in the next period is the price now with the growth factor μ included. Thus, there is no way to predict the price in the future other than extrapolation using the growth rate μ. Any information regarding the market in question and ideas of future scarcity or glut is already factored into the current price. The expected price in period t, $E(P_t)$ and the variance of price are given by the equations below where P_1 is the current, known price.

$$E(P_t) = P_1 e^{\mu t} \tag{6}$$

$$\sigma_t^2 = P_1^2 e^{2\mu t}\left(e^{t\sigma^2} - 1\right) \tag{7}$$

The variance of the price in the future increases, reflecting the idea that periods in the more distant future are less certain. Such a model of prices has the advantage that it strips out any serial correlation and reduces the data requirements of the model. Another advantage of the model is that complications such as storing in times of historically low prices in the expectation of a price rise can be ignored – a complication in the derivation of a rotation length. It has the deficiency that it seems quite reasonable to assume that a good year for lamb birth and survival could depress price in the current year and so would tend to lead to the expectation of negative serial correlation. Increased storage of meat, for example, would go a long way to weaken such linkages. The actual figures employed are presented in the following table and their derivation is explained below.

The figures for sheep are taken from the annual *Farm Business Survey in Wales* (Institute of Rural Studies, various years). This survey is undertaken by the Institute of Rural Studies at the University of Wales, Aberystwyth. The figures from 1985 to 2001 for all flock sizes on Welsh lowland farms show an average lamb price of £35.29. The higher figure used here reflects the extra income earned by farmers in the form of sale of culled ewes, fleeces and the subsidy received. The figure in Table 2 is, more correctly, gross income. The variance of the log-price relative was calculated on the basis of lamb prices alone, the implication being that the other income elements were fixed.

Timber prices are more problematical. In Britain, the market for hardwood timber has tended to be scattered and so there are relatively few published statistics of sufficient length or consistency to allow estimates of time trends or variances. So the following approach was adopted. The Forestry Commission in the United Kingdom publishes a survey of coniferous standing timber prices and these it converts into a Laspere index. The conversion accounts for the change in the composition of the timber offered for sale by the Commission. The variance of the log-price relatives of this index was found and, as it represents the variance of a

Table 2. The price variables in the model.

	Lamb data	*Timber data*
Current price P_1	£46.97 a head	£50 a cubic meter
Average annual growth, μ	0.41%	2.9%
Variance of annual growth, σ^2	0.037	0.015
Cost	£16.96 per lamb	£1.16 per tree

growth term, it was assumed that ash prices showed the same variance of growth. The price now, P_1, is in the range, relative to coniferous prices, suggested by the Forestry Commission (Insley, 1988). Whilst grants might be available for the forestry part of the investment, depending on the planting density, they will not be included in the analysis here.

The implication for prices in this model is shown in Figure 2. There the thicker line shows the expected price over the forty-year study period. The shaded areas show the 95% confidence intervals for the prices. The implication of the model for risk is clear.

The cost figures are adapted from Thomas and Willis (2000) and the Aberystwyth surveys. Note that the sheep costs are on a per lamb basis while the forestry costs are on a per tree basis. This means that high fertility will lead to higher costs for sheep farming. In forestry, the costs of planting are spread over the years of the rotation and the figure in the table is the present value. It is deemed to be constant regardless of final output. This difference causes the equations that follow to have slightly different forms for sheep and forestry – see the Appendix.

4.2. Volume variables

The variance of lamb production was treated in a similar way to that of price. Proportional changes in fertility from year to year were deemed to be drawings from a static distribution with a mean of zero. Although it could be argued that fertility will increase over the study period, it was assumed here not to do so. The average fertility over the 1985 to 2001 period according to the Aberystwyth survey is 1.131

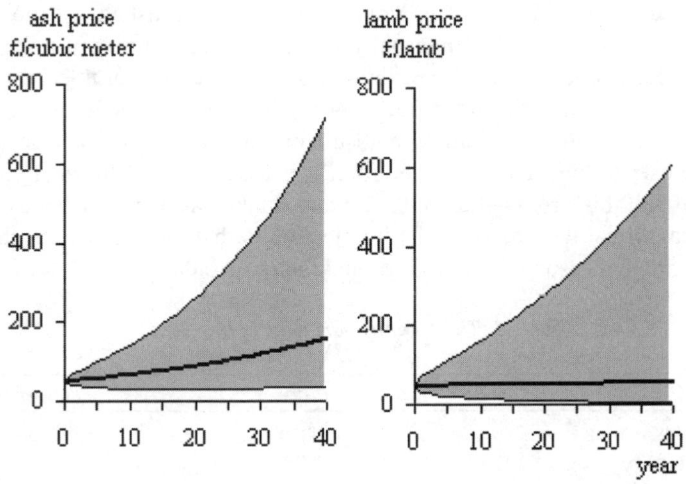

Figure 2. The expected prices and their 95% confidence intervals.

lambs per ewe with a variance of 0.0122. The stocking rate used for monoculture was 12 ewes per hectare. This is the same as Thomas and Willis (2000), similar to Willis, Thomas, and van Slycken (1993) but more than reported by Crowe and McAdam (1999).

For ash, the situation is less straightforward. The volume produced will depend on the rotation and here a forty-year rotation is taken (thereby ignoring the complications that the rotation will depend on the discount rate which will depend on risk). No agroforestry experiments in Britain have yet run to a full rotation and so, in most of the studies, it is assumed that the cumulative yield will be similar to that contained in the Forestry Commission's *Forest Management Tables* (Hamilton & Christie, 1971).

The *Forest Management Tables* were developed when monoculture was seen as being the only method that the Forestry Commission would use. Their underlying idea is that, within a wide range of stocking rates, the cumulative timber yield is constant. The tables are based on so-called *yield classes*, which are denoted by the maximum mean annual increment per hectare for a site. In Thomas and Willis (2000) an ash yield class of 12 is assumed. This generates a cumulative yield of 480 cubic meters per hectare over 40 years. Usually, a substantial portion of this would have been removed as thinnings. However, to keep the analysis here in line with the studies in Britain, it will be assumed that only 25% of the cumulative yield is removed in this way, giving an average final cut of 75% of cumulative yield, or 360 cubic meters. To simplify the analysis, and without too much loss of reality, it will be assumed that thinnings are removed and sold at cost so that the expected value of thinning (and the variance) is zero.

Very little on the variability of volume output is published, so the following approach was adopted. It is not unusual for yield classes to turn out to be one class higher or lower than estimated. In Britain, classes are denoted at intervals of two so it would be possible that the ash productivity was in yield class 10 or 14. However, the plantation thinning would allow adjustments to be made so that the final yield would be closer to the expected. Therefore, it is assumed that the final yield lies in the range of half this, from 440 to 520, and this is taken as being the 95% confidence interval of a normal distribution. The implication is that the variance of the yield is 416 which, when measured as a proportional factor, is 0.0018.

Table 3. The quantity variables in the model.

	Lamb data	*Timber data*
Stocking rate	12 ewes/ha (N_s)	1,800 stems/ha
Average fertility	1.131 $E(F_s)$	n/a
Variance	0.0122 (σ_s^2)	0.0147 (σ_f^2)

4.3. Expected net present values

The expected net present value for forestry is calculated using the formula below where R is the rotation length of 40 years. The discount rate, r, is 5%. The cost, C_f varies according to the number of plants used but is constant for any given planting level. The subscript f refers to forestry and the expected price $E(P_{f,R})$ to the expectation of the price of timber in period R, the expectation being formed now. The expected revenue figure is just the expected price $E(P_{f,R})$ multiplied by expected quantity $E(V_f)$- see the Appendix.

$$E(NPV_f) = \frac{E(P_{f,R})E(V_f)}{(1+r)^R} - C_f \tag{8}$$

The expected price of timber in year 40 is £160.47 and the resulting expected net present value is £6,119 per hectare.

The expected net present value of sheep production is merely the sum of the expected net present values for each period given by equation 9 below.

$$E(NPV_s) = \sum_{t=1}^{t=R} \frac{N_s E(F_s) \times (E(P_{s,t}) - C_s)}{(1+r)^t} \tag{9}$$

The result is £7,665 per hectare.

4.4. The risks

The risk from forestry derives from a single cash flow at the rotation end and this is given by equation 10 – see the Appendix. The variances in the square brackets are those of the growth rates and not those of absolute price or volume. The subscript *pf* refers to forestry price and *vf* to the volume output of forestry.

$$\sigma_f^2 = \frac{E(P_{f,R})^2 E(V_f)^2}{(1+r)^{2R}} \left[\sigma_{vf}^2 + R\sigma_{pf}^2 + R\sigma_{vf}^2\sigma_{pf}^2\right] \tag{10}$$

Note that the variance of price is multiplied by R because of the assumption of timber prices being generated by a growth process. The variance of volume is not influenced in this way. The result is a risk, measured as a standard deviation, of 6,647.

The calculation is similar for sheep. The assumption of price and volume independence through time means that the net present values over the forty years are themselves independent random variables. Thus, the risk can be assumed to be the sum of the variances of the individual net present value terms. For each year the risk, as defined by an equation similar to equation 10 – see Appendix – can be used and the terms summed. The resulting formula is given below. The difference in formulation results from the difference in the treatment of costs.

$$\sigma_s^2 = \sum_{t=1}^{t=R} \frac{E(V_s)^2}{(1+r)^{2t}} \left[E(P_{s,t})^2 \left(t\sigma_{ps}^2 + t\sigma_{vs}^2 \sigma_{ps}^2 \right) + \sigma_{vs}^2 \left(E\left(P_{s,t}^2\right) - C_s^2 \right) \right] \tag{11}$$

The result is a standard deviation of 1,306. Thus, sheep husbandry is considerably less risky than forestry. Although the summation of risks might suggest that the riskiness should increase, the volume of lamb production and the variance of early prices are much lower so the overall risk is reduced.

4.5. End-point solutions – monocultures compared

The results of the analysis are shown as points S (sheep) and F (forestry) in Figure 3. Any rational, risk-averse farmer would choose sheep husbandry over forestry as the expected net present value is greater and the risk is also less.

4.6. Coarse-level integration

The ability to mix trees and sheep on spatially separated parts of the farm means that these ENPVs in equations 8 and 9 can be mixed and any diversification deriving from price can be taken advantage of. However, none of the benefits of intimate

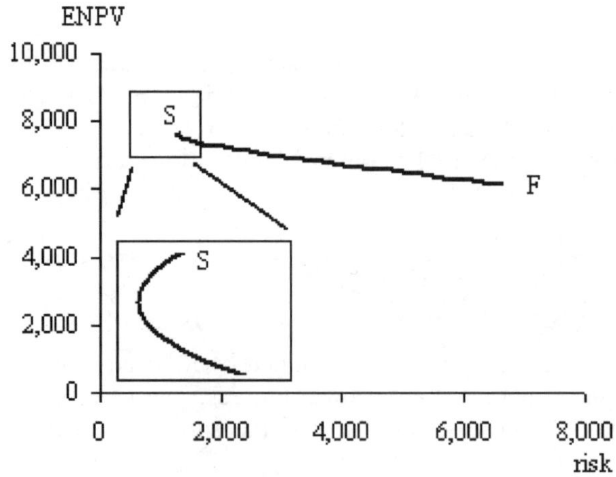

Figure 3. The efficient frontier with coarse-level mixing (box shows magnification of area around point S).

mixing of crops would be available. The relationship between risk and return depends upon the covariance of the expected net present values of forestry and agriculture. These are discussed in more detail in the following section so, as an initial assumption, assume that the price and volume of sheep and timber output are independent so that any covariance term is zero.

Under these assumptions the risk of the two crops together will be given by equation 4 where the covariance elements are zero, the variances are those calculated above and the weights w refer to the proportion of the total area of the farm unit under sheep or trees. The resulting frontier is shown in Figure 3 joining points S and F.

Although the line looks suspiciously linear, it does have a shape similar to that in Figure 1 but inverted. The curve enclosed by the box in Figure 3 shows a magnification of the area of the efficient frontier around point S. Moving from point S to point F involves, to begin with, a movement towards the left, in other words a reduction in risk. There is a combination of sheep and trees that minimizes risk and this is achieved by using 3.7% of the farm area for tree crops and the remaining 96.3% for sheep. This will give an expected net present value of £7,607 and a risk of 1,282.

The graph shows, therefore, that there are few benefits in diversification at the coarse level in this situation. The reasons for this will be discussed later, but first, it is necessary to look at the effect of a true agroforestry system and this means examining in some detail the interrelationships of the variables in the system.

5. A TRUE AGROFORESTRY SYSTEM

There are three sets of relevant figures if portfolio theory is to be applied to a true agroforestry system. The first set consists of the "agroforestry" relationships between the crops themselves. This is where most of the research has been undertaken and, in the terminology of portfolio theory, relates to the expected values of the system. The second is the set of variance terms on which virtually no work has been undertaken. The final set is the covariance element.

5.1. The expected values

The main input in this model, from which all other terms follow, is the initial stocking rate for trees. This ranges from 0 to 1,800. The latter is the stocking rate in the standard Forestry Commission yield tables and so can be seen as "conventional forestry". Not many studies have looked at timber volume under the low stocking rates encountered in agroforestry so, here, the overall cumulative volume of the timber production in the agroforestry system is given by a Gompertz (double exponential) function (Cabenettes, Auclair, & Imam, 1999). This takes the form shown in Figure 4 with the volume per hectare being shown on the right-hand axis.

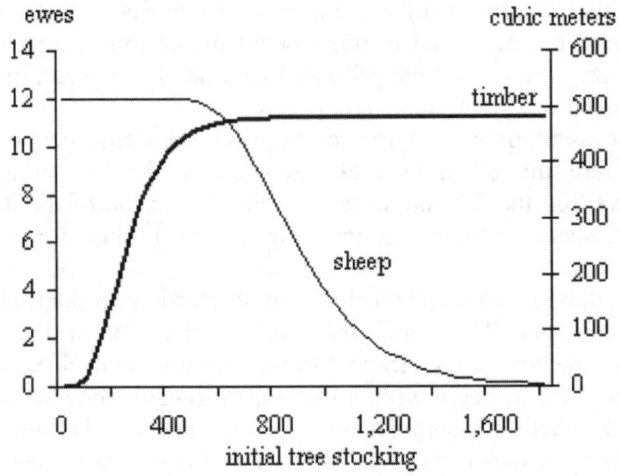

Figure 4. The Gompertz functions applied to timber production and ewe stocking.

The parameters of the function are shown in the Appendix. The rationale behind the curve is the same as that behind the Commission's yield tables. Over a wide range of stocking rates the cumulative output is the same, but at the bottom end the tree crop does not use all resources, so there is an output drop.

The diagram shows that at a stocking rate of 400 stems per hectare the cumulative output is equal to 419 cubic meters. This is less than the volume assumed by Thomas and Willis (2000) and compares with the 480 in the Commission's yield tables. The thinning yield is always 25% of final volume and, as stated earlier, it is assumed to be removed at cost with no stochastic elements and so is ignored in this model.

The stocking rates for ewes is also shown in Figure 4, with the initial flock size on the left-hand axis, the initial stocking rate depending on the tree stocking. The relationship is also modeled by a Gompertz function, and the parameters are included in the Appendix.

Thomas (1991) and Thomas and Willis (2000) hypothesize that there is a reduction in the production of pasture and, therefore, lamb production as the canopy closes. As none of the agroforestry experiments have been running long enough to calibrate this part of the system, Thomas and Willis (2000) use two possible decay functions. One reduces lamb yield by 10% the other by 25%. In both cases the lamb output begins to decline around year ten and stabilizes at the lower around year 24, similar in parts to the results of Teklehaimanot, Jones, and Sinclair (2002). In the model, as in Thomas and Willis's, the difference between the two assumptions is not great given the effect of discounting. Here a 25% figure was used with the parameters of the Gompertz function to generate the relationship shown in the Appendix.

The results of these assumptions are summarized in Figure 5 where the numbers in the graph refer to the initial tree stocking rates in terms of stems per hectare.

These figures are sufficient to produce a graph of the land equivalent ratio defined by equation 5 above. This is shown in Figure 6. The peak in the LER is at an initial planting rate of around 500 stems per hectare. Without an analysis of risk, therefore, the system of Thomas and Willis (2000) using a 400 stem per hectare planting rate would seem to be strong contender for the best agroforestry system.

5.2. The variance terms

There seems to be little or no work on the effect of the *variance* of crop output in agroforestry systems. All of the work is concentrated on finding the average or expected levels of outputs for different crop combinations. Mead and Willey (1980) refer to the "general belief that intercropping yields are more stable" but this, if true, could result from a number of possible combinations of variances and covariances. For example, if the outputs of two crops are independent the variance of their combined output will fall because of the zero covariance term in a suitably redefined equation 2. Indeed, a falling covariance could even be the result of negative correlation but increasing absolute variances.

In the absence of evidence to the contrary, therefore, it will be assumed that the variance terms in Tables 2 and 3 are unaltered under different crop combinations. As they take the form of proportional errors – see the Appendix – the implication is that a crop yield of, say, 5% above the expected is just as probable if the crop is grown alone as it would be if the crop was grown as a small part of an agroforestry system.

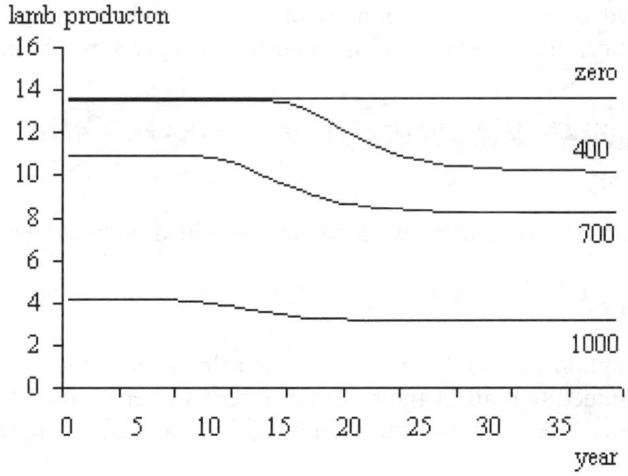

Figure 5. Lamb production under different tree stocking rates.

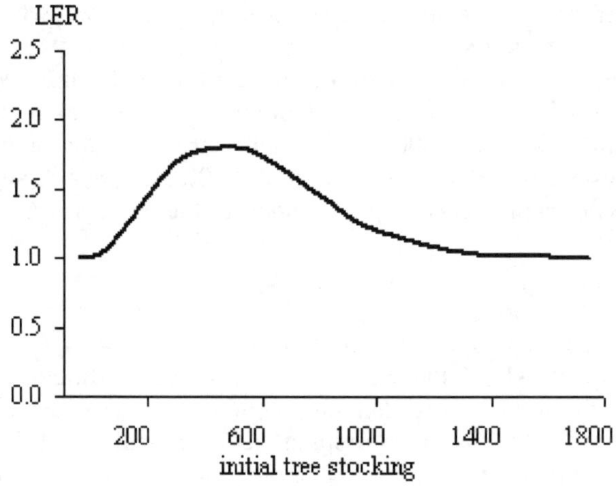

Figure 6. The land equivalent ratio under the assumptions here.

5.3. The covariance terms

The covariance of the net present values is the heart of the portfolio approach. However, the assumption of a lack of serial correlation between volumes and prices allows a great simplification. As there are 41 net present values in the model there should be 820 covariance terms. However, the only one that will be non-zero will be the covariance between the lamb output in year 40 and the timber output. If $\sigma_{vs,f}$ represents the covariance of the volumes of forestry and agriculture and $\sigma_{ps,f}$ the covariance of price, the covariance of the final NPVs is given below in equation 12.

$$\sigma_{s,f} = \frac{1}{(1+r)^{2R}} E(V_f)E(V_s)E(P_{f,R})[(E(P_{s,R})-C_s)\sigma_{vs,f} + E(P_{s,R})\sigma_{ps,f} + E(P_{s,R})\sigma_{vs,f}\sigma_{ps,f}] \quad (12)$$

The derivation of this unattractive formula is outlined in the Appendix.

5.4. The frontier

Under the assumptions above the frontier can be calculated and it is shown in Figure 7. Here it is assumed that all of the covariance terms in equation 12 are zero. This makes the frontier directly comparable with the one in Figure 3 reproduced as the thinner line.

The effects of agroforestry are to increase the ENPV but also to increase risk. The reduction in risk that occurs in financial applications of portfolio theory is, in fact, limited by the very nature of agroforestry. Half of one's investment in a financial asset halves one's returns and similarly reduces risk exposure. Half of

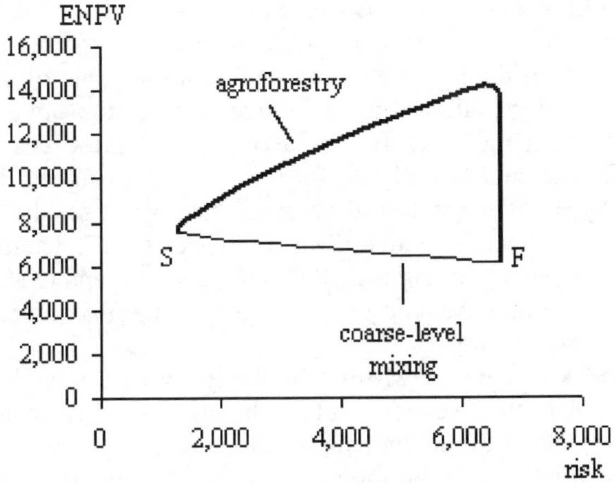

Figure 7. The efficient frontier for sheep/timber agroforestry systems (thicker line) and coarse-level mixing (thinner line).

one's land given over to one of the crops in an intimate mixture does not necessarily halve output and similarly, does not reduce the risk exposure in the same way. Hence, the benefits often ascribed, indeed desired, by agroforestry systems are, at best, a double-edged sword. The returns may well be enhanced, but the risk may also be increased.

The maximum ENPV is given by the peak of the frontier and corresponds to the 500 stems per hectare that maximized the LER in Figure 5. Note that this correspondence is not necessary. The LER is defined in terms of volumes while the ENPV is in value terms.

Recall that there was a slight movement to the 'left' of the coarse-level frontier and this was shown in Figure 3. But apart from this, every point on the coarse-level frontier shown by the thinner line in Figure 7 is dominated by points on the thicker line showing the agroforestry frontier. The implication is that, if mixing of sheep and trees is being considered, it is almost certainly better to mix them in the agroforestry sense than at the coarse-level.

5.5. The Sharpe index

In financial markets, the Sharpe index (SI) often used to measure portfolio performance is adapted for the situation here, and given in equation 13.

$$SI = \frac{E(NPV) - K}{\sigma} \tag{13}$$

The index shows the risk-adjusted, additional benefit compared to fixed costs *K* of undertaking agroforestry. In finance, this is an index that, under certain circumstances, it would be reasonable to expect investors to attempt to maximize. In the agroforestry context, the figure *K* could represent the value of the land and the labor costs required to run the farm, the farmer having the option of giving up agroforestry and selling the land. If it is assumed that these costs are £7,000 a hectare (£6,000 for the land and £1,000 for the labor allowing for the fact that the time horizon is 40 years) the agroforestry system that maximizes the Sharpe index is that with 300 trees per hectare, which gives an index of 1.17. The best that can be achieved under coarse-level mixing is a 100% allocation of land to sheep. This gives an index of 0.51. Note that the land price is less than the maximum ENPV as this latter has risk associated with it.

The Sharpe index can be shown diagrammatically as a straight line from the point *K* on the *y*-axis to the relevant point on the efficient frontier. Maximizing the Sharpe index is equivalent to maximizing the slope of such a line. This is shown in Figure 8 where points S and F are the same as those in Figure 7, point AG is the agroforestry system that maximizes the Sharpe index and M is the point of maximum ENPV.

It follows that the efficient frontier in this example is the line K-AG-M. Any rational, risk-averse farmer would choose systems along that line. The actual point chosen would depend on the risk preferences of the farmer. Diagrammatically, this would be shown by an indifference curve map with the optimum choice being the point on the efficient frontier that coincided with the highest indifference curve. If that point was in segment K-AG the implication would be that the farmer would sell or lease some of the land and put the remainder under the agroforestry system

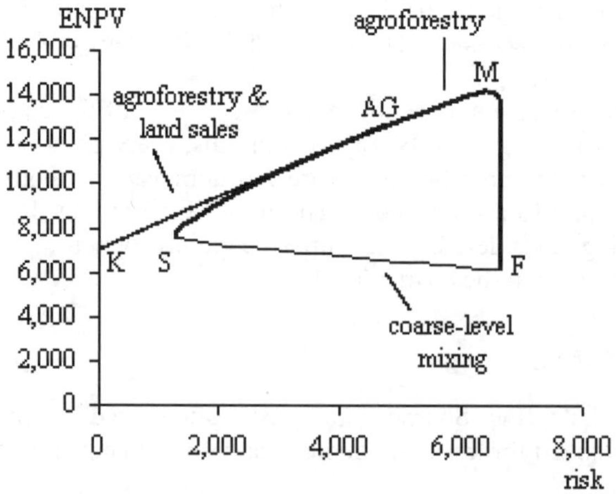

Figure 8. The efficient frontiers for coarse-level mixing and agroforestry with land-sales also allowed.

represented by AG. In the segment AG-M the farmer would opt for an agroforestry system with an initial tree stocking rate of between 300 and 500 stems per hectare. Thus, in this situation, agroforestry rather than coarse mixing of monocultures is always indicated.

In finance the line K-AG can be extended to the right on the assumption of borrowing. Here, such an extension to the line would represent the purchase or renting of extra land and labor at £7,000 a hectare. If such an assumption is allowed the only rational agroforestry system is that shown by AG. Farmers would adjust risk by moving along the extended line K-AG by adjusting the amount of land they farmed. But all farmers would institute system AG on their land.

Altering the assumed correlation terms has virtually no effect on risk in this case. The result is that, for all combinations of correlations between prices and volumes from minus one to plus one, the efficient frontier in Figure 7 remains virtually unaltered. Indeed, no diagram is offered here as, given the scale, the difference is virtually invisible. The reason is that the assumption of serial independence in the price and volume variables means that the only non-zero correlation term relates to year forty. In that year the effect of discounting is at its greatest and so the effect of diversification is minimized. For example, if both the correlations between prices and volumes are plus one, the overall risk of the 500 stems per hectare system is 42,525,259 (as a variance). The covariance element of that is only 1,160,033, or 2.7% of the total risk.

In effect, because of the relatively long rotation of a temperate-zone agroforestry system and the reasonable assumption of economically efficient product markets the diversification effects are very small. Thus, the benefits, or otherwise, of agroforestry derive from the physical/biological effects of intermixing. However, as the following extension shows, this is not the case in shorter rotation systems and, hence, in the application of agroforestry in tropical contexts.

6. A HYPOTHETICAL SHORT-ROTATION AGROFORESTRY SYSTEM

To investigate the effect of shortening the rotation, which according to the discussion above should increase the effects of diversification, the following hypothetical agroforestry system was used. The rotation of the tree crop was reduced to five years and the timber crop in year five was taken to be 60 cubic meters. The value for α for the Gompertz function for yield was, therefore, 60 and not the 480 shown in the table in the Appendix. Thus, a yield class of 12 was still assumed. All of the other variables in the system remained unaltered except for the price of timber. The price assumed was £15 per cubic meter for the current year. In essence, the system can be thought of as a temperate one producing biomass for, say, energy production. Alternatively, it could be viewed as closer to the situation in tropical agroforestry with an annual crop and a short-rotation forest crop.

The result for the efficient frontier is shown in Figure 9 where it is assumed that all correlations are zero. This is a very different frontier from that in Figure 7 describing the previous system. The coarse-level mixture and true agroforestry

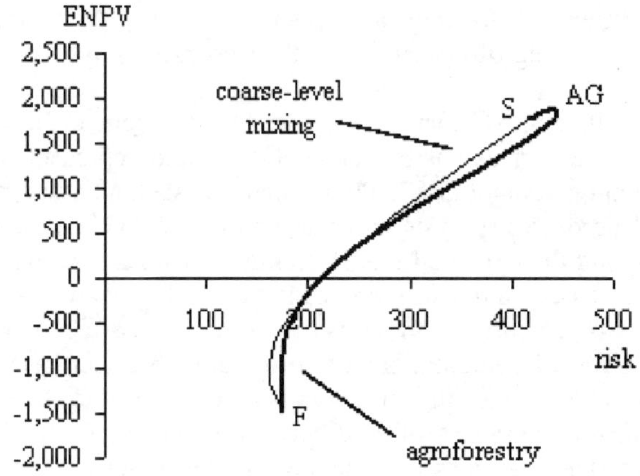

Figure 9. The efficient frontiers for short-rotation sheep/timber agroforestry systems and coarse-level mixing (zero correlation between prices and volumes assumed).

efficient frontiers are nearly coincident and the effect of diversification very apparent. The initial movement in the "top-right" part of the agroforestry frontier, S-AG, reflects the effect of the volume and risk increase that was discussed above, but after that the effect of diversification begins to dominate and the curve is very similar to that shown in the hypothetical example in Figure 1. Note, also, that unlike the frontier in the financial context or the analogous coarse-level mixing frontier, the agroforestry frontier is not necessarily convex from below. The shape depends on the crop interactions.

Although it is not very clear in the diagram, the coarse-level mixing and agroforestry frontiers cross at a positive expected net present value. Thus, for some part of the profitability range coarse-level mixing is better, for some agroforestry dominates.

The result here is interesting because with the price chosen, forestry operates at a loss for planting densities except those between 300 and 500. But despite this apparent lack of profitability, the system with the highest Sharpe index is the agroforestry system using 300 stems to a hectare. Although not clear from the diagram, this is better than any of the coarse-level mixing and better than sheep alone, the slope of the line from the *y*-axis being maximized at point AG. Note that here, because profitability is lower than the previous case, a value for K of £1,000 per hectare has been used. This is the equivalent of about £4,000 per hectare over 40 years, or eight rotations.

If the assumed correlation coefficients are altered the diagrams change noticeably. Now, the covariance term becomes large enough to influence risk significantly. For example, if both correlations are plus one, the overall risk of a 500 stem per hectare system is 266,103 measured as a variance. The covariance part of

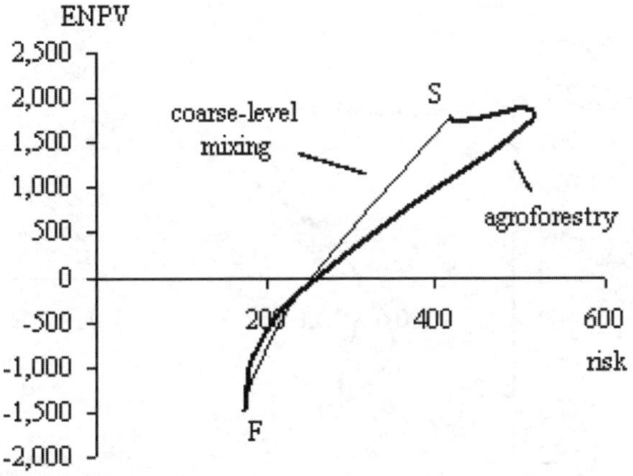

Figure 10. The efficient frontiers for short-rotation sheep/timber agroforestry systems and coarse-level mixing (correlations of plus one assumed between prices and between volumes).

this is 71,139, or 27% of the total. This is ten times greater than the proportion in the forty-year rotation example above.

Figure 10 shows the situation with the worst diversification possibilities. In this case both the correlation of volumes and prices are both taken to be plus one. The agroforestry systems are dominated by the coarse-level mixtures at all positive ENPVs so no agroforestry system would be chosen by a risk-averse rational farmer. The Sharpe index is maximized for the all-sheep option at 1.90.

On the other hand, if the correlation coefficients are set to minus one, the result is shown in the diagram in Figure 11. In this case, the true-agroforestry frontier dominates that of coarse-level mixing at all positive ENPVs and no rational farmer would do anything except intimately mix sheep and trees to gain from the financial and non-financial effects. The Sharpe index is maximized with a tree planting density of around 400 stems at 2.39 indicating that, on that indicator at least, the optimum system would be one that is recognizably agroforestry.

The influence of the correlation coefficients on the relationship of the positions of the two frontiers is shown in Figure 12 below. The y-axis shows the price correlation and the x-axis the volume correlation. At negative correlations of sufficient magnitude agroforestry systems dominate those of coarse-level mixing as shown in Figure 11. At these correlations, agroforestry is the only rational choice. At positive correlations the agroforestry systems are themselves dominated, as in Figure 10, and so no farmer would chose agroforestry over a coarse crop combination.

In the central section of the graph labeled "neither system dominates" the choice will depend on the preferences of the farmer as the frontiers take the form of those shown in Figure 9 with coarse-level mixing dominating in some ranges but not in all.

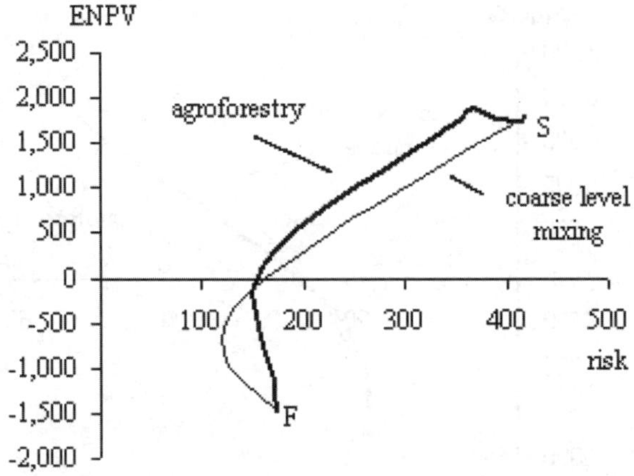

Figure 11. The efficient frontiers for short-rotation sheep/timber agroforestry systems and coarse-level mixing (correlations of minus one assumed between both prices and volumes).

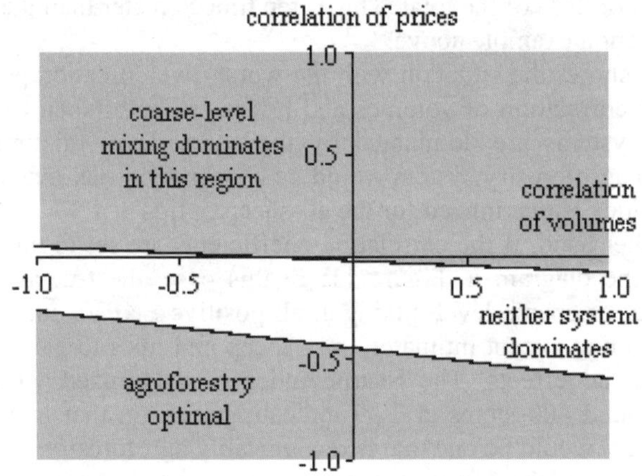

Figure 12. Relationship between optimum land use choice and price and volume correlations.

7. CONCLUDING REMARKS

It should go without saying that the actual results derived here depend upon the figures that have been assumed. Although the actual data used are derived from various sources, many of the interrelationships are speculative. However, the results are intriguing and show that a Markowitz-style analysis of agroforestry adds insights into the analysis of such systems.

A number of points would bear further investigation. For example, are the following statements generally true or are they the results of the specific example chosen here?

- The effects of coarse-level diversification are likely to be very small in temperate-zone agroforestry systems if the rotation length of the forestry element of the system is "long" or the discount rate is high.
- In temperate-zone agroforestry the interactions of the different crop elements are critical to the desirability or otherwise of agroforestry, and price diversification effects can be ignored.
- For shorter rotation crops the correlations of price and volume are critical in both the design and choice of an agroforestry system.
- Paradoxically, therefore, in tropical/subtropical situations it is likely that data for both crop and price interactions are needed – maximum data requirements appear to occur in situations with minimum data availability. This is an observation that is probably true for all operations research-style approaches (Betters, 1988).

Not only are data a problem for tropical/subtropical situations. Poor rural households may approach the issue of risk differently than temperate richer landowners. Here, the analysis assumes that choices can be modeled by smooth indifference curves which, usually by implicit assumption, are sufficiently "flat" to allow the conclusion that optimum solutions will lie somewhere along the efficient frontier and not at a border solution. Thus, it is implicit when discussing diagrams such as Figure 7, that the choice of an optimum system lies somewhere between *S* and the maximum point on the curve. This implies that, while risk is seen as undesirable, it is not catastrophically bad. However, in the tropical arena households are often close to subsistence level. The implication would be that avoiding bad outcomes is a question of survival rather than just a trade-off against the possibility of a higher return.

Technically, the indifference curve map that is implicit in such situations would be represented in the diagrams above by lines that were nearly vertical, showing that a very slight increase in risk would have to be compensated for by a very high increase in potential income.

Thus, in the case of poor land owners, it is possible that corner solutions would become more common. In portfolio terms, and especially in terms of Figure 1, this would seem to mean that monocultures would be indicated unless the efficient frontier "bent back" on itself as it did in the illustration in Figure 1.

However, in such contexts where subsistence farmers are producing crops for their own consumption rather than for the market, decisions are likely to be made in terms of nutritional variables rather than monetary ones. In this case, the relevant figures in the analysis become the expected nutritional output and its variance. Thus, monocultures would be somewhat rarer and the points on the efficient frontier could represent outputs measured in a single nutritional variable that is the result of a mixture of crops.

While this does not invalidate the ideas here it does mean that there are likely to be fruitful areas for work. Even for farmers marketing their crops, ideas of catastrophic risk can be incorporated. For example, there are a number of so-called "safety-first" models that have been proposed in finance. The interested reader is referred, again, to Elton and Gruber's text (1995) where three such models are outlined. One, Roy's criterion, would translate here to minimizing the probability that the ENPV of the agroforestry system was below some predefined subsistence level. Katoka's criterion maximizes the lower limit of the ENPV range subject to the constraint that the probability of achieving this lower limit is not greater than some predetermined value. A third criterion, due to Telser, suggests maximizing the ENPV subject to the constraint that the probability of the actual ENPV being less the subsistence level is not greater than some predetermined figure.

In the context of finance it can be shown that such criteria often lead to the same result as conventional Markowitz-type analysis. However, there is another factor that differentiates subsistence level farming from western finance. In the usual application of these models, the returns from financial assets are symmetrically distributed. Thus, the idea of skewness can conveniently be ignored. However, in many tropical situations it is likely that agricultural outcomes are very skewed and this skewness is likely to be an important decision factor. Thus, in "normal" years output varies according to a usual symmetrical type of distribution but, every now and then, the harvest could be catastrophically low. This downside risk (along with the downside covariances and co-skewnesses) could well figure in the decision-making in subsistence farming but is ignored in the analysis here.

Despite, or maybe because of, these provisos, the portfolio theory approach adds extra insights to agroforestry and can be used in tandem with biologically-based research to define optimum agroforestry systems. If risk in agroforestry is more fully understood and incorporated into evaluations, it is likely that agroforestry will be more attractive to farmers and hence enjoy a higher uptake.

8. APPENDIX

In the derivation of the equations in this chapter, use is made of the following properties of random numbers. First, the expected value of a sum of random numbers is equal to the sum of their expected values. Using the notation $E(x)$ for the expected value of the random variable x, the first result can be written as equation A1.

$$E(x+y)=E(x)+E(y) \tag{A1}$$

If x and y are not correlated, the expected value of the product is equal to the product of the expected values. In other words,

$$E(x \times y)=E(x)\times E(y) \tag{A2}$$

The variance of the sum of two random numbers is one of the principal equations in portfolio theory and is given in equation A3.

$$\sigma_{x+y}^2 = \sigma_x^2 + \sigma_y^2 + 2\sigma_{x,y} \tag{A3}$$

The following results are also employed and can be found in any standard statistical textbook. If n is a constant (or at least is not a stochastic variable) then the following are true.

$$E(nx) = nE(x) \tag{A4}$$

$$\sigma_{nx}^2 = n^2\sigma_x^2 \tag{A5}$$

$$\sigma_{mx,ny} = mn\sigma_{xy} \tag{A6}$$

The basic price and volume model assumes that the random movement can be modeled as a growth element. Thus, if P_1 is the actual price it can be defined in equation A7 where ε is a normally distributed random variable with a mean of zero and variance of σ^2.

$$P_1 = E(P_1)(1+\varepsilon) \tag{A7}$$

A similar equation can be written for the volume term. Note that the assumed growth in prices given by the factor μ is included in the expected price for the next period and not in the error term. Thus, $E(P_2)=E(P_1)(1+\mu)$.

Using equations A2 and A4 above, the expected net present value of forestry, $E(NPV_f)$ can be written as the following where the subscript f refers to forestry, v to volume and p to price, C is the cost that is non-stochastic but varies with the planting density, r is the discount rate and R the rotation. The expected price, $E(P_{f,R})$ is the price of timber in period R the expectation being formed in the current period. This is given by equation 6 in the main text.

$$E(NPV_f) = \frac{E(P_{f,R})E(V_f)}{(1+r)^R} - C_f \tag{A8}$$

The variance of the expected net present value of forestry is somewhat less elegant. The basic form is shown in equation A10 below.

$$\sigma_f^2 = E\left[\left(E(NPV_f) - \frac{E(P_{f,R})(1+\varepsilon_p)E(V_f)(1+\varepsilon_v)}{(1+r)^R} + C_f\right)^2\right] \quad \text{(A9)}$$

If this is expanded and, using the fact that $\varepsilon_v = \varepsilon_p = 0$, the result is given by equation A11.

$$\sigma_f^2 = \frac{E(P_{f,R})^2 E(V_f)^2}{(1+r)^{2R}}\left(\sigma_v^2 + \sigma_p^2 + \sigma_v^2\sigma_p^2\right) \quad \text{(A10)}$$

The variance of the error term in price, σ^2_p is not the same as that given in the text in equation 7. That equation relates to the variance of price itself. Here the variance refers to the multiplicative error term and will be given by $\sigma^2(R)^{0.5}$ where the variance σ^2 is that to be found in Table 2. This is because of the growth in price and increasing variance that this implies. The volume figure is treated differently and so the variance here will be that found in Table 3.

For sheep the expected net present value for production in period t is given by equation A12. This differs slightly from A9 because the cost figure is on a per lamb basis. Higher volume output is likely to lead to more costs for feed, etc. and this is not likely to be the case with extra volume growth in forestry. Also, the volume figure is given by the stocking rate for ewes, taken to be a constant, multiplied by the fertility, $N_sE(F_s)$. It is this latter term to which stochastic variation is attached.

$$E(NPV_s) = \frac{\left[E(P_{s,t}) - C_s\right]E(V_s)}{(1+r)^t} \quad \text{(A11)}$$

The variance of the net present value for sheep for period t is found by the expansion of equation A13 given in A14.

$$\sigma_s^2 = E\left[\left(E(NPV_s) - \frac{\left[E(P_{s,t})(1+\varepsilon_p) - C_s\right]E(V_s)(1+\varepsilon_v)}{(1+r)^t}\right)^2\right] \quad \text{(A12)}$$

$$\sigma_s^2 = \frac{E(V_s)^2}{(1+r)^{2t}}\left(E(P_{s,t})^2\left(\sigma_v^2 + \sigma_p^2 + \sigma_v^2\sigma_p^2\right) - 2E(P_{s,t})C_s\sigma_v^2 + C_s^2\sigma_v^2\right) \quad \text{(A13)}$$

Finally, bearing in mind that serial correlation terms are all zero, the only covariance term is that between the final lamb and timber crops. This is given by equation A15 and its simplification in A16 where $\sigma_{vf,a}$ refers to the covariance of the volume term for forestry and agriculture, etc.

$$\sigma_{s,f} = E\left[\left(E(NPV_s) - \frac{\left[E(P_{s,R})(1+\varepsilon_{p,s}) - C_s\right]E(V_s)(1+\varepsilon_{v,s})}{(1+r)^R}\right)\left(E(NPV_f) - \frac{E(P_{f,R})(1+\varepsilon_{p,f})E(V_f)(1+\varepsilon_{v,f})}{(1+r)^R} + C_f\right)\right] \quad \text{(A14)}$$

$$\sigma_{s,f} = \frac{E(V_s)E(V_f)E(P_{f,R})}{(1+r)^{2R}}\left(E(P_{s,R})(\sigma_{vs,f} + \sigma_{ps,f} + \sigma_{ps,f}\sigma_{vs,f}) - C_s\sigma_{vs,f}\right) \quad \text{(A15)}$$

In the calculations used for the main text, the covariance terms within A16 were calculated using the relationship between standard deviations and correlation coefficients.

Much use is made of the Gompertz function in relating variables in the model. The basic function is given in equation A17 below.

$$y = \alpha e^{\left(-e^{-\beta(x-\chi)}\right)} \quad \text{(A16)}$$

In the formula, α represents the limit to which y tends as x increases, β is a slope variable which defines the rate at which the function "moves" from its lower value to its upper and χ is the point of inflection. The values used in this function in its various guises are given in the table below.

Table A1. The parameters of the Gompertz functions used.

	meaning/value		
y	cumulative volume	initial flock size	lamb output as a percentage of output in year one
x	tree stocking rate	tree stocking rate	year
α	480	12	0.75
β	0.1	0.005	0.3
χ	200	800	See below

The value for χ in the function defining the decay in lamb output was made dependent upon the initial stocking level for the tree part of the system. The argument is that, at higher stocking levels crown cover occurs sooner and so the point at which pasture production begins to decline will also be sooner. The function used is that given below.

$$\chi = \frac{400}{\sqrt{\text{initial tree stocking rate}}} \tag{A17}$$

9. REFERENCES

Betters, D. R. (1988). Planning optimal economic strategies for agroforestry systems. *Agroforestry systems. 7*, 17-31.

Blandon, P. R. (1985). Agroforestry and portfolio theory. *Agroforestry Systems, 3*, 129-249.

Cabanettes, A., Auclair, D., & Imam, W. (1999). Diameter and height growth curves for widely-spaced trees in European agroforestry. *Agroforestry Systems, 43*, 169-181.

Crowe, S. R., & McAdam, H. H. (1999). Silvopastoral practice on farm agroforestry in Northern Ireland. *Scottish Forestry, 53*(1), 33-36.

Does agroforestry pay? (1989). *Agroforestry Today, 1*(3), 7-9.

Elton, E. J., & Gruber, M. J. (1995). *Modern portfolio theory and investment analysis*. New York: John Wiley & Sons.

Hamilton, G. J., & Christie, J. M. (1971). *Forest Management Tables (Metric)*. Forestry Commission Booklet No. 34. London: HMSO.

Insley, H. (Ed.). (1988). *Farm Woodland Planning*. Forestry Commission Bulletin, No. 80. London: HMSO.

Institute of Rural Studies. (various years). Farm Business Survey in Wales. UK: University of Wales, Aberystwyth.

Izac, A.-M. N. (2003). Economic aspects of soil fertility management. In G. Schroth and F. L. Sinclair (Eds.), *Trees, crops and soil fertility* (pp. 13-37). Wallingford, UK: CABI Publishing.

Lilieholm, R. J., & Reeves, L. H. (1991). Incorporating economic risk aversion in agroforestry planning. *Agroforestry Systems, 13*, 63-71.

Markowitz, H. (1959). *Portfolio selection: efficient diversification of investments*. New York: John Wiley & Sons.

Mead, R, & Willey, R. W. (1980). The concept of a "land equivalent ratio" and advantages in yields from intercropping. *Experimental Agriculture, 16*, 217-228.

Moseley, W. G. (1994). An equation for the replacement value of agroforestry. *Agroforestry Systems, 26,* 47-52.

Price, C. (1995). Moderate discount rates and the competitive case for agroforestry. *Agroforestry Systems, 32,* 53-61.

Sanchez, P. A. (1995). Science in agroforestry. *Agroforestry Systems, 30,* 5-55.

Scherr, S. J. & Current, D. (1997). What makes agroforestry profitable for farmers? *Agroforestry Today, 9, No. 4,* 10-15.

Teklehaimanot, Z. Jones, M., & Sinclair, F. L. (2002). Tree and livestock productivity in relations to tree planting configuration in a silvopastoral system in North Wales, UK. *Agroforestry Systems, 56*, 47-55.

Thomas, T. H. (1991). A spreadsheet approach to the economic modeling of agroforestry systems. *Forest Ecology and Management, 45*, 207-235.

Thomas, T. H., & Willis, R. W., (2000). The economics of agroforestry in the UK. In A. M Hislop and J. N. Claridge (Eds.), *Agroforestry in the UK*, Forestry Commission Bulletin No. 122, (pp. 107-125). Edinburgh: HMSO.

Willis, R. W., Thomas, T. H. & van Slycken, J. (1993). Poplar agroforestry: a re-evaluation of its economic potential on arable land in the United Kingdom. *Forestry Ecology and Management, 57*, 85-97.

BEN H. J. DE JONG, SUSANA OCHOA GAONA, SOTERO QUECHULPA MONTALVO, ELSA ESQUIVEL BAZÁN, AND NICOLÁS PÉREZ HERNÁNDEZ

ECONOMICS OF AGROFORESTRY CARBON SEQUESTRATION

A Case Study from Southern Mexico

1. INTRODUCTION

The restoration of tropical lands, degraded by inappropriate anthropogenic practices such as logging, grazing, and agriculture, has the potential to sequester significant amounts of carbon at moderate costs through reforestation and agroforestry activities; the so-called Land Use, Land-Use Change and Forestry (LULUCF) projects. Two international environmental treaties – the Framework Convention on Climate Change (UNFCCC), and the Convention on Biological Diversity – include provisions to make financial resources available to developing countries for global environmental benefits. The Third Conference of the UNFCCC, held in December 1997 in Kyoto, Japan, described two market-based mechanisms that will allow countries to trade in greenhouse gas emission (GHG) reductions:

- Between two Annex 1 countries (countries with binding emission limits), known as Joint Implementation (JI), and
- Between an Annex 1 country and a non-Annex 1 country (countries with no binding emission limits, mainly developing countries), known as the Clean Development Mechanism (CDM).

Under a possible future carbon offset trading program, countries would be most likely to pay for greenhouse gas emission reductions in another country where the cost for reducing emissions is lower. With such a program international carbon emission offsets could become a currency for investing in emission-reducing activities (Tipper & De Jong, 1998). These offsets have the potential to create a system of incentives to develop C-saving projects worldwide, because they stimulate

J. R. R. Alavalapati & D. E. Mercer, Valuing Agroforestry Systems, 123–138.

CO_2 emitters to seek the least expensive emission-reduction measures (Swisher & Masters, 1992).

A strategy for carbon sequestration in the forestry and the agroforestry sector includes the establishment of permanent agroforestry plots to substitute for slash-and-burn agriculture and the conservation of standing old-growth forests as carbon sinks. In addition, carbon sequestration can be enhanced through increased harvesting efficiency in forests and utilizing a higher percentage of total biomass in durable products; improving forest productivity on existing forest lands through management and genetic manipulation; establishing plantations on surplus cropland and pastures; restoring degraded forest ecosystems through natural regeneration and enrichment planting; establishing plantations and agroforestry projects with fast-growing and short rotation tree crops (Sathaye, Makundi, Andrasko, Boer, Ravindranath, & Sudha, 2001); and increasing soil carbon by leaving dead wood, litter, and slash from harvests (De Jong, Soto, Montoya, Nelson, Taylor, & Tipper, 1997).

The potential of agroforestry for carbon sequestration is largely a consequence of the availability of land for tree growing and that tree-growing activities in the tropics are relatively inexpensive. Substantial land areas could be used for mitigation purposes, without displacing current production activities.

Attempts to estimate the global potential for increasing carbon sinks through LULUCF activities are complicated due to the high variability in socio-economic and political factors, such as land availability or the rate of uptake of different options (Read, 2001). IPCC's Third Assessment Report (Kauppi, Sedjo, Apps, Cerri, Fujimori, & Janzen, 2001) supports many of the earlier findings, and provides a broad evaluation of potential carbon pools. For instance, a study by Sathaye and Ravindranath (1998) suggests that 300 million hectares (Mha) may be available for mitigation purposes in ten tropical and temperate countries in Asia. Of these, about 176 Mha is estimated to be available for agroforestry. The cumulative carbon mitigation potential in the Asian countries for all LULUCF options is estimated to be about 26.5 gigatons of carbon (Gt C). Sathaye et al. (2001) examine the sequestration potential in seven major developing countries (Brazil, China, India, Indonesia, Mexico, the Philippines, and Tanzania), which account for about 60% of the forested area in the developing world. The estimated cumulative potential amounts to 1,851 Mt C by 2012 with an average of about 120 Mt C annually.

2. COSTS OF CARBON SEQUESTRATION

Financial analysis of land-use projects are not easily compared as no standard method of analyzing direct, indirect, initial, and recurring costs as well as revenues, has emerged and come into wide use. The cost estimates of land-use projects often include land purchase or rent, land clearing and site preparation, initial planting, maintenance and management, and sometimes data collection and evaluation. The opportunity costs of land are often not included in the cost estimates, nor are the costs associated with project definition, promotion, monitoring, and training of project participants.

Kauppi et al. (2001) distinguishes between three basic ways of estimating the costs of sequestration; point estimates, partial equilibrium estimates, and general equilibrium approaches. Many of the point-estimate studies provide undiscounted private market cost estimates of carbon sequestered in afforestation projects. These types of estimates usually reveal little about how costs might change if the project were to expand. The estimates tend to be biased downwards, partly because the opportunity costs of land are often not included. Partial equilibrium studies provide a more complete estimation of a cost function and includes rising costs associated with increased sequestration activities (Moulton & Richards, 1990). Marginal cost functions generated from such studies suggest that costs usually are higher than those of simple point estimates (De Jong, Tipper, & Montoya-Gomez, 2000a).

Studies of broad geographic/climate regions indicate relatively low costs of carbon sequestration for forest plantations, forest management, and agroforestry. In most of these studies the average storage method has been used. Dixon, Schroeder, and Winjum (1991) find relatively little difference among the boreal, temperate, and tropical regions with respect to the sequestration costs of forest plantations and forest management. Recent studies of individual developing countries provide point-estimates from US$ <0 to 35.10 per ton of carbon (t C), using the annual C flow as the carbon measure (Boer, 2001; Masera, Cerón, & Ordóñez, 2001; Lasco & Pulhin, 2001; Makundi, 2001; Ravindranath, Sudha, & Rao, 2001; Xu, Zhang, & Shi, 2001; Table 1). These estimates are derived from the carbon sequestration potential of forestry and agroforestry projects and the costs of projects and or management options. None of these are based on analysis of on-the-ground projects with carbon sequestration as one of the goals. For example, costs of project design, the time required to explain to farmers the project objectives, carbon related inventories, and the cost of baseline setting are usually not taken into account.

Table 1. Ranges of documented costs of carbon sequestration in the tropics.

Source	*Country*	*Cost Categories*[a]	*Method*[b]	*Costs (US$/t C)*
Xu et al. (2001)	China	Lc Mc	P	0.28 to 5.40
Ravindranath et al. (2001)	India	Lc B	P	0.39 to 18.80
Boer (2001)	Indonesia	Lc B Mc	P	<0 to 1.70
Masera et al. (2001)	Mexico	Lc B Mc	P	0.70 to 35.10
Lasco & Pulhin (2001)	Philippines	Lc B	P	<0 to 1.55
Makundi (2001)	Tanzania	Lc B	P	0.50 to 1.38

[a] Lc = Life Cycle Cost; Mc = Monitoring Costs (Defined as administration, Masera et al, 2001); B = Benefits; [b] P = Point Estimates

De Jong et al. (2000a) estimated the cost of carbon sequestration in Chiapas, Mexico within a range of US$ 2.00 to over US$ 40.00 per t C. Their cost categories project include promotion, training of project participants, and opportunity costs, and they used incremental cost functions to estimate the offset potential of a specific region as a function of carbon offset incentives. They exclude the cost of project design, baseline setting, and monitoring.

In this chapter we focus on the following key questions:

- What are the costs of adopting agroforestry systems for carbon sequestrations?
- What are the costs per unit of carbon that can be sequestered by each agroforestry system?
- What are the costs of developing new agroforestry options within the scope of the Scolel Té project?
- What are the transaction costs associated with the Scolel Té project design?

3. FUNCTIONING OF THE SCOLEL TÉ PROJECT

In Mexico, a carbon-trading program was set up in 1997, known as Scolel Té, providing financial incentives and technical assistance to farmers interested in selling carbon offsets. Farmers (individually or in groups or communities) voluntarily submit a proposal to a local trust fund (Fondo BioClimatico, FBC) in which they present a "land improvement plan", called Plan Vivo (www.planvivo.org; Figure 1). Applications specify the areas and systems to be implemented and the current land use. Feasible proposals are assigned a level of financial and technical support related to the expected amount of carbon sequestration. Financial assistance is spread over time and made contingent upon satisfactory progress towards the development of the new system and its incremental carbon storage (Tipper, Montoya, De Jong, Castillo, March, & Soto, 1997).

The Scolel Té project is used as an example agroforestry project to calculate the costs related to implementing a carbon sequestration project in rural environments dominated by resource-poor small-scale farmers, the cost of designing a project under these conditions within the framework of the Plan Vivo system, and the transaction costs that were required to set up the Plan Vivo carbon management system for this type of project. Our estimates of the cost of the carbon sequestration are based on the present-valued fixed and variable costs of implementing new agroforestry systems, the opportunity cost to farmers of diverting land from current land use, cost of developing Plan Vivos, the cost of project promotion and training, plus the cost of monitoring and internal verification of project performance. Excluded from our calculations are the costs of external verification, certification, and marketing carbon offsets. We include an estimation of the transaction costs required to design new agroforestry options outside the Scolel Té project range, but apply the Plan Vivo system and the impact of these costs on the carbon price as a function of the amount of marketable offsets.

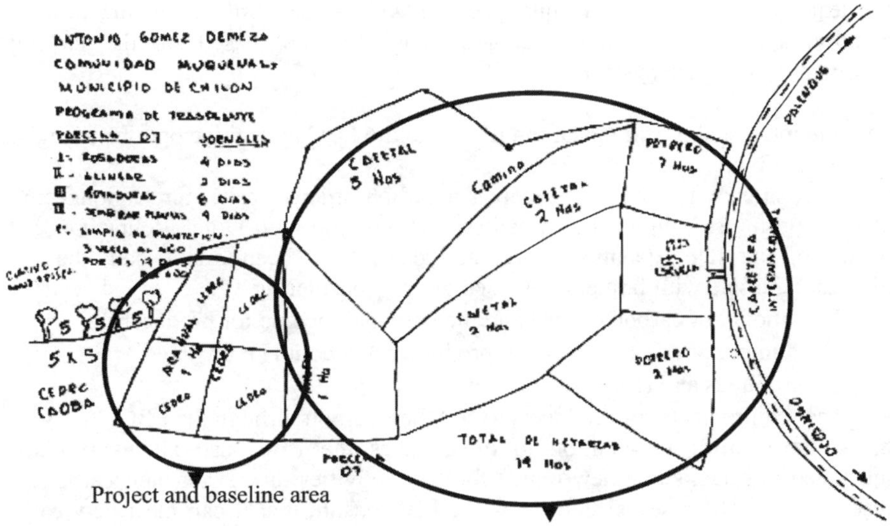

Figure 1. An annotated "Plan Vivo", including the plots where the farmer will plant trees (Project and baseline area), other land he owns, and its use ("Leakage" area).

The Scolel Té project is administered through the Fondo Bioclimático (FBC), a non-profit trust fund registered with a bank. The FBC is currently administered jointly by AMBIO, a rural development and environmental non-governmental organization (NGO) based in San Cristóbal de Las Casas, Chiapas, and the Edinburgh Centre for Carbon Management in the United Kingdom (UK). The FBC acts as an umbrella organization for those interested in selling carbon offset services; it is also a center of support, training, and contact between actors interested in mitigating CO_2 emissions.

The FBC facilitates the sale of carbon-offset services from small-scale rural producers. The FBC is responsible for the administration of finances to secure carbon-offset services from producers registered with the FBC, and for the assessment and monitoring of these activities. The procedures used by the FBC are based on the Plan Vivo System. A Plan Vivo is a simple tool for planning, managing and monitoring (agro-) forestry activities and carbon services.

A Plan Vivo is constructed around the needs and resources of the producer's family or community. The main component is an annotated map of the producer's land showing all the different fields that the producer owns and the land use or vegetation type of each plot. On this map the producer or community marks where he/she will implement the proposed change and provides a work program showing activities, dates, and necessary inputs (in terms of labor and materials). From the information provided in the Plan Vivo the FBC subsequently determines whether the proposal does or does not compete with other land-use activities, if there is potential for leakage of carbon, how much carbon will be sequestered by the proposed activity, and what the implementation cost will be, based on the technical specifications of each system.

Farmers interested in selling carbon offsets to the FBC have three options to register their Plan Vivo:

a. Direct sale: When the FBC receives a carbon order it makes an "announcement of an opportunity", directed to organizations working with target groups. Once the details of the sale (quantity, price, and date) have been established, the FBC discusses the sale with farmers' representatives, considering:

- The amount of carbon needed and the potential income for producers
- The number of communities and producers that could be involved
- Uncertainties and risks associated with the sale
- The requirements for producers to sell their carbon through the FBC

b. Reserve fund: In addition to direct sales the FBC also allows producers interested in establishing new (agro)-forestry activities to register their carbon in a reserve fund. The reserve fund helps the FBC ensure that it can meet new carbon orders quickly and can maintain a diverse portfolio of carbon offset options. For the producers it provides the opportunity to plan activities in advance of confirmed sales.

c. Waiting list: The third means of registering offset activities with the FBC is through the waiting list. In order to register their carbon on the waiting list producers provide details of what type of activity they intend to carry out and what area of land they will use. When new orders are received, producers on the waiting list may be invited to submit full Plan Vivos for assessment by the FBC.

Once a carbon sale is agreed upon between the FBC and the farmers, they sign a collaborative agreement, which clearly defines the rights and responsibilities of all parties involved. This helps to avoid possible misunderstanding and conflicts in the future. All parties sign the agreement and set out a work plan.

In some cases after the collaborative agreement has been signed a small advance payment can be made to new producers/groups to help cover costs incurred while implementing (agro)-forestry activities. The advance payment is in proportion to the implementation costs estimated with the Plan Vivos.

4. COSTS ASSOCIATED WITH THE SCOLEL TÉ PROJECT

4.1. Cost of developing Plan Vivos

To develop a Plan Vivo, the Scolel Té project has developed and tested a cost-efficient participatory methodology. Farmers usually are interested in implementing a variety of agroforestry options adapted to his/her own situation according to land availability, the experience with certain agroforestry related activities, and the particular conditions of his/her farm. This requires a labor-intensive consultation process. Therefore, in each community local technicians are trained in developing a Plan Vivo, through a 2 to 3 day workshop. They learn about the specific goals of carbon sequestration projects, how to calculate carbon benefits, and which options are attractive for the farmers according to the particular conditions of his/her land. Subsequently the trained technicians develop Plan Vivos in their community, either on an individual farmer-to-farmer basis, community-based (such as community forest management or conservation, forest restoration, etc), or group-based proposal. After completion of the Plan Vivos, an assembly between the community and the FBC technicians is called to discuss the proposals. Where necessary, corrections are made and the final versions of the Plan Vivos are presented to the FBC for appraisal.

The total cost of developing one Plan Vivo, thus, is mainly a function of the number of farmers that are interested in participating, the time required to teach the farmers the goals of the project and the land-use options offered by the program, and the traveling distance between the communities and the office of the FBC. Independently of the number of farmers, elaborating a Plan Vivo typically requires about 3 days of training by a professional technician. Salary, transport, and lodging of the technicians are the most important recurring costs for these training sessions, which vary from US$ 400 to 500. A strategy is being developed to make sure that at least 10 Plan Vivos result from these efforts, thus limiting the cost of each Plan Vivo to about US$ 50.

4.2. Cost of establishment and maintenance (first 3 years)

Each farmer or community participating in the Scolel Té project informs the FBC about the labor and material required to establish and maintain his/her proposal. Based on the planned activities, the FBC determines if the proposal is technically and economically feasible. If the proposal passes the test, the FBC offers the farmer a collaborative agreement to be signed by both parties, stipulating the amount of carbon services to be traded, the price of the carbon service and the terms of payment. The cost estimate that the farmer makes for each option reflects his/her level of experience with the specific requirements of planting and maintaining trees as part of his/her productive activities. In general, farmers report the amount of labor required within an acceptable range, varying from about 40 to 50 man days necessary to establish an agroforestry system with 650 to 800 trees per hectare, such as taungya and enriched fallow.

The labor requirements that farmers report for agroforestry activities that require less trees, such as living fences, coffee with shade trees, pasture with dispersed trees, are more or less proportional to the amount of trees to be planted. Some activities require more material and labor, for instance due to the need to protect the trees from grazing animals. The average cost of establishing, monitoring, and maintaining an agroforestry systems with 650 to 800 trees planted is about US$ 890 (Table 2). Taking into account only the establishment cost, this amount would reduce to about half (US$ 430).

Table 2. Average costs of establishment and maintenance of agroforestry plantations (Based on introducing 625 trees/ha).

Category	*Type*	*Cost (US$/ha)*
Labor	Man days (US$ 6/day)	260
Plants	Production and transport	120
Material	Planting, protection	50
Training and design	Workshops	50
Monitoring (first 3 years)	Man days, transport	60
Maintenance (first 3 years)	Man days, plants	350
Total		890

4.3. Opportunity costs

The predisposition of farmers to add the cultivation of trees for timber or other purposes to their current land-use practices, is determined by a mixture of economic, social and cultural factors. The main economic determinant of farmers changing land use will be the opportunity cost, as measured by the net income derived from or subsistence value of the current land use.

To estimate the opportunity costs, Tipper, De Jong, Ochoa-Gaona, Soto-Pinto, Castillo-Santiago, & Montoya-Gomez (1998) calculated the distribution of net income per hectare from maize production. In this study, 69 farmers provided estimates of average yields and costs of production. In the Selva Lacandona, the net income per hectare varied from less than zero to US$ 520 per hectare (Figure 2) and in the Highlands of Chiapas from less than zero to US$ 320 per hectare (Figure 3).

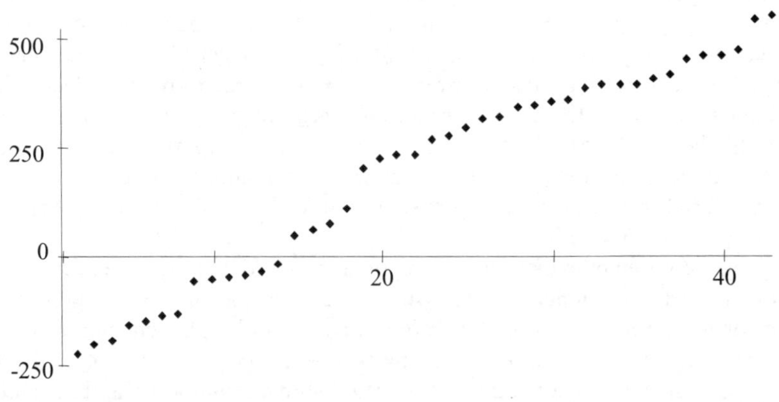

Figure 2. Net income per hectare (in US$) of maize production in the Selva Lacandona region of Mexico (Tipper et al., 1998). Each point represents a farmer interviewed, in ascending order of income per ha.

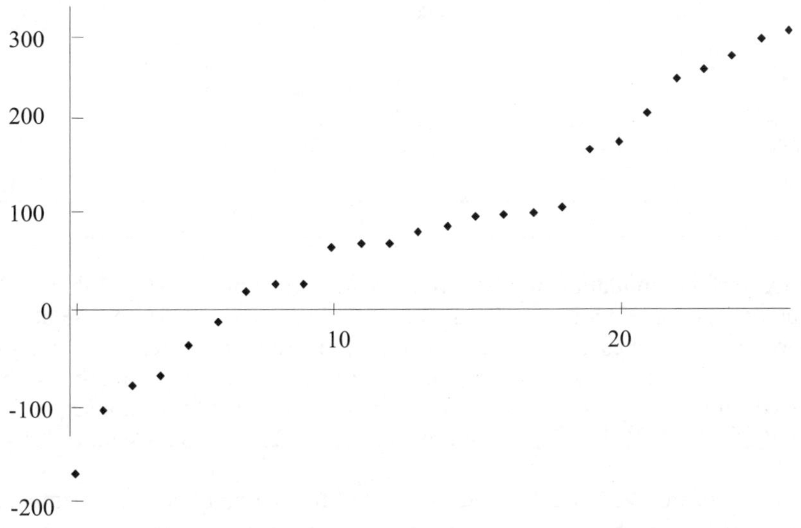

Figure 3. Net income per hectare (in US$) of maize production in the Chiapas Highlands region of Mexico (Tipper et al., 1998). Each point represents a farmer interviewed, in ascending order of income per ha.

Tipper et al. (1998) also estimated that the opportunity costs associated with transferring land from cattle ranching to forestry would vary between US $ 65 and 260 and converting fallow into forestry between US $ 6.50 and 195, all per hectare and per year. De Jong et al (2000a) used these incremental curves of opportunity costs to estimate the carbon sequestration potential for the Highlands of Chiapas. In this chapter we apply the median opportunity cost to calculate the cost per unit of carbon to be sequestered.

4.4. Cost per unit of carbon sequestered

A carbon sequestration estimation profile has been developed for each option, based on the proposed activities, such as selected tree species to be planted, planting distance, local ecological conditions, etc. These profiles are based on empirical data collected on tree growth in each ecological zone, soil carbon under various land-use practices, and data on carbon dynamics of other pools (e.g., De Jong, Cairns, Ramírez-Marcial, Ochoa-Gaona, Mendoza-Vega, & Haggerty, 1999; De Jong, Ochoa-Gaona, Castillo-Santiago, Ramírez-Marcial, & Cairns, 2000b). We used the

Table 3. Labor requirements (man days) to implement 1 ha of agroforestry option. in Southern Mexico (Living fence: 200 trees in 400 m. of fence).

Activity	*Living fence*	*Improved fallow*	*Plantation in Pasture*
Transport of plants	2	1.2	1.7
Cleaning	3	10.2	9.0
Digging holes	4	8.6	7.0
Planting	4	6.0	7.0
Weeding	4	14.8	25.1
Total	17	40.8	49.8

CO2FIX carbon simulation model (Mohren & Goldewijk, 1990; Mohren, Garza Caligaris, Masera, Kanninen, Karjalainen, & Pussinen, 1999) to calculate the expected carbon fluxes in the systems, and converted the outcome to long-term average increase in stock (De Jong et al, 2000a). According to our estimates, improved fallow and plantations in current pastureland would have a sequestration potential of 128 (± 25.6) t C per hectare, and living fences would sequester 54.1 (± 10.9) t C.

Some costs are fixed and thus independent of the option to be implemented, such as training the farmers and monitoring the project performance. The labor requirements to establish these systems vary according to the number of trees planted, and amount of cleaning and weeding of the plots (Table 3).

Taking into account these fixed and variable costs, the cost per unit of carbon varies between US$ 7.92 and 9.73 for each t C to be sequestered (Table 4). As the current Scolel Té project revenue amounts to US $ 13.00 / t C, this gives the FBC a margin of between US $ 3.27 and 5.08 for each t C traded, to cover the costs of project administration, marketing the carbon offsets and, eventually, the development of new options.

Table 4. Cost of sequestering 1 t C for three agroforestry systems ($US/t C), based on a case study carried out in Tenosique, Tabasco, Mexico.

Activity	*Living fence*	*Improved fallow*	*Plantation in Pasture*
Training	50	50	50
Establishment	214	454	608
Maintenance (3 years)	150	300	350
Opportunity cost	0	150	177
Monitoring	60	60	60
Total	474	1.014	1,245
C-accumulation (t C / ha)	54.1	128	128
Cost per t C ($US / t C)	8.76	7.92	9.73

5. COSTS OF DEVELOPING NEW OPTIONS

When a new agroforestry option is being proposed or farmers want to participate from regions where no carbon data are available, a new option profile needs to be designed. This will require significant additional funding, either from external sources, such as government institutes, or from former revenues of carbon-offset sales. The latter will have an impact on the financial assets of the project, the level of which will depend on the potential of additional carbon sales from the new option.

The project design cost estimate presented here includes the fixed costs incurred in calculating the potential for carbon sequestration of the option (baseline development, biomass and carbon stock flow measurements, option design), and variable costs such as promotion in farmer communities and training of farmers, among others. The curve of project design cost in relation to the carbon offset potential of the proposal would thus give an indication of the minimum amount of carbon that has to be sold in order to maintain the relative cost of project design within acceptable limits. We used data from a case study, carried out between November 2002 and August 2003, complemented with information from a feasibility study carried out in 1995 (De Jong et al., 1997). The costs were categorized into five activities. Fixed-cost activities include the cost of estimating the carbon sequestration potential of the agroforestry options within acceptable confidence limits, the development of a regional baseline, training of the technicians, and data analysis. Variable-cost activities are those associated with the development of the Plan Vivos in each community, such as the cost of community involvement, (Table 5).

The cost of developing a Plan Vivo are more or less fixed and only vary slightly according to the number of communities participating (effort of training a small versus large number of technicians), the size of each community (number of technicians to be trained in relation to the number of Plan Vivos to be developed), and the educational background of the farmers to be trained (time required to train a technician) and interviewed (time required to develop each Plan Vivo).

Table 5. Costs related to developing new agroforestry options.

Activity	*Nature*	*Type*	*Cost (US$)*
Community involvement	Workshops	Variable	4,000
Training of community technicians	Field courses		
Carbon estimation	*3 technicians, 1 week*	Fixed	2,000
Plan vivo development	*8 technicians, 4 days*	Variable	2,400
Data collection	Fieldwork		
Carbon estimation	*Inventories, land use maps*	Fixed	20,000
Plan vivo development	*Discussions with farmers*	Variable	8,000
Data analysis	Laboratory, computer	Fixed	6,000
Final conclusions	Workshops	Variable	3,000
Total			45,400

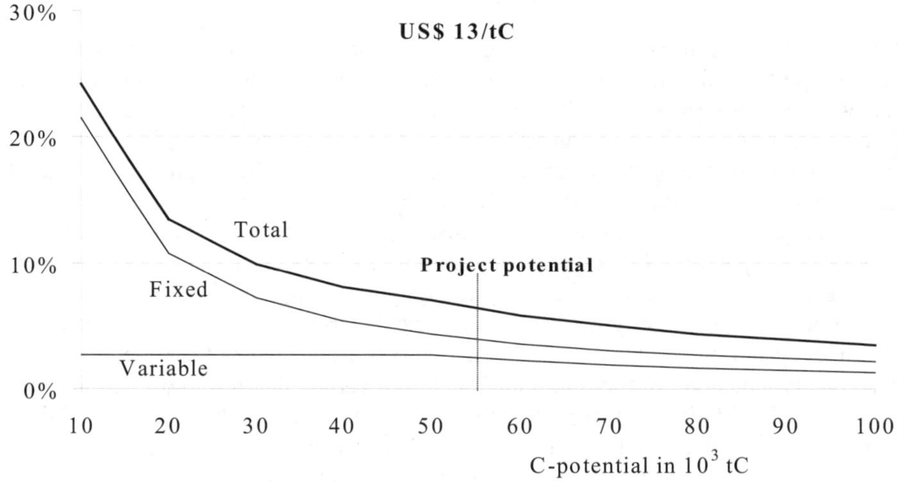

Figure 4. Variable, fixed and total cost of project design as percentage of hypothetical carbon revenues.

Weighing the cost of project design against the potential revenue of future carbon sales shows that, if the project were to result in a sale of 10,000 t C at a price of US$ 13/t C (the current price of the carbon offset), the cost of project design would represent almost 25% of the revenue (Figure 4).

In contrast, if the project is able to sell the amount of carbon equivalent to the sequestration potential of all submitted Plan Vivos (about 55,000 t C), this cost item would represent about 7% of total revenue. If a 10% contribution to the expected revenue is used as an upper limit, the new option needs to sell at least 30,000 t C.

Presently, developing new types of activities are outside the financial capacity of the Scolel Té project, as current sales of the offsets are not enough to finance these activities. As such, external funding offered by the Mexican government has so far given the opportunity to carry out these type of studies. Offset sales to be developed through the Scolel Té program will benefit from the database created by the study. In turn, this could generate an additional financial asset for the Scolel Té project, when marketing the technology to other initiatives. Future sales of Scolel Té within a range that covers the new options now could be made more cost-effectively, since these new initiatives only have to deal with the costs to develop Plan Vivos, which represent a more or less fixed cost per plan.

6. TRANSACTION COSTS TO DESIGN THE PLAN VIVO SYSTEM

To set up the Scolel Té project, various institutions assisted financially to cover the initial transaction costs, predominantly through specific projects to answer certain research questions and to set up the Plan Vivo Management System, all with the ultimate goal of establishing a prototype carbon-offset project that can be scaled-up to a regional or national level or that can be used in other countries. The Mexican government started the process with financing a feasibility study to assess the willingness of resource-poor small-scale farmers to participate in a project with C-sequestration as the main objective, which land-use options would be the most attractive activities, both from the point of view of the farmers and the C-sequestration perspective, and what would be the C-sequestration potential of each option (De Jong, Montoya-Gómez, Nelson, Soto-Pinto, Taylor, & Tipper, 1995). Various agroforestry systems were considered economically, technically, and socially attractive as opportunities to offset carbon as well as converting the current land-use activities into sustainable systems. Farmers were interested in developing a variety of agroforestry alternatives ranging from low-intensity interventions, such as living fences, to systems (almost) completely converting the current land use to a new one, such as taungya and fallow-enrichment planting (De Jong et al., 1997).

Carbon inventories and regional land-use data that were required to develop a regional baseline and to estimate the C-sequestration potential of each option were also financed with external funding (De Jong et al., 1999, 2000b). At the time that the project started to trade carbon offsets in 1997, external funding was available to set up the Plan Vivo system, to assist in the capacity building of local representatives (social advisors, farmer representatives and individual farmers), and to set up a network of permanent sample plots. This funding allowed the project to develop a prototype system for carbon offset trading, to develop a set of technical specifications of the most attractive agroforestry options for carbon sequestration, and to establish a baseline for an area of about 2,700,000 ha. In Table 6 we present an outline of the costs of each of the activities required to design the Plan Vivo management system.

From 2000 onward, all transaction costs related to project administration; technical support to communities for planning and implementing forestry activities, evaluating management plans and monitoring activities have been covered directly by the carbon-offset revenues. The use of community technicians for training and monitoring activities has helped to increase community involvement and reduce

Table 6. Brief outline of transactions costs required designing the Plan Vivo Management System, financed with external funding.

Activities	*Period*	*Costs (US$)*
Feasibility study	1995	80,000
Carbon inventories and land-use analysis	1994 to 1997	400,000
Development of Plan Vivo system	1997 to 2000	210,000
Establishment of permanent sample plots	1999	20,000
Development of regional baseline (2,700,000 ha)	1999 to 2002	350,000
Total	1995 to 2002	1,060,000

operational cost. External transaction costs, such as improving regional baselines, extending the project to other regions, and project verification are not yet covered by the revenues and are still dependent on external funding.

Research on measuring the offsets is ongoing, with continual improvements on the originally conservative estimates. On the other hand, the Plan Vivo Management System is currently applied in a project in India and tested in some countries in Africa (www.planvivo.org).

7. DISCUSSION

Cost curves could yield information that is not available in point estimates or ranges of cost per unit of carbon. These curves are especially important to estimate the potential of carbon sequestration at a regional level, as a function of the amount of carbon offset incentives (De Jong et al., 2000a). They also give an indication of which land-use alternatives are the most attractive, either from the point of view of rural farmers as part of a rural development strategy, or from the point of view of the global climate mitigation economy. It would be interesting if cost-benefit curves could also be developed for other ecological services, such as biodiversity, water and soil conservation, and incorporate these in a model to calculate additional co-benefits of the land-use alternatives.

Our assessment of opportunity costs associated with diverting land from current land use to forestry is based on limited samples. Experience gained from the Scolel Té pilot project indicates that the participating families have sufficient resources available to initiate afforestation activities on 0.5 to 1.5 ha, without a significant drain on the labor resource for current subsistence and cash crop agriculture. Participating communities agreed to set aside between 50 and 100 ha of communal land for reforestation or forest conservation, through the community decision-making process. To date, carbon offsets market through the Scolel Té project have been very limited due to lack of selling opportunities. This means that most of the farmers or communities that entered the program had relatively low opportunity costs, making the decision to enter the program relatively easy. Once the project expands the sales of carbon offsets significantly, it may be increasingly more difficult to find farmers willing to participate due to the increasing opportunity costs.

Funding to develop new options eventually has to be covered by the carbon-offset sales. It will be critical therefore for the project to first invest in marketing offsets. Increasing the budget will decrease the cost of administration per unit of carbon, which currently takes up almost 30% of the total budget.

Technical constraints for the project include, among others, the lack of high quality planting material in many areas. Government and private tree nurseries are relatively few and rather poorly stocked. If the Scolel Té project were to produce its own planting material, this would require additional investments for a period of at least 3 to 4 years to collect and propagate sufficient high quality stock in readily accessible nurseries.

The agroforestry options from which farmers select are specifically chosen measures expected to be economically self-sustainable in the long-term, assuming a stable price of timber and agricultural products. Depending on the terms and conditions within any sequestration agreement, the responsibility for assuring the conservation of sequestered carbon stocks in perpetuity currently rests with the individual farmers. The state of the future market products from the agroforestry systems will clearly be a crucial factor in determining the sustainability of the new systems. The quality of technical and organizational support available now and in the future to the new farmer enterprises will also play a role. Critical factors include the selection of appropriate local genetic material for specific climatic and soil conditions, consideration of the multiple uses that agroforestry systems have for local people, design and implementation of appropriate measures for fire and pest control, tenure rights on land associated with the program, inclusion of relevant stakeholders in the design and management of the new agroforestry systems, and the development of local management capacities (in both individuals and institutions).

8. REFERENCES

Boer, R. (2001). Economic assessment of mitigation options for enhancing and maintaining carbon sink capacity in Indonesia. *Mitigation and Adaptation Strategies for Global Change, 6*, 3-4, 257-290.

De Jong, B. H. J., Montoya-Gómez, G., Nelson, K., Soto-Pinto, L., Taylor J., & Tipper; R. (1995). Community forest management and carbon sequestration: a feasibility study from Chiapas, Mexico. *Interciencia 20*, 6, 409-416.

De Jong, B., Soto, L., Montoya, G., Nelson, K., Taylor, J., & Tipper, R. (1997). Forestry and agroforestry alternatives for carbon mitigation: an analysis from Chiapas, Mexico. In W. Adger, D. Pettenella, & M. C. Whitby (Eds.). *Climate Change Mitigation and European Land Use Policies* (pp. 263-284). Wallingford, United Kingdom: CAB International.

De Jong, B. H. J., Cairns, M. A., Ramírez-Marcial, N., Ochoa-Gaona, S., Mendoza-Vega, J., & Haggerty, P. K. (1999). Land-use change and carbon flux between the 1970s and 1990s in the Central Highlands of Chiapas, Mexico. *Environmental Management 23*, 3, 373-385.

De Jong, B. H. J., Tipper, R., & Montoya-Gómez, G. (2000a). An economic analysis of the potential for carbon sequestration by forests: evidence from southern Mexico. *Ecological Economics 33*, 2, 313-327.

De Jong, B. H. J., Ochoa-Gaona, S., Castillo-Santiago, M. A., Ramírez-Marcial, N., & Cairns, M. (2000b). Land-use/land-cover change and carbon flux in the Selva Lacandona, Mexico. *Ambio 29*, 8 504-511.

Dixon, R. K., Schroeder, P. E., & Winjum, J. K. (Eds.). 1991. Assessment of promising forest management practices and technologies for enhancing the conservation and sequestration of atmospheric carbon and their costs at the site level. [Gen. Tech. Rep. EPA/600/3-91/067]. Washington, D.C.: United States Environmental Protection Agency.

Kauppi, P., Sedjo, R., Apps, M., Cerri, C., Fujimori, T., & Janzen, T. (2001). Technical and economic potential options to enhance, maintain and manage biological carbon reservoirs and geo-engineering, In O. Davidson, B. Metz, & R. Swart (Eds.). *Climate Change 2001: Mitigation* (pp 301-344). Cambridge, United Kingdom: Cambridge University Press.

Lasco, R. D. & Pulhin, F. B. (2001). Climate change mitigation activities in the Philippine forestry sector: Application of the COMAP model. *Mitigation and Adaptation Strategies for Global Change, 6*, 3-4, 313-334.

Makundi, W. R. (2001). Potential and cost of carbon sequestration in the Tanzanian forest sector. *Mitigation and Adaptation Strategies for Global Change, 6*, 3-4, 335-353.

Masera, O., Cerón, A. D., & Ordóñez, A. (2001). Forestry mitigation options for México: Finding synergies between national sustainable development priorities and global concerns. *Mitigation and Adaptation Strategies for Global Change, 6*, 3-4, 291-312.

Mohren, G. M. J., Garza Caligaris, J. F., Masera, O., Kanninen, M., Karjalainen, T., & Pussinen, A. (1999). CO2FIX for Windows: a dynamic model of the CO2-fixation in forests, Version 1.2. (IBN Research Report 99/3). The Netherlands: Wageningen.

Mohren, G. M. J., & Goldewijk, C. G. M. K. (1990). CO2FIX: a dynamic model of the CO2-fixation in forest stands. (Report No. 624[35]). Wageningen, The Netherlands: De Dorschkamp, Research Institute for Forestry and Urban Ecology.

Moulton, R. J., & Richards, K. R. (1990). Costs of sequestering carbon through tree planting and forest management in the United States. (Gen. Tech. Rep. WO-58). Washington, DC: USDA-FS.

Ravindranath, N., Sudha, P., & Rao, S. (2001). Forestry for sustainable biomass production and carbon sequestration in India. *Mitigation and Adaptation Strategies for Global Change, 6*, 3-4, 233-256.

Read, D. (2001). The role of land carbon sinks in mitigating global climate change. (Policy Document 10/01). Retrieved February 17, 2004, from The Royal Society of London Website, http://www.royalsoc.ac.uk/files/statfiles/document-150.pdf.

Sathaye, J. A., & N. H. Ravindranath, (1998). Climate change mitigation in the energy and forestry sectors of developing countries. *Annual Review of Energy and Environment 23*, 387-437.

Sathaye, J. A., Makundi, W. R., Andrasko, K., Boer, R., Ravindranath, N. H., & Sudha, P. (2001). Carbon mitigation potential and costs of forestry options in Brazil, China, India, Indonesia, Mexico, the Philippines, and Tanzania. *Mitigation and Adaptation Strategies for Global Change, 6*, 185-211.

Swisher, J., & Masters, G. (1992). A mechanism to reconcile equity and efficiency in global climate protection: international carbon emissions offsets. *Ambio, 21*, 2 154-159.

Tipper, R., & De Jong, B. H. (1998). Quantification and regulation of carbon offsets from forestry: comparison of alternative methodologies, with special reference to Chiapas, Mexico. *Commonwealth Forestry Review 77*, 3 219-227.

Tipper, R., De Jong, B. H. J., Ochoa-Gaona, S., Soto-Pinto, M. L., Castillo-Santiago, M. A., & Montoya-Gómez, G. (1998). Assessment of the cost of large-scale forestry for CO2 sequestration: Evidence from Chiapas, Mexico. UK: International Energy Agency R & D Programme.

Tipper, R., Montoya, G., De Jong, B. H., Castillo, M. A., March, I., & Soto, L. (1997). *Assessing the cost of large-scale forestry for CO2 sequestration in Southern Mexico: some preliminary results.* In P. W. F. Riemer, A. Y. Smith, & K. V. Thambimuthu (Eds.). Greenhouse Gas Mitigation: Technologies for Activities Implemented Jointly (pp. 177-185). Amsterdam, The Netherlands: Pergamon.

Xu, D., Zhang, X. Q., & Shi, Z. (2001). Mitigation potential for carbon sequestration through forestry activities in southern and eastern China. *Mitigation and Adaptation Strategies for Global Change, 6*, 3-4, 213-232.

OSCAR CACHO AND ROBYN HEAN

DYNAMIC OPTIMIZATION FOR EVALUATING EXTERNALITIES IN AGROFORESTRY SYSTEMS:

An example from Australia

1. INTRODUCTION

Integrating trees in cropping and grazing systems – agroforestry – can provide many benefits in the Australian agricultural context. These include the production of timber and non-timber products such as oils and flowers, fodder, windbreak protection, shade and shelter, wildlife habitat, flood mitigation, soil-erosion control, improved water quality, and reduced dryland-salinity emergence. Prinsley (1992), Cleugh, Prinsley, Bird, Brooks, Carberry, and Crawford et al. (2002) and other authors in the same volumes present some examples for Australia. Many of the benefits from trees are off-farm environmental services, which are public goods or externalities and which landholders may not take into account. The social benefits from trees may therefore exceed the private benefits, and given such a divergence, landholders may conserve and plant too few trees from a social perspective. Without perfect information, landholders may even underestimate their private benefits from trees and further exacerbate this divergence. Concern for this issue is not new in Australia (see Tisdell, 1985).

Market failure due to externalities and imperfect information provides a rationale for government intervention to encourage landholders to invest in vegetation management and reforestation. Regulatory, extension, and market-based approaches are all being used to this end by governments in Australia.

State Governments have established regulatory controls on land clearing of private native vegetation (Walpole, 1999; Stoneham, Chaudhri, Ha, & Strappazzon, 2003), such as the Native Vegetation Conservation Act in New South Wales (NSW) and The Planning and Environment Act in Victoria. State and Federal Governments

J. R. R. Alavalapati & D. E. Mercer, Valuing Agroforestry Systems, 139–163.

have also implemented extension programs to provide funds and/or assistance to landholders and community groups to manage native vegetation on private land. At the State level, these programs include Land for Wildlife and Trust for Nature in Victoria, and the Voluntary Conservation Agreement Program in Queensland. At the Federal level, there is Landcare, One Billion Trees, Save the Bush, Bushcare, National Heritage Trust, and the National Action Plan for Salinity and Water Quality. Stoneham et al. (2003) provides an overview of some of these programs.

State and Federal governments are also trialing market-based approaches to managing environmental externalities. For example, the NSW Government is trialing an Environmental Services Scheme where landholders can apply for funding to change their land-use practices to improve any of six environmental services (carbon-sequestration, biodiversity, salinity, soil, water, and acid-sulfate benefits). In Victoria, the Government is trialing a BushTender initiative through which landholders 'bid' to conserve areas of private native vegetation for biodiversity enhancement (Stoneham et al., 2003). The Commonwealth, State, and Territory Governments are also jointly funding a National Market-Based Instruments Pilots Program to trial the use of trading instruments and offset schemes to change the behavior of landholders towards important natural-resource management issues, such as biodiversity, dryland and irrigation salinity, and water quality.

An evaluation of the economic efficiency of these three approaches is beyond the scope of this chapter. We emphasize here that, whichever approach is used, it is important to understand the environmental benefits that can be provided by trees. Lack of information about the environmental benefits from trees is one of the obstacles to restoring landscape vegetation to address land-degradation issues in Australia. When making tree/crop decisions, landholders tend to focus on the relative input and output prices of crops and trees, as well as risk considerations. Moreover, the long time lag between investment and returns in the forestry enterprise generally makes this activity unattractive compared with alternative land uses when not all the benefits are considered.

In this chapter, we demonstrate a general technique for evaluating externalities provided by trees for crops through improvements in land productivity from mitigation of land degradation. This extends the analysis presented by Cacho (2001) for an agroforestry system in a catchment affected by dryland salinity. In the analysis presented here we also include payments for carbon sequestration by the forestry enterprise. A general model of an agroforestry system comprising an annual crop and a tree crop is initially developed for a catchment represented by a homogenous plot of land under the control of a single-decision maker or managed as common property. An application is then presented, using a dynamic optimization approach to estimate optimal land allocation between forestry and agriculture.

2. MODEL AND METHODOLOGY

The optimal management of an agroforestry system in the presence of land degradation is addressed by the decision question: what is the optimal area of trees to plant in an agroforestry system and how long should the trees be kept before harvest? In essence, the question boils down to estimating the direct (timber) and indirect (prevention of land degradation, carbon sequestration) net benefits provided by trees and comparing them to the net benefits provided by crops. The decision is made in a dynamic setting, where decisions taken today have a bearing on the range of decisions available in the future.

The problem of the socially-optimal forest rotation in the presence of non-timber values in forests was first studied by Hartman (1976). He derived the decision rule (with an infinite planning horizon) for the optimal rotation when the standing forest provides positive externalities. He showed that the optimal harvest age is that at which the growth rate of the forest equals the value of the discounted stream of non-timber benefits relative to timber benefits up to the time of harvest. Hartman's work has been extended by authors such as Bowes and Krutilla (1985, 1989), Englin and Klan (1990) and Swallow, Parks, and Wear (1990).

Bowes and Krutilla (1985) focus on the multiple-use management of public forestlands, where land managers must consider not only the value of timber harvested, but also non-market benefits such as recreation, water flow, and wildlife. They introduce multiple forest stands of different ages and the manager's decision is whether to harvest some or all of them at a given point in time. Bowes and Krutilla (1989) present a detailed review of previous models as well as an applied forest-management model where the aesthetic, water flow, and amenity values are expressed as functions of the age of the forest stand. They also show how to apply linear programming to solve multiple-use forest management problems.

In the case of agroforestry, timber production and non-timber benefits may occur in different areas of land, where the area planted to crops benefits from the area planted to trees. This results in an additional decision variable, not considered in the papers discussed above; the proportion of the area available that should be planted to trees, with the remaining area planted to crops.

2.1. General model

Consider a homogeneous plot of land that can be planted to any combination of trees and crops, and where trees provide land conservation services. The benefit obtained from a hectare of land over a single forest rotation of length T is:

$$NPV(k,T) = (1-k)\sum_{t=1}^{T} a_t(s_t,k)\cdot(1+r)^{-t} + k\sum_{t=1}^{T} f_t(s_t,k)\cdot(1+r)^{-t} - k\cdot c_E\,; \quad \text{with } 0 \le k \le 1 \qquad (1)$$

where s_t is the state of the land in year t, k is the proportion of the plot planted to forest and c_E is the cost of forest establishment. The state variable, s_t, represents a quality indicator, such as soil depth, soil-carbon content and soil fertility; or it may be defined to represent a negative quality, such as soil salinity or sodicity. Note that this equation considers only establishment costs per hectare, so the cost of forest establishment increases linearly with the proportion of land under trees. In other words, no fixed costs that may cause economies of scale are considered.

The net present value (*NPV*) of the flow of benefits obtained over a single forest rotation (equation 1) consists of the accumulation of direct monetary benefits, $a(\cdot)$, provided by an annual agricultural crop, and the benefits provided by a forestry operation, $f(\cdot)$. The discount rate is represented by r.

The dynamics of the state variable are given by:

$$s_{t+1} = s_t + \Delta s_t(s_t, k) \tag{2}$$

with s_0 given.

The expected signs of the key derivatives are:

$$\frac{\partial a_t}{\partial s_t}, \frac{\partial f_t}{\partial s_t} \begin{cases} < 0 \text{ if } s_t \text{ measures land degradation} \\ > 0 \text{ if } s_t \text{ measures land quality} \end{cases}$$

$$\frac{\partial a_t}{\partial k} \begin{cases} < 0 \text{ if trees compete with crops for resources} \\ = 0 \text{ if the only effect of trees on crops is through } \Delta s_t \text{ (equation 2)} \\ > 0 \text{ if trees benefit crops directly (i.e., other than through } \Delta s_t\text{)} \end{cases}$$

$$\frac{\partial f_t}{\partial k} \begin{cases} < 0 \text{ unlikely} \\ = 0 \text{ if } k \text{ does not affect forest growth (other than through } \Delta s_t\text{)} \\ > 0 \text{ if there are economies of scale in forest establishment and growth} \end{cases}$$

Since, s_t is the state variable, which represents some measure of land quality, if s_t refers to a 'negative quality' such as salinity then an increase in s_t will lead to a decrease in returns from agriculture and forestry. Conversely, if the measure refers to a 'positive quality' such as soil fertility, then an increase in s_t will lead to an increase in returns from agriculture and forestry.

The reasoning behind the signs of the derivatives with respect to k is as follows: trees may compete with crops for light, water, and nutrients, so a larger area of trees may result in lower crop yields per unit area ($\partial a/\partial k < 0$); but trees may also benefit crops directly, by providing services such as shelter, nitrogen fixation, and pest control ($\partial a/\partial k > 0$). If both effects are present then the ultimate direction of change depends on the balance between positive and negative interactions. Important indirect services are also provided by trees, such as prevention of soil erosion and salinity, in these cases it is possible to have $\partial a/\partial k = 0$. This means that the effects of trees on crops occur indirectly, through changes in land quality (equation 2) and the resulting changes in crop yields, as represented in equation 4 below.

The relative area of trees in an agroforestry system may also affect the net benefits obtained from the forestry enterprise. Although changes in tree density can affect the growth rate of individual trees, this does not apply in our case because we assume constant planting density; so the case where $\partial f/\partial k < 0$ is unlikely to occur in the context of the problem studied here[1]. A larger area of trees may be associated with larger returns from forestry ($\partial f/\partial k > 0$) if there are economies of scale. In the case study presented later we assume that both derivatives with respect to k are zero, so the area planted to trees affects land quality only indirectly (through equations 2, 4 and 6).

The net monetary benefits obtained from agriculture in any given year t are:

$$a_t = p^a y_t^a - c^a \tag{3}$$

where p^a is the price of the crop and c^a is the cost of crop production per hectare. y_t^a is crop yield obtained in year t and is affected by the productivity of the land as follows:

$$y_t^a = \bar{y}^a \, Q_t^a(s_t) \tag{4}$$

where $\bar{y}^a$ is expected yield under 'normal' land productivity and Q_t^a is the land-productivity function for agriculture.

The annual net monetary benefits of the forestry operation are given by:

$$f_t = p^f y_t^f - c_t^f \tag{5}$$

where the price of forest outputs (p^f) is assumed to be constant irrespective of the age of the forest, while costs (c_t^f) and yields (y_t^f) do depend on forest age. Yields are also affected by land productivity:

$$y_t^f = \bar{y}_t^f \, Q_t^f(s_t) \qquad (6)$$

where $\bar{y}_t^f$ is expected yield in year t and Q_t^f is the land-productivity function for forestry. Because the model is solved in discrete time, c_t^f can be conveniently represented as a vector of known annual values rather than as an explicit function of time. The annual yields of the forestry operation, y_t^f , can also be represented as a vector (Cacho, 2001). In the numerical model developed below, a forest-growth function is used to estimate both annual forest outputs and final harvest. The model allows carbon-sequestration services (the annual forest output) to produce revenues. The forest-growth function is based on Cacho, Hean, and Wise (2003).

In the numerical analysis presented later, the economic model (1) is first solved in simulation mode, where the NPV is estimated for any arbitrary combination of input values, this is useful to gain a general understanding of how expected profits are affected by the decision variables and the initial state of the land. The model can also be solved in optimizing mode, by finding the values of k and T that maximize the value of (1) subject to the constraints imposed by equations (2) to (6).

Cacho (2001) showed the derivation of the first-order conditions for maximization of equation 1 with respect to k, while keeping T constant at the recommended tree-harvest age. Some of the relationships in the model are nonlinear and have complex derivatives. This means that it is more convenient to undertake direct numerical maximization of the objective function, using any of a number of constrained optimization techniques, than to use the first order conditions which require derivatives to be estimated.

The *NPV* function (1) represents the first forestry cycle only, so it does not take account of the opportunity cost of keeping trees on the ground rather than harvesting them. Cacho et al. (2003) used an infinite-rotation model to deal with this problem in their carbon-sequestration model. This common approach to including the opportunity cost of not harvesting is not applicable here, because land quality changes over time. This means that subsequent forestry cycles are not identical to the first cycle and optimal cycle length and optimal forest area may change between cycles. So, with forestry cycles denoted by n, the optimization problem is:

$$\max_{k_n, T_n} V_0 = \sum_{n=1}^{\infty} NPV(s_n, k_n, T_n) \cdot (1+r)^{-(T_0+\ldots+T_{n-1})} \qquad (7)$$

subject to equations (1) to (6) and with $T_0 = 0$. This is a very complex problem with an infinite number of decision variables. Fortunately, thanks to Bellman's principle of optimality, it can be simplified to a two-decision variable problem to be solved recursively. So we convert the multiple-rotation problem into a dynamic programming (DP) problem, where the stages are forestry cycles of length T_n. The DP recursive equation is:

$$V_n(s_n) = \max_{(k_n, T_n)} \left(NPV(s_n, k_n, T_n) + V_{n+1}(s_{n+1}) \cdot (1+r)^{-T_n} \right). \tag{8}$$

subject to equations (1) to (6). The problem is solved for an infinite planning horizon, by backward induction (Kennedy, 1986), until convergence in V_n is achieved. This involves combining the DP algorithm with the numerical model described below.

2.2. Numerical model

Dryland salinity is a major land degradation problem in Australia. It has been caused by replacing perennial native vegetation with farming and grazing systems that allow a larger proportion of rain to recharge groundwater systems, and is evidenced by high and rising saline water tables in low-lying, discharge areas of catchments. Greiner and Cacho (2001) provide an overview of the problem.

An issue that has received some attention recently, although it has not been debated in the formal literature, is the possibility of payments for carbon-sequestration services to contribute to salinity mitigation (see Hean, Cacho, & Menz, 2003). Although the main focus in the global-warming debate is on emissions (sources), sinks, such as carbon-sequestration in trees, have a role to play. Trees remove carbon dioxide from the atmosphere during photosynthesis and store the carbon in wood, leaves and roots; while the oxygen is released back into the atmosphere. At the international level, the Kyoto Protocol has provided much of the impetus for the policy debate, but other policies exist at different government levels. In Australia, both Victoria and NSW have enacted legislation to allow for the separate ownership of land, trees, and carbon-sequestration rights to facilitate carbon-credit exchanges, and they have also implemented schemes to investigate the potential for carbon-credit markets.

The numerical model below is based on dryland-salinity emergence in Australia in the presence of carbon-sequestration payments. Control of the land-degradation problem is driven by forest growth, defined as:

$$\Delta b_t = \left(\alpha_G \cdot b_t^{\beta_G} - \gamma_G \cdot b_t \right) \cdot Q_t^f \tag{9}$$

where b_t is aboveground forest biomass, Δb_t is the annual biomass increment, and Q_t^f is the land-productivity function for forestry as previously defined. α_G, β_G and γ_G are parameters determined by tree species and site characteristics, in this study we use parameter values for *Eucalyptus nitens* in southern Australia (Table 1). Biomass is measured in metric tons per hectare (t/ha).

The forestry operation is assumed to have two types of outputs: annual outputs (carbon-sequestration services) that depend on forest biomass accumulation, Δb_t, and a final harvest (timber) that depends on stemwood volume, v_T, measured in cubic meters per hectare (m^3/ha). Assuming 50 percent of forest biomass is carbon

Table 1. Parameter values used in the numerical model. (Currency in Australian\$).

Parameter	Value	Units	Description	Equation
Biophysical				
α_G	4.189	1/yr	forest growth parameter	9, 15
β_G	0.681	*	forest growth parameter	9, 15
γ_G	0.595	1/yr	forest growth parameter	9, 15
θ_G	453.976	t/ha	maximum forest biomass	14, 15
d	0.7	t/m^3	wood density	14
α_a	3.684	*	land-quality parameter, crop	16
β_a	2.608	*	land-quality parameter, crop	16
α_f	1.5	*	land-quality parameter, tree	16
β_f	2.608	*	land-quality parameter, tree	16
α_{RA}	80	mm/yr	crop recharge intercept	18
β_{RA}	40	mm/yr	crop recharge slope	18
α_{RF}	55	mm/yr	tree recharge intercept	19
β_{RF}	0.7	mm/yr	tree recharge slope	19
γ_R	160	mm/m	recharge conversion factor	17
$\bar{y}^a$	4.5	t/ha	maximum crop yield	4, 18
Economic				
r	6, 12	%	discount rate	1, 8
p^a	180	\$/t	price of crop output	3
p_v	30	\$/m^3	price of timber	11
p_c	20	\$/t	price of carbon	10, 11
c^a	140	\$/ha	variable cost of crop	3
c_E	2000	\$/ha	forest establishment cost	1
c^f_t	50	\$/ha	forest maintenance cost	5, 10

**coefficient has dimension 1.*

(Brown, 1997; Hamburg, 2000), annual forest output is proportional to carbon-sequestration rate. Hence, equations (5) and (6) are together replaced by:

$$f_t = p_c \cdot 0.5 \cdot \Delta b_t - c_t^f \text{ ; for } 0 < t < T \quad (10)$$

for annual forest outputs, and by:

$$f_T = p_c \cdot 0.5 \cdot \Delta b_t + p_v \cdot v_T - p_c \cdot 0.5 \cdot b_T \quad (11)$$

for final harvest. p_c and p_v are the prices of carbon (Australian$/ton of Carbon [$/t C]) and timber (Australian$/cubic meter [$/m^3]) respectively. In the second term in equation 11, which represents the value of the timber harvest, p_v is assumed to be constant, although more realistically it would depend on the stemwood diameter of the trees at harvest (see Cacho et al., 2003).

The first term in both equations (10) and (11) represents the annual payments from carbon sequestered in the interval (0, …, *T*), while the last term in equation 11 represents the assumption that carbon credits received during forest growth have to be fully redeemed upon harvest. Full debit at harvest means that the total amount of carbon credits received during the life of the forest must be paid back to the investor by the landholder at harvest. As pointed out by Cacho et al. (2003), this implicitly assumes that the contract ends as the sequestered carbon is no longer under the control of the landholder. In other words, the contract between an investor (e.g., a power company) and a landholder to capture and maintain a given amount of carbon out of the atmosphere expires when the forest is harvested, and the landholder cannot guarantee that the terms of the contract will continue to be fulfilled. Once the contract expires, the investor would have to find an alternative sequestration project, or pay a carbon tax. This scheme is equivalent to the rental carbon market proposed by Marland, Fruit, and Sedjo (2001).

Biomass and timber volume are estimated by numerical integration of the model as follows:

$$b_{t+1} = b_t + \Delta b_t \quad (12)$$

$$v_{t+1} = v_t + \Delta v_t \quad (13)$$

with b_0 and v_0 given. Increments in timber volume are estimated by:

$$\Delta v_t = \frac{0.7 \cdot \Delta b_t \cdot \left(\dfrac{b_t}{\theta_G} \right)^{0.2}}{d} \quad (14)$$

where d is wood density (t/m^3) and θ_G is the maximum biomass achievable by the forest (t/ha):

$$\theta_G = \left(\frac{\alpha_G}{\gamma_G}\right)^{1/(1-\beta_G)} \tag{15}$$

Equation 14 assumes that an increasing proportion of new growth is stemwood. When trees are young, they generally have more branches and foliage relative to stem than old trees. By maturity, approximately 70 percent of biomass is stemwood.

Where dryland-salinity emergence is a problem, trees can be strategically placed in recharge areas to reverse trends in rising water tables. In this example, land quality (s_t) in equation 2 is represented by the depth of the water table (w_t) measured in meters (m) below the soil surface. Hence, the land-productivity function is defined as:

$$Q_t^j = 1 - \alpha_j \cdot \exp\left(-\beta_j \cdot w_t\right) \text{ for } j{=}a{,}f \tag{16}$$

where the parameters α_j and β_j depend on land and plant characteristics. This function implicitly accounts for the relationship between the water table and salinity, and the effect of salinity on crop yields and tree growth (see Cacho, 2001 for an interpretation of this function).

Annual changes in the water-table depth are given by:

$$\Delta w_t = -\frac{(1-k)R_t^a + k\,R_t^f}{\gamma_R} \tag{17}$$

where R_t^j represents the amount of recharge associated with activity j (j=a,f) in year t and γ_R converts total recharge (in mm/m^2) to water-table depth changes (in m) over the watershed. The value of γ_R depends on characteristics of the aquifer and the nature of water movements below the soil surface, and can be adjusted to represent areas with different levels of propensity to dryland-salinity emergence.

Recharge rates depend on the amount of rainfall received, the amount of runoff to streams and the amount of water taken up by plants. Young trees do not eliminate deep water, but as they grow larger their roots reach deeper into the water table and eliminate large volumes of water through evapotranspiration, so R_{tf} becomes negative as trees grow. The recharge rates associated with agriculture and forestry are given by:

$$R_t^a = \alpha_{RA} - \beta_{RA} \cdot \frac{y_t^a}{\overline{y}^a} \quad (18)$$

$$R_t^f = \alpha_{RF} - \beta_{RF} \cdot b_t \quad (19)$$

where α_{RA} is the recharge (millimeters/ year [mm/yr]) that would occur under fallow land and α_{RF} is the recharge under newly planted tree seedlings. β_{RA} and β_{RF} are parameters, and the remaining variables are as previously defined.

The model just described can be simulated for any given tree area and forestry cycle-length by solving equations (9) to (19) while numerically integrating the water table as follows:

$$w_{t+1} = w_t + \Delta w_t(w_t, k) \quad (20)$$

Equation 20 replaces equation 2 in this example, since land quality (s_t) is represented by the depth of the water table.

The model was simulated for the base-case parameter values shown in Table 1. These parameters represent good-quality land in Australia. The tree-growth parameters are for *Eucalyptus nitens* (Cacho et al., 2003), and the salinity parameters are largely based on Cacho (2001), but the expected crop yield, $\overline{y}^a$, is higher to represent good-quality land (i.e., 4.5 t/ha). Currency ($) is measured in Australian dollars, and biomass and carbon (C) weights are measured in metric tons per hectare (t/ha).

Four scenarios (Table 2) were simulated, including the base case, to explore the effects of carbon payments and discount rates on optimal solutions. There is a lot of uncertainty about the price of carbon that may emerge from a competitive carbon market. The carbon price used here has been used in other studies (see Cacho et al., 2003) and is conservative.

Table 2. Scenarios simulated. (Currency in Australian$).

Scenario	*Carbon price ($/t)*	*Discount rate (%)*
1 (base)	0	6
2	20	6
3	0	12
4	20	12

2.3. Implementation of the DP model

The application of dynamic programming (DP) models to forestry is not new. Kennedy (1986) devotes a chapter to the derivation of optimal forestry-rotation rules in a DP context. He describes both deterministic and stochastic models, where optimal thinning and harvesting rules are derived based on the age of the forest. In the simplest version (no thinning), the decision variable is binary: keep the trees, or harvest and replant. When thinning is introduced the decision variable becomes continuous, representing the proportion of the forest harvested at a given age, so total harvest is represented by a value of one and no thinning has a value of zero. Although the decision variable is continuous within the interval (0-1), it is generally expressed as a set of discrete values in DP models. Because of problems with dimensionality, these models are usually solved at time steps of 5 or 10 years. Kennedy (1986) presents a summary of DP applications to forestry management published between 1966 and 1982.

Solution of a DP model consists of solving a recursive equation that maximizes the reward obtained from managing the system. The reward in this case is the net present value of the agroforestry system, and the decision variables are the area of trees to plant (k) and the rotation length (T). The recursive equation is:

$$V_n(s_n) = \max_{(k_n,T_n)}\left(NPV(w_n,k_n,T_n) + V_{n+1}(w_{n+1})\cdot(1+r)^{-T_n}\right) \tag{21}$$

The DP model for the salinity problem was implemented by solving this equation recursively, and backwards in time, starting from time period T+1. The model is solved for a discrete set of values of k, T and w and an optimal decision rule is derived. The simulation model represented by equations (1), (3), (4), and (9) to (20) was solved for values of k ranging from 0 to 1 at increments of 0.01, T from 1 to 50 years at increments of one year and w from 1.8m (shallow) to 4.0m (deep) at increments of 0.1m. So there are 22 states (w), resulting in 484 (or 22×22) possible state transitions, $w_t \rightarrow w_{t+1}$, to be controlled by 101 possible values of k and 50 possible values of T.

The recursive equation 21 can be solved by interacting directly with the simulation model. This involves running a simulation for each possible combination of k and T and finding the (k,T) values that result in each of the 484 possible state transitions. However, the direct-simulation approach is not efficient, because the same set of simulations is run at every stage n of the DP algorithm as the recursive equation is solved backwards in time.

We follow a more efficient approach, consisting of creating and saving all the relevant matrices once, by running 101×22×22 simulations, each representing 50 years of forest growth. The saved matrices can be reloaded at any time to solve the DP algorithm. Using this approach it took about 1/6th of the time to solve each DP problem as compared with direct simulation.

The DP approach has the advantage that it yields not only an optimal solution for a particular scenario, as other dynamic optimization methods do, but it also produces an optimal decision rule based on the state of the system at any time. The optimal decision rule can be repeatedly applied to derive optimal (state and decision) paths through time, without having to solve a new optimization problem for each initial state.

3. RESULTS AND DISCUSSION

This section presents the results of both simulation and optimization. A typical simulation run is implemented by solving equation 1 for a set of arbitrary values of the decision variables k and T and for a given initial value of the state variable (w_0). The simulation run produces a state trajectory (w_t) for the given assumptions. Useful insights into the workings of the agroforestry system are obtained by running simulations. However, it is virtually impossible to find the optimal decision trajectory by trial and error with the simulation model. So the DP algorithm described in the previous section is also solved, to obtain a set of sequential values of k_n and T_n that maximize the present value of the agroforestry system for the given initial conditions. The optimization model was solved for an infinite planning horizon, but only 100 years of results are presented, sufficient time to determine whether the system is sustainable.

3.1. Simulation

Figure 1 shows simulation results for a single forestry cycle for selected combinations of initial water-table depth (w_0) and forest area (k). With no trees (k=0), the water table rises over time to reach the soil surface (w_t decreases to zero) by year 15 (Figure 1A). This trend in w_t results in decreasing crop yields, particularly after year 10 (Figure 1B). By year 15 crop yields drop to zero as the land becomes irreversibly lost to conventional agriculture.

Planting 20 percent of the area to trees (k = 0.2) causes the water table to stabilize at around 4m depth by year 30 (Figure 1C); this results in stable crop yields through time (Figure 1D). When the initial water table is close to the soil surface (2m), a tree area of 0.2 is not enough to achieve sustainable crop yields; so it is necessary to set k=0.5 to control salinity emergence (Figure 1E). In this case, crop yields decrease initially (Figure 1F), as the water table approaches the surface (as w_t approaches zero), but they recover after year 9 to regain their original value (4.5 t/ha) by year 15.

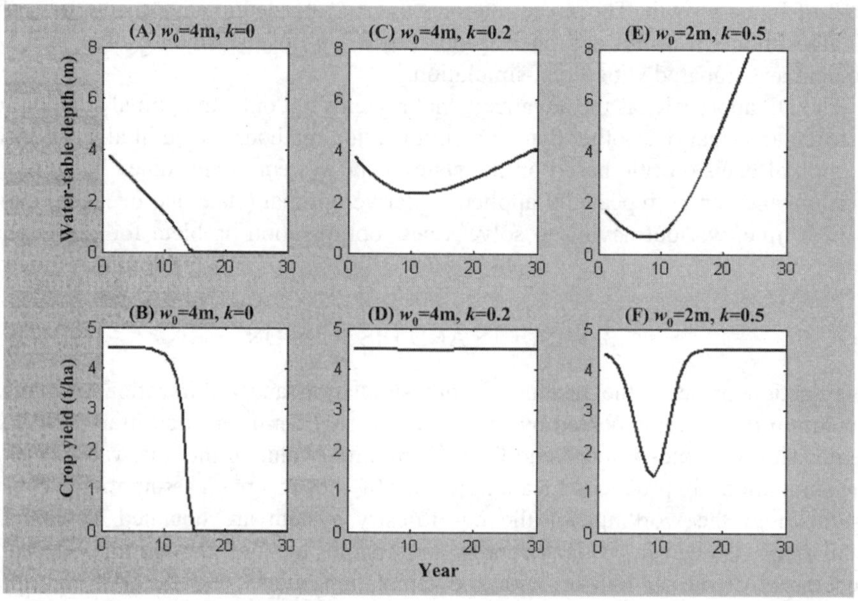

Figure 1. Simulation results: trajectory of the water-table depth (A, C and E) at three different combinations of initial conditions (w_0) and tree area (k), and the corresponding trajectories of crop yields (B, D and F).

The benefits provided by trees take time to become evident (Figure 2). The maximum *NPV* at base parameter values is obtained with $k = 0.15$ in year 50 (Figure 2A). If one were constrained to a planning horizon of five years or less, crop monoculture would be more profitable than agroforestry. This is clear from the negative slope of *NPV* with respect to k (in Figure 2A) during the early years. Carbon payments help somewhat in making forestry more profitable and make it more attractive to keep trees longer (Figure 2B), but the maximum *NPV* occurs at the same k as with no carbon payments.

As the water table approaches the soil surface (as w_t approaches zero), an interesting twist is introduced (Figures 2C and 2D). With $w_0 = 2$m, the value of k at which maximum *NPV* occurs is now 0.5 (compared with 0.15 in the base case). Furthermore, at values of k less than 0.4, *NPV* decreases if the planning horizon is extended beyond five years as crop yields are reduced by salinity (Figure 4C). This effect is similar but more pronounced with carbon payments (Figure 4D). The abrupt increase in *NPV* as k increases between 0.4 and 0.6, in Figures 4C and 4D, is where the positive externality provided by trees for crops is more pronounced. In this region, marginal increases in tree area reverse the salinization process to a level that maintains crop yields.

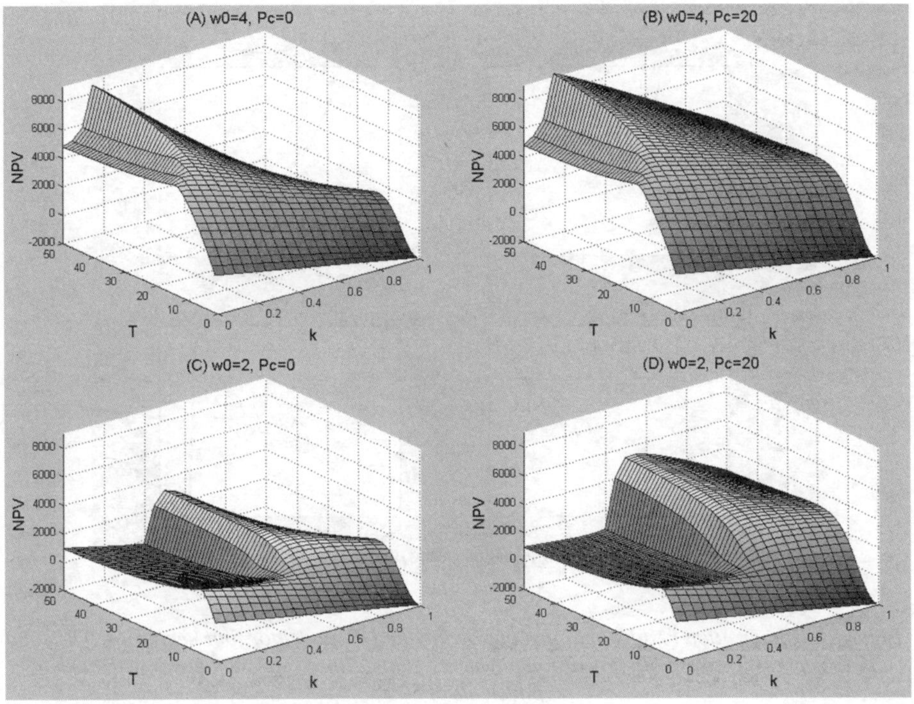

Figure 2. Simulation results: net present value (equation 1) of the agroforestry operation with respect to tree area (k) and cycle length (T). Two initial water-table depths (w_0 = 4 m and w_0 = 2 m) and two carbon prices (p_c = \$0/t and p_c = \$20/t) are shown.

These simulation results are interesting and help us understand the economics of the salinization problem; but they are based on a single forestry cycle and, as discussed before, do not take into account the opportunity cost of keeping trees on the ground. The multiple-cycle problem is addressed by solving the DP problem (equation 21) for an infinite planning horizon in the following section.

3.2. Optimization

The DP model was solved for the base parameters in Table 1 and the four scenarios in Table 2. As explained earlier, these matrices were saved and later used to solve the DP model and perform post-optimality analysis.

The DP solutions for the four scenarios considered are presented in Figure 3. The optimal decision rule (k^*,T^*) is shown for discount rates of 6 percent (Figures 3A and 3B) and 12 percent (Figures 3C and 3D).

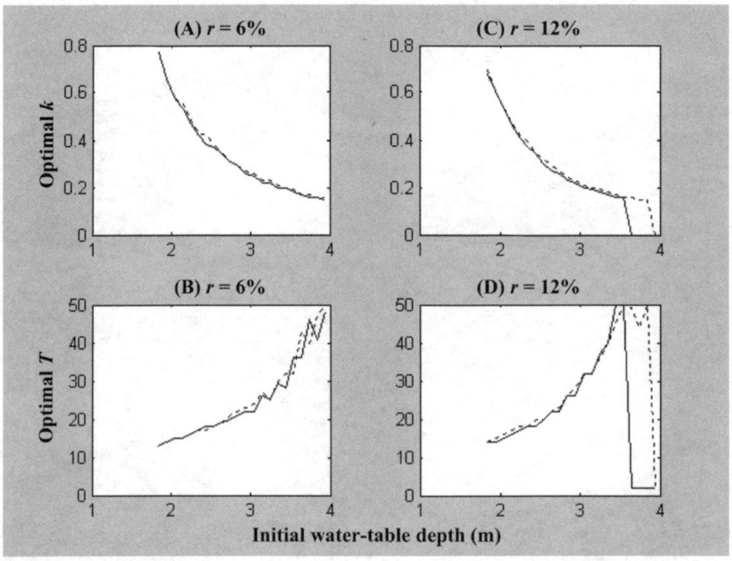

Figure 3. Optimization results obtained by solving the DP model for four scenarios. The solid lines represent cases with no carbon payments, dotted lines represent cases with carbon payments of $20/t.

At low values of w_t (near the soil surface), it is optimal to plant most of the land to forestry (Figure 3A) with a relative short cycle (Figure 3B); at w_t = 2m, for example, the optimal control $(k^*,T^*) = (0.61, 14)$. But this changes as the water-table depth increases below the soil surface; so at w_t= 3.5m, $(k^*,T^*) = (0.2, 32)$.

A similar pattern for the optimal control (k^*,T^*) is observed at the high discount rate (Figures 3C and 3D), but with slightly lower areas planted to forestry (k^*= 0.57 and 0.16 for w_t = 2m and 3.5m respectively). Interestingly, increasing the discount rate does not affect the optimal cycle length at low values of w_t (T^* remains at 14 years), but causes it to increase at high values of w_t (T^* = 50 years when w_t= 3.5m in Figure 3D, compared to 32 years in Figure 3B). An important effect of the high discount rate is that forestry drops out of the optimal solution at water-table depths further than 3.5m below the soil surface, i.e., at w_t > 3.5m, $(k^*,T^*) = (0, 0)$ in Figures 3C and 3D.

The optimal state transitions, given by the changes in the water-table depth ($w_t \rightarrow w_{t+1}$) resulting from applying the optimal decision rule, are shown in Figure 4.

The optimal state transitions (Figures 4A and 4B) do not result in smooth lines with respect to w_t, because of the discrete nature of the solution procedure and the consequent introduction of rounding and truncation errors; so optimal state transitions are presented as points for the relevant set of w_t values in the state vector. Both sets of points show a distinctive downward trend over the relevant interval

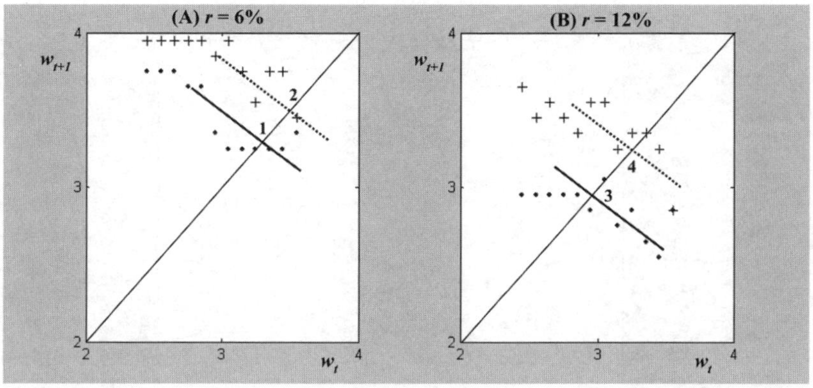

Figure 4. Optimal state transitions for the water table depth at two different discount rates and with (dotted line) or without (solid line) carbon payments. The 45° line starting from the origin represents the steady state.

of w_t values. These trends are indicated by negatively-sloped segments, one for each scenario. The 45-degree lines starting at the origin represent the steady state, where Δw_t = 0m. The intersection between the optimal state transition and the steady-state line marks the optimal 'target' water-table depth in the long run. These target w_t values are: 3.2m, 3.5m, 2.9m, and 3.2m, for scenarios 1, 2, 3, and 4, respectively (Figures 4A and 4B). Hence the optimal equilibrium water-table depth is positively related to carbon price and negatively related to the discount rate.

The optimal value function (V in Figure 5) is the present value of the current water-table depth when the system is managed according to the optimal decision rule in perpetuity. The value of V increases at a decreasing rate as w_0 increases in depth below the soil surface. The pattern is the same for all four scenarios, but the position of the curve changes depending on the discount rate and price of carbon.

The shadow price of the water table is a useful measure, because it indicates the dollar value of an improvement in land quality (deeper water table). This value is in present-value terms, so it takes account of changes in the productive capacity of the land in perpetuity. The shadow price is equivalent to the costate variable (λ) used in the discrete version of the maximum principle in optimal control theory (see Kennedy, 1988). The shadow price is defined as $\lambda_0 = dV/dw_0$. This derivative was estimated numerically from the DP results shown in Figure 5. As would be expected, the value of λ_0 is high at low values of w_0 (poor land quality), implying that the marginal benefit of an improvement in land quality is high (over \$3,000/ha at w_0 = 2.0m). The value of λ_0 decreases as land quality improves, to reach about \$500/ha at w_0 = 3.5m.

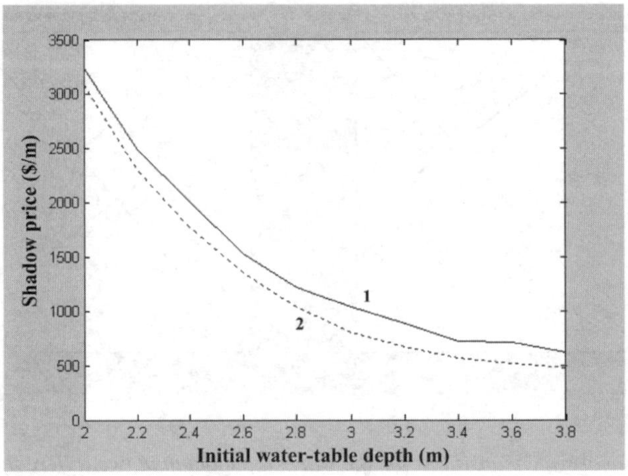

Figure 5. Optimal value function for the four scenarios.

The shadow prices in Figure 6 are not affected by the discount rate within the range tested, but they decrease when carbon payments are introduced. This means that the marginal value of land improvement is lower when carbon payments are available. The reason for this is not immediately obvious, but will become clear later when carbon stocks are discussed.

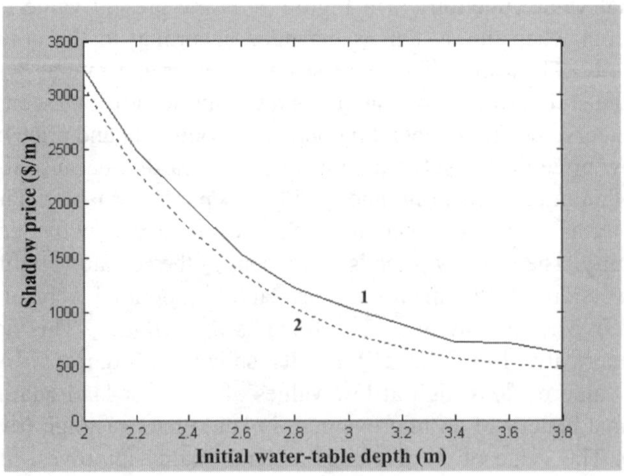

Figure 6. Shadow price of the water table for scenarios 1 and 2.

Further insight into optimal results can be gained by analyzing the optimal paths for selected initial conditions. The optimal paths discussed below were derived by applying the optimal decision rules (Figure 3) and simulating the system as explained in the methodology section.

3.3. Optimal Paths

Figure 7 shows the water-table paths that result from applying the optimal decision rule for each scenario over a period of 100 years. Two different initial states (w_0 = 2m and w_0 = 4m) are presented. The target w_t values, derived in Figures 4A and 4B above, correspond to the ends of forestry cycles. Each cycle starts with a decrease in w_t (as the water table approaches the surface), followed by an increase (as the water table becomes deeper) as trees grow and absorb more water. Each forestry cycle ends at a peak w_t (at values of between 2.6m and 4m depending on the discount rate and carbon price). These peaks represent the target w_t values discussed above.

With w_0 = 2m, the presence of carbon payments makes the second and subsequent cycles longer (dotted lines in Figures 7A and 7C); but the first cycle is not affected by carbon price. This is because the priority in the first cycle is to fix the shallow water-table problem before it is too late, so a large area of forest is planted regardless of whether carbon payments are received.

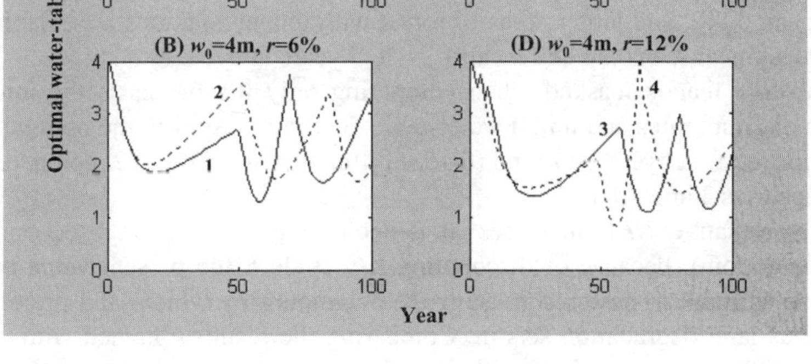

Figure 7. Optimal state paths at two different discount rates and two carbon prices. Numbers next to the curves represent the four scenarios considered.

With w_0 = 4m, the situation is different (Figures 7B and 7D). With a deeper water table (4m), there is no pressure to fix the problem immediately. So it is optimal to run down the system for the first 20 years. After year 20, preventive action starts having an effect on land quality and w_t starts increasing in depth for the rest of the first forestry cycle. Carbon payments have an interesting effect at the low discount rate (Figure 7B). In this case carbon payments do provide an incentive for land conservation, and the optimal value of w_t at the end of the first cycle (around year 50) is 3.5m below the soil surface with carbon payments (scenario 2) as compared to 2.5m deep without carbon payments (scenario 1). This result suggests that, in situations where salinity emergence is not imminent, carbon payments do make a difference in encouraging early preventive action. This occurs only at the low discount rate.

With w_0 = 4m and the higher discount rate (Figure 7D), the story changes. Carbon payments (scenario 4) do not have much of an effect on w_t during the first 40 years as compared with the case where no carbon payments are made (scenario 3). The higher discount rate (12 percent) gives more weight to the value of the early crops and discounts future productivity losses more heavily, so it is optimal to run down the system for a little longer before taking preventive action.

A striking result, which becomes apparent when comparing Figures 7A and 7C against 7B and 7D, is the strong path dependency on initial conditions that the problem exhibits during the first 100 years. The initial water-table depth dramatically influences the optimal state path. Eventually though, the optimal paths obtained for different initial states will tend to converge towards the target w_t explained in connection with Figures 4A and 4B.

The optimal control paths that cause the state paths depicted in Figure 7 are presented in Table 3. With w_0 = 2m, the general trend consists of a large area of forestry (mean 0.55) and a short rotation (mean 15.3 years) established in the first cycle; followed by smaller forest areas (between 0.22 and 0.26 on average) with longer rotations (between 25 and 31 years on average) in subsequent cycles. With w_0 = 4m (Table 3), the trend is different; a small area of trees is planted in the first cycle (mean 0.08), and larger areas of forest with longer rotations are established in later cycles (mean k^* = 0.24 and mean T^* = 26.5 years in cycle 6).

Something that is masked when comparing only the means is the interaction between discount rates and initial water-table depth. At w_0 = 4m, the optimal control (k^*, T^*) for the first cycle is around (0.15, 50) for scenarios 1 and 2, compared to (0, 2) for scenarios 3 and 4.

The cumulative *NPV* increases at a decreasing rate with subsequent cycles (Table 3, bottom), because of discounting. By cycle 6 the present value of future cycles is negligible. These results suggest the amount by which land prices should decrease as land degradation sets in. Comparing the results obtained with w_0 = 2m ($6,173) against those obtained with w_0 = 4m ($8,940) suggests that, under perfect competition in a deterministic world, the better quality land should be worth about $2,667 more per hectare.

Table 3. Optimal control paths and the resulting cumulative profits over six forestry cycles for four scenarios at two initial water-table depths.

Cycle	w_0=2m					w_0=4m				
	Optimal tree area (k^*)									
	Scenario					Scenario				
	1	2	3	4	Mean	1	2	3	4	Mean
1	0.57	0.57	0.53	0.53	0.55	0.15	0.16	0.00	0.00	0.08
2	0.25	0.26	0.30	0.21	0.26	0.34	0.19	0.00	0.15	0.17
3	0.25	0.17	0.30	0.18	0.23	0.20	0.23	0.16	0.53	0.28
4	0.25	0.23	0.30	0.19	0.24	0.22	0.21	0.27	0.17	0.22
5	0.26	0.21	0.26	0.19	0.23	0.25	0.19	0.26	0.20	0.23
6	0.25	0.19	0.27	0.18	0.22	0.26	0.21	0.27	0.20	0.24
	Optimal cycle length (T^*, years)									
	Scenario					Scenario				
	1	2	3	4	Mean	1	2	3	4	Mean
1	15	15	15	16	15.3	48	50	2	2	25.5
2	22	24	20	32	24.5	19	32	2	44	24.3
3	22	40	20	40	30.5	29	25	50	16	30.0
4	22	25	20	36	25.8	26	30	22	45	30.8
5	22	30	22	36	27.5	22	32	22	32	27.0
6	22	32	22	40	29.0	22	30	22	32	26.5
	Cumulative NPV (AUS\$/ha)									
	Scenario					Scenario				
	1	2	3	4		1	2	3	4	
1	2,748	3,548	1,133	2,032		8,470	9,190	1,132	1,132	
2	5,230	6,398	1,624	2,728		8,769	9,626	2,035	4,914	
3	5,917	7,293	1,676	2,748		8,910	9,686	4,661	4,926	
4	6,107	7,364	1,681	2,748		8,934	9,701	4,667	4,930	
5	6,158	7,382	1,682	2,748		8,939	9,704	4,668	4,930	
6	6,173	7,385	1,682	2,748		8,940	9,704	4,668	4,930	

3.4. Carbon stocks

One remaining question in regard to the optimal paths is what happens to carbon stocks. The optimal water-table paths discussed above are associated with optimal biomass paths as trees grow during a forestry cycle. These biomass paths can be expressed in terms of carbon (Figure 8) by multiplying biomass by 0.5.

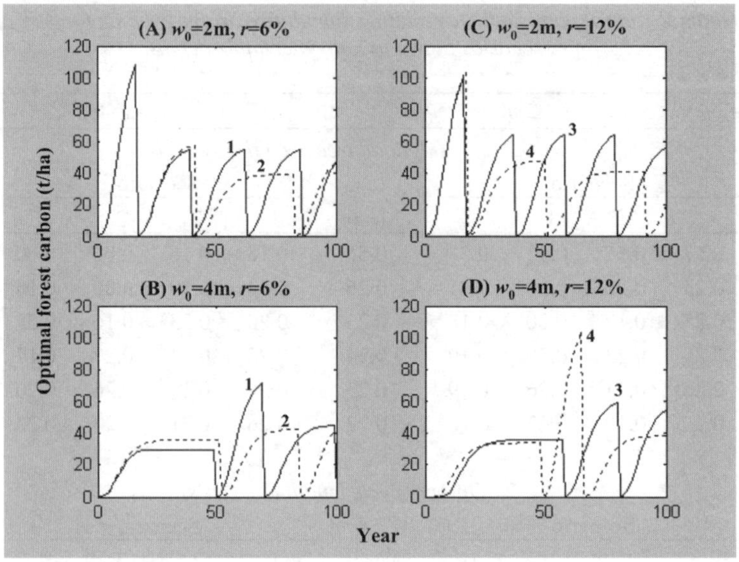

Figure 8. Optimal carbon stock paths.
Numbers next to the curves represent the four scenarios considered.

The optimal carbon path (Figure 8) reinforces the findings in the previous section, that with w_0 = 2m, land restoration practices (a large area of trees) are adopted early (Figures 8A and 8C); whereas with w_0 = 4m, the system is exploited during the first cycle to produce a larger area of crops, with only a small area of trees being planted (Figures 8B and 8D). The early exploitation of the system leads to the need to restore land quality by the second cycle, requiring a larger area of trees to be planted in year 50 (solid line in Figure 8B).

The average carbon stocks (or time-averaged carbon) can be used to assess whether carbon payments influence the amount of carbon that is sequestered by the agroforestry system. The average amount of carbon fixed in the forest over the first 100 years is presented in Table 4 for each scenario. Carbon payments and discount rates have only a small effect on the average amount of carbon stocks. Optimal carbon stocks range between 30.3 and 32.2 tC/ha, on average, for the four scenarios. In contrast, initial water-table depth has more of an effect on carbon stocks, with averages of 33.7 and 28.6 tC/ha for the cases with w_0 = 2m and w_0 = 4m respectively. This is a 15 percent drop in optimal carbon stocks as the initial water-table depth increases from 2 to 4 meters below the soil surface. This explains the result obtained in Figure 6, where the shadow prices of w_0 were lower in the presence of carbon payments than in their absence. Higher values of w_0 result in lower optimal carbon stocks (Table 4), because there is less need for land restoration. Hence, in the presence of carbon payments, the benefit of lower salinity

Table 4. Optimal forestry carbon stocks (tC/ha) for eight scenarios. Only aboveground biomass of trees is considered.

	r=6%		*r=12%*		
w_0	p_c=0	p_c=20	p_c=0	p_c=20	*Mean*
2	33.1	32.9	36.1	32.5	33.7
4	28.3	27.8	28.3	29.9	28.6
Mean	30.7	30.3	32.2	31.2	

is slightly reduced by a decrease in carbon payments received, because of reductions in the biomass carbon stock.

The foregoing analysis is based on land of excellent quality by Australian standards. Land of lower quality can be represented by reducing expected crop yields (parameter $\bar{y}^a$) and tree-growth parameters (α_G, β_G, γ_G). This was not done here, as it would have added little to the analysis, given that the purpose of this chapter is to demonstrate the application of the techniques available to analyze externalities in agroforestry systems, and not to provide policy prescriptions.

Finally, it must be pointed out that, although the model developed in this chapter has been applied to the problem of dryland salinity in Australia, its applicability is much broader. The analytical techniques illustrated here can be applied to virtually any land degradation problem that involves trees and crops. The key to adapting the model to other problems resides in estimating two equations: i) the state-transition as a function of the control variables k and T; and ii) the land productivity function $Q(s_t)$ which describes the effect of land quality on crop and tree growth. In our example, these functions are represented by equations 2 and 16. Equation 16 implicitly represents the effect of salinity on yields; the shape of this function probably holds for many land-degradation problems, but different functional forms may be appropriate in some cases (Swallow et al., 1990; Swallow, Talukdar, & Wear, 1997).

4. CONCLUDING COMMENTS

In this chapter we have developed a model of an agroforestry system where trees and crops interact. The model was used within a dynamic programming algorithm to determine the optimal area of trees to plant and the optimal length of forest cycles in an Australian example. A numerical model of dryland-salinity emergence was solved for a homogeneous plot of land under the management of a single decision-maker. Therefore the only externalities considered were those that trees provide for crops, through mitigation of land degradation. The negative externalities that upstream landholders may impose on downstream landholders were not considered.

An important finding of our numerical analysis was the optimal decision rule, whereby forest area and cycle length depend on initial land quality. When initial land quality is poor, it is optimal to plant a large area of forest and follow a short cycle. As land quality improves, it becomes optimal to plant a smaller area of trees and keep the trees for a longer period. The main driver of this decision is salinity emergence. In general, payments for carbon-sequestration services had little influence on the optimal solution. Although they increased landholder profits, they did not tend to increase the optimal amount of carbon stored in the forest. It must be pointed out, however, that only aboveground biomass carbon was considered in the payment scheme, soil carbon was not included.

The model developed here was applied to a salinity problem in Australia, but the general approach is widely applicable to other land-degradation problems and locations. The two main limitations of our model are that it is deterministic and that it assumes a homogeneous plot of land under single management. The former limitation is important, because of the variability of rainfall, which is the main driver of water-table recharge that leads to salinity emergence. Also, given the possible irreversibility of severe salinity, it is important to account for the probability that high-recharge events will occur. The latter limitation becomes relevant only if we wish to extend the analysis to account for interactions among landholders, with heterogeneous land, who may participate in market-based solutions to land-degradation problems.

Our dynamic-programming approach has advantages for making the model stochastic. In this regard, the state-transition matrix can be replaced by a set of transition-probability matrices associated with sets of decisions. The DP algorithm can then be used to estimate expected profits. So the main problem is to obtain sufficient data to estimate probabilities that certain events (such as rainfall, yields, and prices) will occur.

Relaxing the assumption that the land is homogeneous would require a more complex model; one that includes different types of land as well as the interaction between upstream and downstream plots. This may cause a dimensionality problem that is difficult to handle with DP, particularly if the model is also stochastic.

5. NOTES

[1] A reviewer pointed out that low-density planting may lead to lower growth and tree quality and hence to lower tree values, particularly if trees are intimately mixed. These effects are not considered in our model.

6. REFERENCES

Bowes, M. D., & Krutilla, J. V. (1985). Multiple use management of public forestlands. In A. V. Kneese & J. L. Sweeney (Eds.), *Handbook of natural resources and energy economics*, Vol. 2 (pp. 531-569). Amsterdam: Elsevier Science Publishers.

Bowes, M. D., & Krutilla, J. V. (1989). *Multiple management: The economics of public forestlands*. Washington, D.C.: Resources for the Future.

Brown, S. (1997). *Estimating biomass and biomass change in tropical forests: A primer*. Rome: Food and Agriculture Organization of the United Nations.

Cacho, O. J. (2001). An analysis of externalities in agroforestry systems in the presence of land degradation. *Ecological Economics, 39*, 131-143.

Cacho, O. J., Hean, R. L., & Wise, R. (2003). Carbon-accounting methods and reforestation incentives. *Australian Journal of Agricultural and Resource Economics, 47*, 153-179.

Cleugh, H. A., Prinsley, R., Bird, P. R., Brooks, S. J., Carberry, P. S., & Crawford, M. C. (2002). The Australian National Windbreaks Program: overview and summary of results. *Australian Journal of Experimental Agriculture, 42*(6), 649-664.

Englin, J. E., & Klan, M. S. (1990). Optimal taxation, timber and externalities. *Journal of Environmental Economics and Management, 18*, 263-275.

Greiner, R., Cacho, O. J. (2001). On the efficient use of a catchment's land and water resources: dryland salinization in Australia. *Ecological Economics, 38*, 441-458.

Hamburg, S. P. (2000). Simple rules for measuring changes in ecosystem carbon in forestry-offset projects. *Mitigation and Adaptation Strategies for Global Change, 5*, 25-37.

Hartman, R. (1976). The harvesting decision when a standing forest has value. *Economic Inquiry, 14*, 52-58.

Hean, R. L., Cacho, O. J., & Menz, K. (2003). Farm forestry, carbon-sequestration credits and discount rates. In T. W. Graham, D. J. Pannell, & B. White (Eds.), *Dryland Salinity: Economic Issues at Farm, Catchment and Policy Levels* (pp. 133-144). Cooperative Research Centre for Plant-Based Management of Dryland Salinity, University of Western Australia, Perth.

Kennedy, J. O. S. (1986). *Dynamic programming: applications to agriculture and natural resources*. London: Elsevier Applied Science.

Kennedy, J. O. S. (1988). Principles of dynamic optimization in resource management. *Agricultural Economics, 2*, 57-72.

Marland, G., Fruit, K., & Sedjo, R. (2001). Accounting for sequestered carbon: the question of permanence. *Environmental Science and Policy, 4*, 259-268.

Prinsley, R. T. (1992) The role of trees in sustainable agriculture – an overview. *Agroforestry Systems, 20*, 87-115.

Stoneham, G., Chaudhri, V., Ha, A., & Strappazzon, L. (2003). Auctions for conservation contracts: an empirical examination of Victoria's BushTender trial. *Australian Journal of Agricultural and Resource Economics, 47*(4), 477-500.

Swallow, S. K., Parks, P. J., & Wear, D. N. (1990). Policy-relevant nonconvexities in the production of multiple forest benefits. *Journal of Environmental Economics and Management, 19*, 264-280.

Swallow, S. K., Talukdar, P., & Wear, D. N. (1997). Spatial and temporal specialization in forest ecosystem management under sole ownership. *American Journal of Agricultural Economics, 79*, 311-326.

Tisdell, C. A. (1985). Conserving and planting trees on farms: lessons from Australian cases. *Review of Marketing and Agricultural Economics, 53*, 185-194.

Walpole, S. C. (1999). Assessment of the economic and ecological impacts of remnant vegetation on pasture productivity. *Pacific Conservation Biology, 5*, 28-35.

SUBHRENDU K. PATTANAYAK AND BROOKS M. DEPRO

ENVIRONMENTAL SERVICES FROM AGROFORESTRY

Economics of Soil and Forest Conservation in Manggarai, Indonesia

1. ENVIRONMENTAL SERVICES FROM AGROFORESTRY

This chapter develops and tests an economic model of environmental services from agroforestry. Agroforestry includes a spectrum of land uses in which trees are combined with crops across time or space (Lundgren & Raintree, 1982). Thus, agroforestry in general, or tree planting on farms in particular, has the potential to generate socially valuable environmental services such as soil conservation, forest protection, biodiversity conservation, carbon sequestration, and protection of water quality and quantity. Through the provision of these services, as well as tree-based products such as construction wood, fuelwood, and fodder, agroforestry can improve socio-economic welfare. Given this promise, we might expect that agroforestry would be widely embraced by farmers, and would improve livelihoods and environmental conditions where adopted. Despite some impressive technological and scientific advances over the years, however, adoption rates have been low and dis-adoption is not unusual. We contend that this is partly because the claim that agroforestry generates environmental services is largely untested and overly influenced by controlled field trials and laboratory experiments of natural scientists.

We believe that a social science or socio-economic perspective on environmental services from agroforestry is important for four reasons. First, in contrast to controlled experiments, farmers in the real world can produce sub-optimal levels of environmental services (or returns to agroforestry) by biophysical experimental standards because they face income, production and information constraints, all of which are insufficiently incorporated into experiments and trials. Second, environmental services are often times "externalities" because they do not directly benefit farmers/adopters and, therefore, are not produced even at socially optimal levels, let alone ecologically optimal levels. Third, even "internalized"

J. R. R. Alavalapati & D. E. Mercer, Valuing Agroforestry Systems, 165–182.

environmental services from agroforestry (i.e., those considered by farmers) cannot be viewed exclusively through an "ecological or agronomical production" lens because agroforestry and its associated services can impact explicit market prices and implicit shadow prices of non-market resources, causing farmers to adjust and adapt their production and consumption choices, including the production of environmental services. Fourth, markets for products, including "environmental services," are often incomplete or missing causing household demographic and consumption characteristics to play an important role in the production of environmental services. This is especially so in most rural areas of developing countries that has been the focus for most agroforestry projects.

Unfortunately, the literature is noticeably thin with regards to economic modeling of environmental services in general and from agroforestry in particular. Consider, for example, one of the most prominent environmental services – hydrological and soil benefits from watershed protection. There is general consensus that forests can play a protective role in various watershed processes (Lal, 1993) and policy makers have provided sustained attention and public funds to promote watershed services (Tomich, van Noorwijk, & Thomas, in press). However, the benefits associated with environmental services are imprecise and rarely quantified (Dixon, 1997; Georgiou, Whittington, Pearce, & Moran, 1997). Tomich et al. (in press) go as far as to say that research to date has provided policymakers with "surprisingly little useful information" about watershed services. Thus, one of the key challenges for agroforestry proponents is to identify and demonstrate the benefits of agroforestry practices. We use a case study from Indonesia to address this challenge with respect to two important environmental services frequently associated with agroforestry: erosion of on-farm soils and collection of fuelwood from adjoining public forests.

Soil erosion poses an economic and environmental concern in many parts of the world where farming is an important and expanding activity, such as in the tropical uplands (Gill, 1995; Dixon, 1997). This problem can be controlled by tree planting on farms, particularly with agroforestry systems such as contour hedgerow farming. Other potential soil related benefits of trees include: maintenance or increase in organic matter and diversity, nitrogen fixation, and enhancement of soil physical properties (Nair, 1993). Fuelwood consumption has similar economic and environmental concerns because it accounts for about 15% of the primary energy supply in developing countries and provide up to 80% of total energy in some countries (World Resources Institute (WRI), 2000). In addition, fuelwood collection is also responsible for much local forest cover loss in parts of Asia, Africa, and Latin America (WRI, 2000). Agroforestry practices can provide an attractive on-site substitute for fuelwood and reduce its collection from public forests.

The remainder of the chapter is organized as follows. In Section 2, we use household production theory to develop a stylized model of environmental services from agroforestry and generate testable hypotheses. Section 3 presents a brief overview of the data set to empirically estimate our model. Results from non-parametric statistical tests and parameteric regression models in Section 4 provide mixed evidence in support of environmental services from agroforestry. We conclude with a brief summary and discussion of our methods, data, and findings in Section 5.

2. HOUSEHOLD PRODUCTION OF ENVIRONMENTAL SERVICES

Consider a stylized characterization of a farmer's problem, often described in the literature as the farm household model (Chayanov, 1966; Sen, 1966) or more recently as the agricultural household model (Singh, Squire, & Strauss, 1986). This model modifies Pattanayak, Abt, and Holmes' (2003a) household model of joint production of timber and amenities that considers the case of farmers producing income earning crops and environmental services by using labor, land, and materials under agroforestry. A typical farmer maximizes utility comprising income (**π**) and environmental services (**e**).[1] Including income in the utility function is equivalent to including a composite consumption commodity, whose price is normalized to 1, because the farmer will presumably buy a weighted basket of consumption with any income earned through farming and other activities. The farmer may consume this environmental service directly, because of aesthetic or cultural value, or indirectly, by consuming utility yielding services such as cooked food and heating produced by services such as fuelwood. Environmental services can also directly affect the "production sector" of the household as discussed below. Acknowledging nonseparability of production and consumption spheres of the household decision making, presumably due to incomplete markets for environmental services, utility will be conditioned by preference parameters θ that measure the shape of the utility curve and account for risk preferences.

Utility is maximized subject to two constraints. The first constraint is a multi-input, multi-output production function (G) that is twice differentiable, continuous, and convex. Subscripts **y** and **e** are vectors of crop and environmental services, and **a** is the agroforestry technology that represents the joint investments of labor and materials, which will be collectively embodied in the amount of land dedicated to agroforestry. Agroforestry can therefore be conceived as one among many sets of coordinated investments to enhance overall productivity, which will be conditioned by ecological factors, Z. The shape of the production function, including cross effects, is intuitive and summarized in equation 1.

$$G_y \geq 0,\ G_{yy} \leq 0,\ G_e \geq 0,\ G_{ee} \leq 0,\ G_{ye}\,?,\ G_a \geq 0,\ G_{aa} \leq 0 \qquad (1)$$

The cross partial G_{ye} merits further discussion. To the extent that the environmental service enters the production of crops, by definition a "service" such as soil conservation will enhance crop production by extending the production possibility frontier. On the other hand, the process of producing crops could reduce and deplete the soil. Like all assumptions, we can only confirm the validity of this assumption through empirical analysis.[2]

The second constraint is a budget constraint such that the sum of the cost of inputs into agroforestry technology (**r·a**) and consumed income (**π**) are no greater than the sum of exogenous (**π**$_e$) and crop income (**p·y**). Thus, **r** reflects the weighted costs of labor, land, and material investments into agroforestry. In this stylized model, there are no practical benefits of decomposing constituent inputs to agroforestry to account for different types of practices and productivity.

The Lagrangian for this problem, in which μ and λ are the Lagrangian multipliers, is presented above. μ is the marginal utility of jointly produced crop and amenities and λ is the marginal utility of income.

$$\ell_{\pi, a, y, e} = U(\pi, e, \theta) - \mu[G(y, e, a; Z)] + \lambda[\pi_e + py - ra - \pi] \quad (2)$$

The first order conditions of this utility maximization are presented in equation 3. Simultaneous solution of these first-order conditions determines optimal allocation of resources (labor, land, and materials) and consumption levels (environmental services and crop income).

$$\begin{bmatrix} \ell_\pi \\ \ell_e \\ \ell_y \\ \ell_a \end{bmatrix} = \begin{bmatrix} 0 \\ 0 \\ 0 \\ 0 \end{bmatrix} = \begin{bmatrix} U_\pi + \lambda \\ U_e - \mu \bullet G_e \\ \lambda \bullet p - \mu \bullet G_y \\ -\lambda \bullet r - \mu \bullet G_a \end{bmatrix} \quad (3)$$

Essentially, resources are allocated such that marginal opportunity costs are equal to marginal utility of consumption generated by that resource. For example, by manipulating the first three conditions in equation 3 we get

$$U_e = -U_\pi p_y \bullet \frac{G_e}{G_y} \quad (4)$$

This implies that the marginal utility of environmental services is equal to the marginal utility of the crop income forgone to self-produce environmental services. In other words, at the margin amenity benefits equal the cost of self-producing environmental services. See Pattanayak et al. (2003a) for details of the type of comparative statistics that can be conducted within this framework.

There are three important considerations in establishing an empirical environmental services supply model from these first order conditions. First, because all consumption and production commodities are linked to the budget constraint either directly or indirectly (i.e., linked to market commodities through a joint production technology), all prices influence all consumption and production allocations including self-produced/consumed amenities. Second, all production allocations including supply of crops and demand for inputs will depend on the optimally chosen level of environmental services because there is no price for environmental services. Third, the optimal level of environmental services depends on the discrete decision to invest land, labor, and materials in agroforestry, in addition to the levels of investment in agroforestry, which in turn depend on the relative input and output prices. That is, in a situation including some households who have adopted agroforestry and others who have not adopted agroforestry, we might expect that the "adoption dummy variable" will impact the level of environmental service produced. The decision to adopt will of course depend on several factors in addition to or separate from the factors that affect the level of investment. Pattanayak, Mercer, Sills, and Yang (2003b) present a review of agroforestry adoption factors, including market prices, biophysical characteristics, socio-demographic factors, and policy variables among others.

Collectively, these points suggest that the optimally chosen household level of environmental services depends on prices of consumption commodities and production inputs, factors affecting preferences and production possibilities, and discrete adoption choice. We have presented a reduced form characterization of environmental services production, instead of a structural representation, because the choice of functional form for a structural representation of all ecological and economic functions would be arbitrary. Additionally, the resulting analytical expressions would be sufficiently complex that the signs of most partial derivatives would be indeterminate without specific information, not only about the general functional forms, but also about the magnitudes of all of the parameters.

3. AGROFORESTRY IN MANGGARAI, INDONESIA

We consider the case of environmental services from agroforestry in the Manggarai region of Indonesia. The data for our case study are drawn from a project concerning the economics of biodiversity conservation in the Manggarai region on the island of Flores in eastern Indonesia (Kramer, Pattanayak, Sills, & Simanjuntak, 1997; Kramer, Sills, & Pattanayak, 2003; Pattanayak & Butry, 2003; Pattanayak & Kramer, 2001; Pattanayak, Sills, & Kramer, in press). We use data from a survey of 494 households in 47 desas (villages) in the buffer zone of Ruteng Park in February 1996. Households were chosen from a sampling frame of about 13,500 households in the buffer zone on the basis of stratified random sampling in which the weights reflected the population densities of the desas. The survey was administered by 16 Indonesian undergraduate agronomy students who spoke the local Manggarai dialect. The interviewers received three days of training and were monitored during data collection. The survey was comprised of five sections: demographic characteristics, environmental profile of farm land, farm and home production budgets, labor and financial allocations, and contingent valuation. It was designed using focus groups, pretests, and expert opinions (Kramer et al., 1997).

Survey results show that the average Manggarai household—described in terms of household with the mean or modal characteristic—exhibits a heavy reliance on agriculture, with about 90% of the local population engaged in agriculture.[3] They farm approximately an average of 1.4 hectares of land, distributed across three parcels of approximately 0.4 hectares each. Only 8% plan to increase their landholdings in the near future. Parcels are typically passed on from one generation to another with approximately 80% of the parcels acquired through inheritance and used for about 15 years. For about 45% of the parcels, soils can be characterized as rocky or sandy. Only 23% of the parcels are on flat land, and only 15% receive rain-fed irrigation. Farmers primarily grow coffee and rice, keeping chickens and pigs. Labor and fertilizers are the key agricultural inputs. Manggarai households do have some off-farm employment opportunities, including positions with the local government, non-governmental organizations, kiosks, and logging crews.

On average, households have six members, with 37% males and 38% children. The average age across members is 34 years. Approximately 35% of household members have experienced at least one morbidity event from malaria, dysentery, tuberculosis, respiratory infection, typhoid, or others in the 12 months preceding the survey. Most households have little or no schooling with the typical household head having completed 6 years of schooling.

Considering a few community characteristics, we find that credit opportunities exist in approximately 80% of the villages. However, only about 30% of these villages have any "cooperatives" (community-based farmer support groups) and only 15% have been visited by extension agents of Ruteng Park. On average, these villages are about 15 km from the district capital (Ruteng) and 7 km from the nearest paved road. Given this profile of Manggarai households, it is not surprising that they struggle economically. The average household earns US$700 per year in terms of the cash value of its agricultural production plus any wage income.[4] This typical household owns one "modern" asset from a list including radio, television, electricity, wall clock, wrist watch, kerosene stove, and motorcycle.

3.1. Agroforestry Data

Among the various agricultural practices, our focus is primarily on the planting of nitrogen-fixing trees on land parcels and, secondarily, on contour farming. These two practices are conceptually closest to the notion of agroforestry that includes farmer investment that enhances long-term returns to land through conservation of soil, water, and forest resources. The survey shows that approximately 62% of the parcels use at least one of these two agroforestry technologies, with an even distribution of 31% between nitrogen-fixing and contour farming practices. In nearly all of these cases, households only use one technology.

As discussed in Section 1, we consider two environmental services that could result from either or both of these agroforestry practices—conservation of soils and forest fuelwood. Soil conservation is measured as on-farm (on-site) erosion that could be reduced by either tree planting or contour farming on the parcel. Forest conservation is measured as fuelwood collection from local public forests by households that again could be reduced by tree-planting on the parcels. Agroforestry may "directly" conserve soil on-farm in contrast to "indirectly" conserving the forest at off-farm locations. This "directness," due to the on-site (on-farm) occurrence of the service, could be a mixed blessing for any empirical work attempting to investigate the impact of agroforestry on environmental services because of the potential for "reverse causality."

The survey asked farmers about erosion on their land parcels in the form of a 5-level Likert scale: 1 = very low; 2 = low; 3 = medium; 4 = high; 5 = very high. The distribution of erosion levels is described in Figure 1A. It has the expected normal distribution with a large concentration of values at the medium levels of erosion. Nevertheless, about 30% of the sample show a "high" or "very high" levels of erosion, and the empirical question is whether these are more likely on parcels that have not adopted agroforestry.

Data on fuelwood collection were gathered using the notion of a "typical trip" for collecting forest products (c.f., Pattanayak & Sills, 2001; Pattanayak, Mehta, Sills, & Kramer, 2003c; Pattanayak et al., in press,). The survey asked households to describe typical collection trips for forest fuelwood (identified through focus groups and pre-testing) in terms of (a) frequency of collection trips and (b) quantity of forest products collected on each trip. Quantity of forest fuelwood was thus calculated as the product of (a) "number of trips" and (b) "yield per trip." There is some imprecision in the reported measures of both these variables, particularly the"yield" measure, in addition to the fact that a sizeable percentage of households do not collect any fuelwood from public forests. Therefore, we categorize the

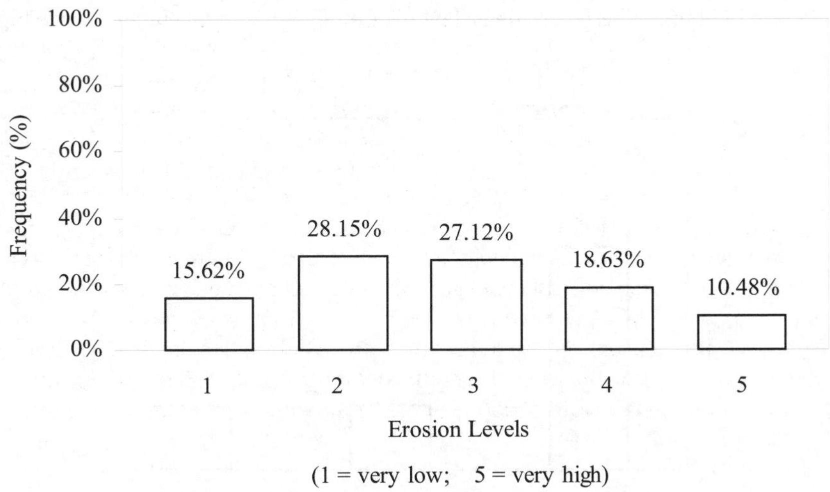

Figure 1A. Distribution of parcel-level erosion: Thirty-percent of the sample experience "high" or "very high" levels of erosion.

collection of fuelwood from local public forests into three broad categories: 0 = no collection; 1 = some collection; 2 = significant collection. The distribution of this variable is reported in Figure 1B. The empirical question is whether households who did not plant trees are more likely to collect "significant" amounts of forest fuelwood, whereas household who planted trees are more likely to "not collect." We also test the robustness of our finding by analyzing the "trip number" and "trip yield" variables independently and checking if we get the same or similar results regarding the impact of agroforestry.

4. EMPIRICAL MODEL RESULTS

We investigated the impacts of agroforestry on soil and forest conservation using a two-level approach. First, we conducted non-parametric analysis by comparing the distribution of the soil and fuelwood variables for "adopting" and "non-adopting" households. Specifically, we used rank-sum tests for ordered categorical variables to compare the distribution. Second, we conducted parametric regression analysis to control for other influences on soil conservation and fuelwood collection (see Edmunds, 2002; Jalan & Ravallion, 2003 for the use of "regression controls"). We estimated ordered probit models of soil erosion and fuelwood collection for this purpose.

4.1. Non-parametric Analysis of Environmental Services

Starting with soil erosion, Figure 2A shows that parcels with nitrogen-fixing trees are no less likely to experience lower levels of erosion and may even produce higher

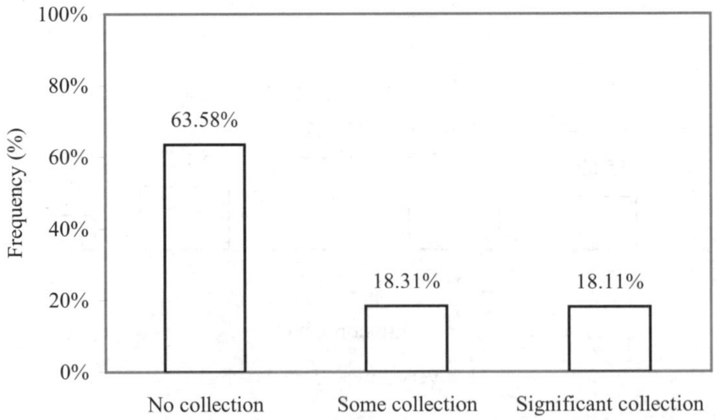

Figure 1B. Distribution of household-level fuelwood collection: Forty-percent of the sample collect fuelwood from forests

levels of erosion. We compare the two samples using the Wilcoxon/Mann-Whitney test (Wilcoxon, 1945; Mann & Whitney, 1947). The Wilcoxon Z statistic of –2.63 suggests that "adopting" parcels experience higher levels of erosion. Figure 2B shows that parcels with contour farming are likely to experience less erosion, a finding that is confirmed with by the Wilcoxon Z statistic of 2.85. Turning to forest fuelwood collection, Figure 2C shows that households that have planted nitrogen-fixing trees on their parcels are much less likely to collect fuelwood from forests. Again, this finding is confirmed by a Wilcoxon Z statistic of 3.99.[5]

Although these tests are an important first step in our analysis, we need additional evidence to establish that agroforestry provides these services. These results would be sufficient to evaluate the environmental benefits of agroforestry if and only if the adopting and non-adopting parcels were identical in all other respects. That is, agroforestry was the result of a randomized experiment. Unfortunately, that is not the case. As Table 1 shows, parcels that have nitrogen-fixing trees on them, and households who have planted these trees, are significantly different (based on t or z statistics) with respect to several plot, household, market, and policy characteristics, which have been identified as critical determinants of agroforestry in a recent meta-analysis of agroforestry adoption (Pattanayak et al., 2003b). We find that the similarity of adopters and non-adopters (identified as "cannot reject" in last column of Table 1) is limited to average use, size, and inheritance of the parcel; the existence of village cooperatives; rice price; ratio of males; and household wealth. Thus, we next conducted multivariate regression analysis to control for all remaining factors and evaluate impacts of agroforestry on environmental services *ceteris paribus*.

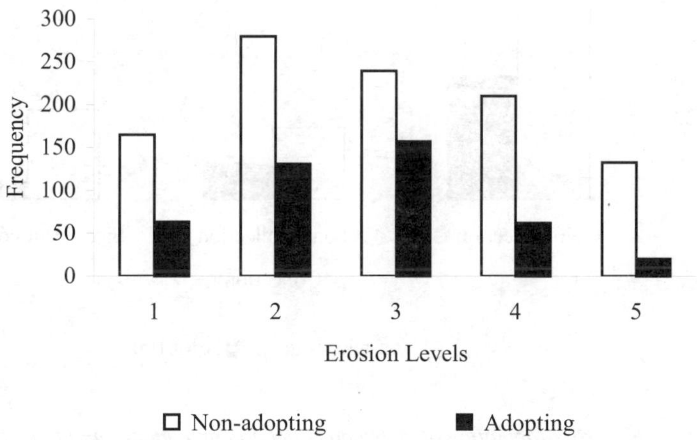

Figure 2A. Cross-tabulation of erosion and adoption of tree planting: Parcels with nitrogen-fixing trees do not necessarily experience less erosion

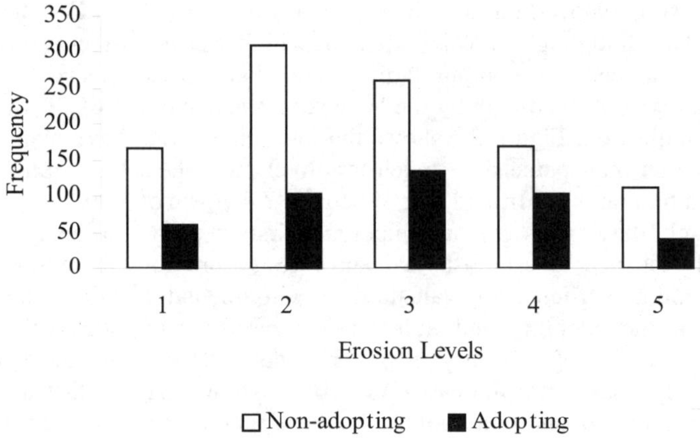

Figure 2B. Cross-tabulation of erosion and adoption of contour farming: Parcels with contour farming are likely to experience less erosion.

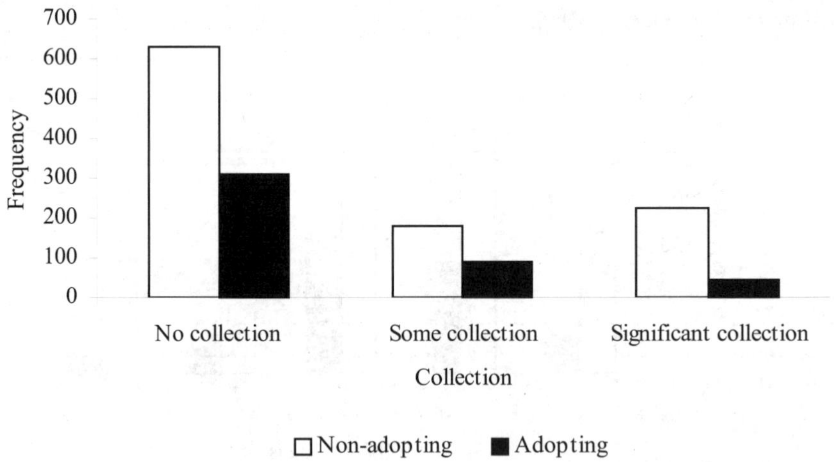

Figure 2C. Cross-tabulation of fuelwood collection and adoption of tree planting: Households that planted nitrogen-fixing trees typically do not collect fuelwood from forests.

Table 1. Comparing adopters and non-adopters in terms of plot, household, market, and policy characteristics

Variable	*Description*	*Adopt*	*Non-adopt*	*t-test or z-test*	*p-value*	*Difference in means equal to zero?*
Luas	plot size	0.446	0.409	–0.786	0.432	Cannot reject
Sandrock	sandy or rocky plot dummy	0.493	0.440	–1.878*	0.060	Reject
Dflat	flat plot dummy	0.180	0.244	2.708*	0.007	Reject
Dirig	irrigated plot	0.011	0.164	8.341*	0.000	Reject
Tahun2	years used	15.325	14.946	–0.486	0.627	Cannot reject
Drent	rented plot	0.045	0.038	–0.667*	0.505	Cannot reject
Plabhour	hourly wage rate	244.140	248.059	1.439	0.150	Reject
Pfert	fertilizer price	0.186	0.181	–1.901	0.058	Cannot reject
Ppadi	rice price	0.180	0.175	–1.385	0.166	Cannot reject
Pkopi	coffee price	1.770	1.816	2.841	0.005	Reject
Illrate	illness rate	0.200	0.171	–2.293	0.022	Reject
Chadpct	ratio of children to family size	0.357	0.395	3.223	0.001	Reject
Mfpct	ratio of males to family size	0.368	0.363	–0.470	0.638	Cannot reject
Aveduc	average education level	2.133	1.997	–3.781	0.000	Reject
Belong	count of assets	0.793	0.880	1.230	0.219	Cannot reject
Linc	index of income	5.755	5.666	–0.936	0.349	Cannot reject
Dincfarm	plans to increase farm size	0.144	0.056	–5.709*	0.000	Reject
Staffmn	index of park staff visits	0.191	0.142	–4.923	0.000	Reject
Dcoop	village cooperative dummy	0.297	0.321	0.916*	0.360	Cannot reject
Dcredit	village credit facility dummy	0.811	0.852	1.997*	0.046	Reject

**Z-statistic for two-sample Wilcoxon rank-sum (Mann-Whitney) test.*

4.2. Parametric Ordered-Probit Models of Environmental Services from Agroforestry

Because both soil erosion and fuelwood collection variables are categorical data with a pre-specified ordering, we used ordered-probit regression models to control for other confounding influences (Greene, 2003; Sills & Pattanayak, 2003). Specifically, we included biophysical plot characteristics, market prices, socio-demographic household influences, and policy variables that are shown in Table 1 to be sufficiently different between adopting and non-adopting households. Essentially, we estimated equation 5:

$$\begin{aligned}
&\text{Prob}(y = 0 \mid x) = \Phi(-x'\beta)\\
&\text{Prob}(y = 1 \mid x) = \Phi(\alpha_1 - x'\beta) - \Phi(-x'\beta)\\
&\text{Prob}(y = 2 \mid x) = \Phi(\alpha_2 - x'\beta) - \Phi(\mu_1 - x'\beta)\\
&\vdots\\
&\text{Prob}(y = J \mid x) = 1 - \Phi(\alpha_{J-1} - x'\beta)
\end{aligned} \tag{5}$$

where the x is a vector of control variables, and α and β are vectors of parameters to be estimated. Note we estimate J-1 additional α parameters or level-specific thresholds for a dependent variable with J levels, unlike a probit which includes just one constant. Thus, there are 4 and 2 α parameters for the soil erosion and fuelwood collection models, respectively.

The regression results are reported in Tables 2 through 4, corresponding to the two soil erosion and one fuelwood collection models. The fuelwood collection model is a weighted regression to account for the fact that collection is reported at the farm (household) level, and not at the parcel level.

For each regression model, we report the results of three specifications—model 1 includes only agroforestry; model 2 includes only biophysical plot controls; and model 3 includes the full set of regression controls. We do not discuss any of the models and/or variables in detail because of our primary interest in the impact of agroforestry on conservation of on-farm soils and off-farm forests. Suffice it to say, coefficients of other regression controls meet our expectations with respect to the sign, size, and significance for most parts. Additionally, all models are statistically significant, with the fuller specifications exhibiting considerably greater statistical power.

Turning to the primary result, in Table 2 we see that parcel level soil erosion (*tekn2*) is not correlated with tree planting. That is, when regression controls are introduced into the analysis in the "biophysical plot" and the "full" specification, we do not find any statistical relation between agroforestry adoption and soil erosion. In contrast, Table 3 shows that on-farm soil erosion is negatively correlated with contour farming (*tekcfarm*).

Table 2. Ordered probit regression models of parcel-level soil erosion as a function of tree planting

	Agroforestry Only		*Biophysical Controls*		*All Controls*	
Variable	*Coefficient*	*P-valuel*	*Coefficient*	*P-value*	*Coefficient*	*P-value*
tekn2	0.131	0.07	0.063	0.398	0.105	0.154
Sandrock			0.045	0.489	–0.002	0.971
Dflat			–1.108	0.000	–1.085	0.000
Dirig			–0.082	0.323	–0.038	0.661
Plabhour					–0.002	0.036
Pkopi					–0.109	0.400
Pfert					1.951	0.016
Ppadi					0.893	0.263
Illrate					0.521	0.100
Chadpct					–0.015	0.937
Aveduc					–0.258	0.000
Dincfarm					–0.131	0.384
Dcredit					0.101	0.336
Staffmn					–0.338	0.119
α_1	–0.973		–1.343		–1.962	
α_2	–0.117		–0.357		–0.964	
α_3	0.592		0.409		–0.174	
α_4	1.296		1.143		0.606	
Observations	1,460		1,460		1,454	
Log likelihood	–2,272.3		–2,145.8		–2,093.2	
Pseudo R^2	0.001		0.057		0.077	

This result is robust to the specification at the 5% significance level. Similarly, Table 4 shows that collection of fuelwood from off-farm forests is negatively correlated with on-farm tree planting (*tekn2*).

Again, this result is highly robust to alternative specifications, even at the 1% level of significance. Although we do not report the regression results in this chapter, we find that this general result holds for the "trip number" and "trip yield" variables as well. All models show a reduction in size of the coefficient on the adoption variable by as much as 20% as more variables are added into the model. Collectively, these models present mixed evidence regarding the effectiveness of tree planting and contour farming on conservation of on-farm soils and off-farm forests.

Table 3. Ordered probit regression models of parcel-level soil erosion as a function of contour farming

	Agroforestry Only		Biophysical Controls		All Controls	
Variable	Coefficient	P-value	Coefficient	P-value	Coefficient	P-value
tekcfarm	–0.177	0.007	–0.228	0.003	–0.148	0.043
sandrock			0.045	0.489	0.000	0.994
dflat			–1.127	0.000	–1.106	0.000
dirig			0.017	0.850	0.001	0.988
plabhour					–0.002	0.088
pkopi					–0.118	0.363
pfert					2.025	0.014
ppadi					1.050	0.184
illrate					0.552	0.085
chadpct					–0.003	0.988
aveduc					–0.241	0.000
dincfarm					–0.110	0.461
dcredit					0.107	0.311
staffmn					–0.361	0.095
α_1	–1.065		–1.425		–1.880	
α_2	–0.213		–0.440		–0.884	
α_3	0.497		0.329		–0.094	
α_4	1.209		1.072		0.690	
Observations	1,460		1,460		1,454	
Log likelihood	–2,270.3		–2,140.2		–2,092.2	
Pseudo R^2	0.002		0.059		0.077	

5. SUMMARY AND DISCUSSION

Agroforestry in general, or tree planting on farms in particular, can generate various socially valuable environmental services such as soil conservation, forest protection, biodiversity conservation, carbon sequestration, water "production," and water quality protection. Given this promise, researchers and policymakers alike have been surprised to discover that agroforestry has not been widely adopted. To date, researchers have relied on controlled field trials and laboratory experiments to support claims that agroforestry provides environmental services. We contend this evidence is far from convincing and has ultimately provided only limited incentives

Table 4. Ordered probit regression models of fuelwood collection from public forests as a function of on-farm tree planting

	Agroforestry Only		*Biophysical Controls*		*All Controls*	
Variable	*Coefficient*	*P-value*	*Coefficient*	*P-value*	*Coefficient*	*P-value*
tekn2	–0.290	0.001	–0.281	0.002	–0.275	0.003
sandrock			0.414	0.000	0.425	0.000
dflat			0.157	0.104	0.130	0.188
dirig			0.082	0.447	0.064	0.549
plabhour					–0.001	0.572
pkopi					0.108	0.605
pfert					–2.451	0.053
ppadi					1.504	0.187
illrate					0.891	0.030
chadpct					0.557	0.062
aveduc					–0.042	0.642
Dincfarm					–0.204	0.327
Dcredit					–0.040	0.806
Staffmn					–0.032	0.918
α_1	0.301		0.543		0.595	
α_2	0.816		1.071		1.136	
Observations	486		486		484	
Log likelihood	–430.0		–422.0		–415.0	
Pseudo R^2	0.006		0.024		0.039	

to adopt agroforestry practices. We attempt to address the gap and offer a social science perspective by developing and testing a socio-economic model of environmental services from agroforestry.

We draw on household production theory to conceptualize agroforestry as a bundled investment of land, labor, and money that potentially generates on-site and off-site environmental services. Our model suggests that the optimal household production of environmental services depends on the prices of consumption commodities and production inputs, factors affecting preferences and production possibilities, and the discrete adoption choice. Most critically, it highlights the importance of empirical analysis to evaluate this claim. We used data from a survey of households in Manggarai region of eastern Indonesia case study to consider conservation of on-farm soils and off-farm forests. Specifically, we tested if agroforestry reduces soil erosion and forest fuelwood collection. Our non-parametric analyses support the hypothesis that agroforestry reduces the collection of forest fuelwood and suggest that conservation of on-site erosion depends on the type of agroforestry practice. Multivariate ordered probit regressions, which allow us to control for potentially confounding factors, confirmed these findings. These regressions also suggest other factors and variables that could influence the production of environmental services.

Consider two qualifications to our empirical findings. First, we might not have sufficiently accounted for the endogeneity of the adoption choice or controlled for all confounding factors. Unfortunately, the use of instrumental variables or more sophisticated matching methods to address this issue, such as those proposed by Edmunds (2002) and Jalan and Ravallion (2003), is beyond the scope of this chapter. Second, cross-sectional data might not approximate long-run choices and farmer self-reports of biophysical quality and quantity may not be correlated with more objective measures. We would certainly welcome the opportunity to use longitudinal or panel data, including independently collected on-farm biophysical data by agronomists, to evaluate dynamic "ecological" production processes and economic choices such soil conservation.

In summary, despite the quantitative approach in this chapter, our numerical results should be treated as indicative of the impacts, and not as precise estimates. More generally, our goal is to illustrate a methodology for evaluating environmental benefits of agroforestry. Specifically, we believe that the household production logic offers a structured approach for viewing landscape level environmental services as the outcome of choices made by optimizing farmers in the face of biophysical constraints, market signals, socio-demographic influences, and policy forces. Unless these factors are collectively considered in the design and evaluation of agroforestry policies, agroforestry will fall short of its promises.

6. NOTES

[1] The utility function is assumed to be concave, continuous, and twice-differentiable with the following properties: $U_a > 0$, $U_\pi > 0$, $U_{aa} < 0$, $U_{\pi\pi} < 0$, and $U_{\pi a} > 0$. The first four properties are usual assumptions. The last condition cannot be similarly extracted from standard assumptions. It says that a landowner will value amenities more at higher incomes than at lower incomes—the normal good argument for amenities. Like all assumptions, the validity of these can only be determined empirically.

[2] Our focus on household choices considers off-site environmental services only as externalities. That is, households will not consider services that do not directly benefit them. Nevertheless, some of their private choices can create public benefits (and costs) off-farm or downstream.

[3] All means and modes discussed are parcel-weighted such that households who own more parcels have a greater influence on the average characteristics.

[4] Given the difficulties with measuring incomes for rural households, we follow Yang and An (2002) to measure "aggregate product or value-added" as the sum of agricultural revenues and off-farm wage income. At the time of the survey, the exchange rate was 2,200 Indonesian *rupiahs* per US dollar.

[5] Using this data and examining frequency distributions, we found that the average number of trips for households that plant nitrogen-fixing trees is lower than non-adopting households. Moreover, the yield or average fuelwood collection per trip is also lower for adopters vs. non-adopters.

7. REFERENCES

Chayanov, A. V. (1966). Peasant farm organization (Trans.). In D. Thorner, B. Kerblay, & R. Smith (Eds.), *A.V. Chayanov: The theory of peasant economy*, Homewood, Illinois: Richard Irwin. (Reprinted from *Peasant farm organization*, by A. V. Chayanov, 1925, Moscow: Cooperative Publishing House).

Dixon, J. (1997). Analysis and management of watersheds. In P. Dasgupta & K. Goran-Mäler (Eds.), *The environment and emerging development issue* (Vol. I, pp. 371-398). Oxford: Clarendon Press.

Edmunds, E. (2002). Government initiated community resource management and local resource extraction from Nepal's forests. *Journal of Development Economics, 68*(1), 89-115.

Georgiou, S., Whittington, D., Pearce, D., & Moran, D. (1997). *Economic values and the environment in the developing world.* Cheltenham, United Kingdom: Edward Elgar.

Gill, G. (1995). *Major natural resource management concerns in South Asia.* (Food, agriculture, and the environment discussion, Paper 8). Washington, DC: International Food Policy Research Institute.

Greene, W. H. (2003). *Econometric analysis* (5th ed.). New Jersey: Prentice Hall.

Jalan, J., & Ravallion, M. (2003). Does piped water reduce diarrhea for children in rural India? *Journal of Econometrics 111*(1), 153-173.

Kramer, R., Pattanayak, S. K., Sills, E., & Simanjuntak, S. (1997). *The economics of Siberut and Ruteng protected areas.* Indonesia: Directorate General of Forest Protection and Nature Conservation.

Kramer, R., Sills, E. O., & Pattanayak, S. K. (2003). *National parks as development and conservation projects: Gauging local support.* Durham, North Carolina: Duke University.

Lal, R. (1993). Challenges in agriculture and forest hydrology in the humid tropics. In M. Bonnell, M. Hufschmidt, & J. Gladwell (Eds.), *Hydrology and water management in the humid tropics* (pp. 395-404) New York: UNESCO/Cambridge University Press.

Lundgren, B. O., & Raintree, J. B. (1982). Sustained agroforestry. In B. Nestel (Ed.) *Agricultural research for development: potential and challenges in Asia* (pp. 37-49). The Hague, Netherlands: ISNAR.

Mann, H. B., & Whitney, D. R. (1947). On a test of whether one of two random variables is stochastically larger than the other. *Annals of Mathematical Statistics, 18,* 50-60.

Nair, P. (1993). *An introduction to agroforestry.* Boston, Massachusetts: Kluwer.

Pattanayak, S. K., & Kramer, R. (2001). Worth of watersheds: A producer surplus approach for valuing drought control in Eastern Indonesia. *Environment and Development Economics, 6*(1), 123-45.

Pattanayak, S. K., & Sills, E. (2001). Do tropical forests provide natural insurance? The microeconomics of non-timber forest products collection in the Brazilian Amazon. *Land Economics, 77*(4), 595-612.

Pattanayak, S., & D. Butry. (2003). Forest ecosystem services as production inputs. In E. Sills and K. Abt (Eds.), *Forests in a market economy* (pp. 361-379), Forestry Sciences Series, Volume 72, Dordrecht: Kluwer Academic Publishers.

Pattanayak, S., Abt, K., & Holmes, T. (2003a). Timber and amenities on non-industrial private lands. In E. Sills and K. Abt (Eds.), *Forests in a market economy* (pp. 243-258), Forestry Sciences Series, Volume 72, Dordrecht: Kluwer Academic Publishers.

Pattanayak, S. K., Mercer, D. E., Sills, E., & Yang, J. C. (2003b). Taking stock of the agroforestry adoption studies. *Agroforestry Systems, 57*(3), 173-186.

Pattanayak, S. K., Mehta, A. D., Sills, E. O., & Kramer, R. A. (2003c). Local uses of parks: Uncovering patterns of household production from the forests of Siberut, Indonesia. *Conservation and Society, 1*(2), 209-222.

Pattanayak, S. K., Sills, E., & Kramer, R. (in press). Seeing the forests for the fuel. *Environment and Development Economics.*

Sen, A. K. (1966). Peasants and dualism with and without surplus labor. *Journal of Political Economy 74,* 425-450.

Sills, E. O., & S. K. Pattanayak. (2003). *Markets, modernization and Mentawai: Explaining differences in forest dependence.* Raleigh, North Carolina: North Carolina State University.

Singh, I., L. Squire, & J. Strauss, (Eds.). (1986). *Agricultural Household Models.* Baltimore, Maryland: Johns Hopkins University Press.

Tomich, T. P., van Noordwijk, M., & Thomas, D. E. (in press). Environmental services and land use change: Bridging the gap between policy and research in Southeast Asia. Special issue of *Agriculture Ecosystems and Environment*.

Wilcoxon, F. (1945). Individual comparisons by ranking methods. *Biometrics, 1,* 8-8.

World Resources Institute (WRI). (2000). World resources 2000-2001: People and ecosystems: The fraying web of life. Available at http://pubs.wri.org/pubs_description.cfm?PubID=3027. Accessed November 19, 2003.

Yang, D. T., & An, M. Y. (2002). Human capital, entrepreneurship, and farm household earnings. *Journal of Development Economics, 68*, 65-88.

8. AUTHOR'S NOTE

The data used in the case study are drawn from a larger project led by Randy Kramer and funded by the Asian Development Bank, Government of Indonesia, Duke University, Howard Gilman Foundation, and Joshua Trent Foundation. This study was made possible in part by the assistance of Erin Sills, Frans Dabukke, Sastrawan Manullang, Mariyanti Hendro, and Nining N.P.

RAM K. SHRESTHA AND JANAKI R. R. ALAVALAPATI

ESTIMATING RANCHERS' COST OF AGROFORESTRY ADOPTION

A Contingent Valuation Approach

1. INTRODUCTION

Agroforestry practices enable landowners to maintain trees and vegetation cover on farms and ranches. These practices enhance environmental benefits on private lands by controlling soil erosion, reducing pollution runoff, sequestering atmospheric carbon dioxide, and protecting wildlife habitat (Alavalapati & Nair, 2001; Garrett, Rietveld, & Fisher, 2000). Some of these environmental benefits may have positive relationships with agriculture outputs and thus enhance the productivity of the land. Conserving soil and preventing landslides, for instance, will have a lasting positive impact on agricultural productivity (Pimentel, Harvey, Resosudarmo, Sinclair, Kurz, & McNair et al., 1995). Other environmental benefits, habitat conservation for example, can only be realized at the expense of some agricultural outputs. If environmental benefits and agricultural productivity are competing, farmers are less likely to adopt agroforestry practices.

Environmental benefits provided by agroforestry are examples of positive externalities that provide so called "public goods." Such goods and services are, by their nature, provided to the benefits of all society. Consumers of public goods, in this case all members of society, receive benefits without incurring any expense. Producers, on the other hand, must incur the cost of providing such goods. For example, atmospheric carbon dioxide sequestration through agroforestry would improve air quality and benefit society as a whole but only landowners would incur the costs of practicing agroforestry. The non-rival and nonexclusive nature of public goods and services may not, therefore, provide sufficient incentive for landowners to adopt agroforestry practices at a level that society considers optimal (Nicholson, 1998; Randall, 1987; Sugden, 1999; Varian, 1992).

J. R. R. Alavalapati & D. E. Mercer, Valuing Agroforestry Systems, 183–199.

An optimal supply of environmental services may require encouraging landowners to adopt conservation practices. Various federal and state farmland conservation programs in the U.S., such as Conservation Compliance, Sodbuster, Swampbuster, and Conservation Reserve Programs (CRP), provide financial incentives to encourage conservation practices on private farmlands (Feather, Hellerstein, & Hansen, 1999; Lant, 1991). In order to provide environmental services through agroforestry practices, landowners may need incentive payments to compensate for the additional costs associated with adoption. Designing incentive policies for agroforestry adoption, however, requires information on the benefits and costs of the policies or programs. More specifically, it requires information on ranchers' expected cost of agroforestry adoption relative to conventional practices.

Several environmental economic valuation methods and techniques can be used to estimate opportunity costs of agroforestry adoption in ranch lands. Although, these methods are extensively used in agricultural, natural resource, and environmental economics, they are relatively new to agroforestry. This chapter provides an overview of a contingent valuation methodology and its application to an agroforestry system. We estimate Florida cattle ranchers' willingness to accept (WTA) incentive payments to adopt silvopasture practices. The ranchers' WTA will reflect overall opportunity costs of silvopasture adoption, which enhances environmental conservation in ranchlands.

The next two sections illustrate environmental valuation methodology and an application of the dichotomous choice contingent valuation method using silvopasture adoption as a case study. Section 4 details the potential extensions in our valuation methods. Finally, in Section 5 we discuss the implications of the results and draw some conclusions from the study.

2. ENVIRONMENTAL VALUATION METHODS

Valuation of environmental goods and services involves estimating economic costs and benefits of environmental policies using appropriate methodologies. The alternative methodologies to valuation are largely based on either observed or hypothetical behavior measures or a combination of both (Freeman, 1993; Mitchell & Carson, 1989). Methods that derive value estimates using market information are often referred to as "revealed preference" techniques. These methods capture individual's preferences for environmental goods and services based on the participant's past behavior in a market. Methods that determine values using hypothetical or simulated markets are referred to as "stated preference" techniques. These methods elicit values by asking questions regarding a specific improvement in an environmental attribute such as, "how much would you pay or accept in compensation for the improvement?" or "would you vote yes or no for the improvement?"

The indirect revealed preference methods that rely on observed behavior are widely used in environmental valuation. For example, travel costs of recreational visitors to agroforestry lands can be used as a proxy to value recreational benefits

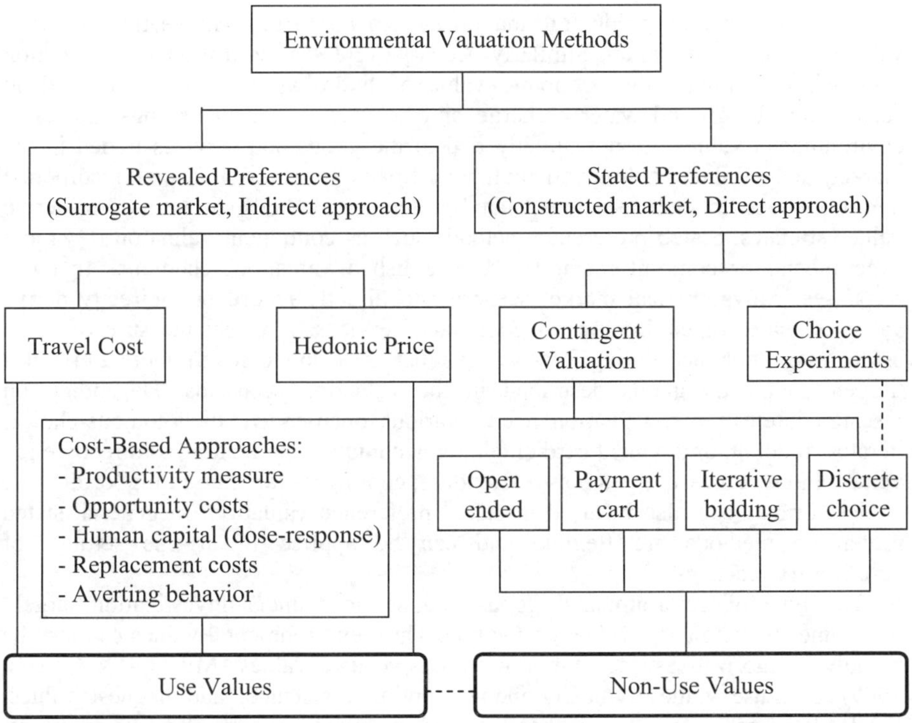

Figure 1: Valuation methods to estimate environmental benefits of agroforestry.

associated with agroforestry practices. Similarly, property price differentials between agroforestry and other land uses can be attributable to the value of environmental services and amenities. The values associated with these amenities can be estimated using the private property values in a hedonic analysis. Furthermore, random utility models may be used to determine values of environmental attributes using observed behavior such as purchases of organic products or "green beef" produced from environmentally friendly silvopasture practices. On the other hand, there are various cost-based, revealed preference approaches to value environmental services (Garrod & Willis, 1999). For example, the market value of lost soil productivity compared to a conventional practice can be a valid measure of the soil conservation benefits associated with agroforestry practices.

Revealed preference methods rely on the notion that complementary or weak-complementary relationships exist between market goods and environmental services. For example, Pattanayak and Butry (2003) estimated the value of drought mitigation in a watershed using labor inputs as weakly complementary to ecosystem services, such that increased ecosystem services raise the value of the marginal product of farm labor. The authors used a combination of market information, survey data, and geographic information system (GIS) to estimate the economic value of drought mitigation. Similarly, Acharya (2000) used a production function approach to estimate the economic value of hydrological services of wetland ecosystems for ground water recharge or discharge. In many settings, however, environmental values are not directly tied to the goods and services traded in the market, and production function or input demand functions may not incorporate non-use values. In such situations, stated preference methods can be used to derive value estimates. Stated preference methods such as contingent valuation (CV) and experimental or conjoint methods are established valuation techniques that use responses to hypothetical market scenarios to directly record or indirectly derive value estimates of environmental goods and services. A direct measure of value estimates are obtained through CV surveys that ask respondents to quote a price for the environmental goods described in the valuation scenarios. Depending on research interests and valuation needs, various open-ended, dichotomous-choice, iterative bidding, or payment card elicitation techniques can be used to directly elicit values. Indirect measures such as referendum, contingent ranking, rating, or discrete choice formats can also be used in stated preference valuation. In general, stated preference methods are flexible and can be applied in diverse settings of agroforestry practices.

The type of environmental goods and services under investigation largely determines the choice of valuation method. Thus, environmental values can also be broadly categorized as use or non-use (passive use) values (Mitchell & Carson, 1989). Non-use values typically include option, existence, and bequest values. While stated preference approaches are used to estimate any or all of these values, revealed preference methods can only be applied to estimate use values. Since markets provide no information about non-use values, revealed preference approaches are clearly inappropriate for determining non-use values of agroforestry practices.

In the following section, we illustrate an application of the contingent valuation methodology. The application involves estimating Florida cattle ranchers' willingness to accept incentive payments to adopt silvopasture practices.

3. RANCHERS' WILLINGNESS TO ADOPT SILVOPASTURE IN FLORIDA

Cattle ranching is a dominant land-use practice in Florida with nearly 60% of the land area under improved and semi-improved pasture (Boggess, Flaig, & Fluck, 1995; Florida Agriculture Statistics Service [FAS], 2002)[1] Florida ranks 10th in the United States and 3rd in states east of the Mississippi River for beef cattle herd size (FAS, 2002). While 98% of Florida's ranches are relatively small and are entirely dependent on cultivated grasses and legumes for grazing, the remaining 2% (located primarily in central and south Florida) comprise about 75% of the pastureland and support 48% of Florida's cattle. These large ranches consist of more than 750 cows each (Wade, Minton, & Delargy, 2001). Cattle ranching has long been a significant environmental concern in this region since phosphorus runoff from cattle ranches can cause significant ecological degradation to the environment.

Silvopasture, which combines forage and livestock with trees, is expected to reduce phosphorus runoff and generate other environmental benefits such as soil conservation, carbon sequestration, wildlife habitat protection, and aesthetics (Clason & Sharrow, 2000). Specifically, trees and other vegetation help filter surface runoff and absorb surplus nutrients before they reach streams and lakes (Environmental Protection Agency [EPA], 1995). Added tree cover in silvopasture would also sequester atmospheric carbon dioxide and, thus, can be considered for carbon credits under the Kyoto Protocol (Cannell, 1999; Sedjo, 2001). Ranch lands are known to provide habitat for wildlife including many threatened and endangered species (Morrison & Humphrey, 2001), habitat that is further enhanced through silvopasture. For these reasons, land-use policies and programs can be designed to provide incentives for ranchers to encourage and adopt silvopasture

In this paper, cattle ranchers' willingness to adopt silvopasture and the optimal level of incentive payments required for a silvopasture program are evaluated using a contingent valuation method (CVM). We applied the dichotomous-choice (DC) CVM to elicit ranchers' WTA for silvopasture adoption in Florida.

3.1. Dichotomous Choice Contingent Valuation method

The dichotomous-choice CVM is considered incentive compatible and is often preferred in eliciting values of environmental goods (Arrow, Solow, Portney, Leamer, Radner, & Schuman, 1993; Mitchell & Carson, 1989). In DC CVM respondents are asked to provide only a 'yes' or 'no' response to whether a randomly assigned dollar bid adequately represents the value of the desired change in the quantity or quality of environmental goods or services. Thus, DC CVM modeling accounts for the probability that the respondent's minimum WTA is less than or equal to the offered amount. A detailed discussion on the theory and estimation techniques used in DC CVM is provided in the appendix section of this chapter.

Eliciting WTA instead of WTP (willingness to pay) is a more relevant method for our silvopasture environmental valuation assuming that ranchers would be willing to adopt silvopasture with appropriate incentive payments.[2] These incentive payments could be annual direct payments per acre of land under silvopasture management or premium prices on beef produced under silvopasture. The total payments to silvopasture adopters could be justified as a proxy for the societal value of the environmental benefits supplied by the additional land in silvopasture. In essence, the public would indirectly encourage ranchers to produce these goods through incentive payments.

3.2. Study design and implementation

The DC CVM survey questionnaire includes three main sections: introduction and description of the good, valuation scenario and elicitation questions, and socioeconomic information questions (Mitchell, 2002; Mitchell & Carson, 1989). We elaborate each of these sections here in the context of our silvopasture case study.

First, to introduce the survey, we explained its objective and confirmed that this survey will not track confidential information. Then, the respondents were asked to provide information on the natural attributes of their ranch including forest cover, ranch size, location, cattle population, and land uses. Second, we presented a concise description of the valuation scenario with proposed changes in current land use to silvopasture and details of its benefits and costs. The valuation scenario included a description of the goods to be valued (description), explanation of how the good will be provided (provision), and a valuation question (elicitation) (Mitchell, 2002). The development of this scenario is the critical part of the CVM survey. Based on the feedback from pre-tests, the term 'tree-cattle' pasture was used throughout the survey to represent silvopasture systems. In particular, the proposed changes included 20% pastureland with forest or brush cover, 60 foot (ft) streams, 12 ft grass buffer strips and restoration of wetlands, as applicable (Figure 2). Valuation questions followed the scenario description. Finally, demographics, household income and occupation questions were presented.

Choices of payment type and bid amounts are an integral part of DC CVM survey design. While the researchers have full control on the determination of the mode of payment and number of bid amounts to be used in the survey a careful decision is required to avoid biases (Mitchell & Carson, 1989). Based on the feedback from agricultural professionals and focus groups we used a premium price on beef produced from silvopasture adoption as payment type. The prevailing market price of cattle at farm-gate $0.38/ pound (lb) was used as mid point of the bid and remaining bids were determined through linear extrapolations. A total of 11 bid amounts, ranging from $0.02 to $0.92 per lb of beef, were used in the survey.

Information from focus group meetings and pre-testing was used to revise the questionnaire. Prior to distributing the survey, we published a brief informational article in the Florida Cattlemen Association (FCA) magazine about the upcoming survey. Respondents were drawn from the FCA membership directory, and as information about FCA members is confidential, the FCA provided assistance in sample selection and mailing. A sample of 1,000 respondents was drawn randomly from member lists, and the survey packet with a questionnaire and cover letter was mailed to each respondent in the first week of May 2002. Following Salant and Dillman (1994), a three stage mail survey protocol was developed to increase the response rate, whereby the initial questionnaire was followed by a postcard reminder and a subsequent mailing of the questionnaire to non-respondents each spaced approximately 10 days apart. We were contacted by a number of respondents indicating that they were not ranchers although they had membership in the FCA. These allied members of this organization accounted for about 10% of the total members. After the second mailing and excluding allied members from the sample, we received a total of 421 survey responses, resulting in a 47% response rate.

Tree-Cattle Pasture, a new management strategy:

Adding forest cover, strips of vegetation (buffer strips) and wetlands:

- Controls phosphorus run-off substantially. The buffer strips below reduce phosphorus runoff by 65 – 90%.
- Generates income from timber.
- Provides habitats for wildlife.
- May increase recreation revenues.
- Costs more and may reduce cattle output, thus necessitating a higher beef price.

Imagine you are asked to change your management practices from Current Pasture (no change in current management) to Tree-Cattle Pasture with the changes described below. Please answer the following question in the box.

Practices	Current Pasture	Tree-Cattle Pasture
Bush/ Forest Cover	No Change	20% of land
Buffer Strips	No Change	60-ft stream buffer 12-ft grass strips
Wetland(s)	No Change	Create or restore as appropriate

Q. Would you choose to change to Tree-Cattle Pasture if an average of $_______ per lb. is added to the price you receive for your beef?
(Check answer)

____ YES
____ NO

Figure 2: Valuation scenario of the silvopasture adoption survey.

3.3. Respondent characteristics and empirical model

Descriptive analyses of the survey responses indicate that cattle ranches in Florida provide various use and non-use values to the public. On average, ranchers noticed that nearly five species of birds or animals are present on their ranches with a maximum of 19 species. About 40% to 50% of ranchers reported that some form of marsh or wetland, creek/stream, and some hardwood trees exist on their ranches. Hunting, fishing, and horseback riding are noted as popular recreation activities on many ranches. About 10% of ranchers responded that they have commercial hunting leases on their land.

Ranchers' accepted more than half (59.1%) of the times the proposed incentive payments. There was a sharp increase in ranchers' positive responses (% Yes) to the proposed incentive (bid offer) at each increment in lower bid amounts. The positive

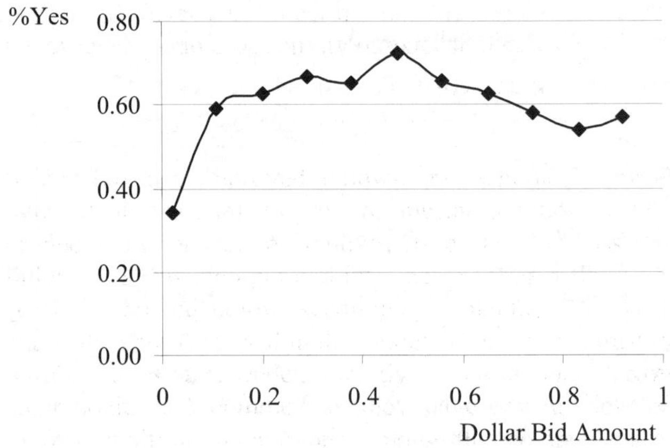

Figure 3. Distribution of positive responses to incentive payment bids.

responses leveled off near the mean bid amounts. The positive responses slightly decreased after the mean bid but ended with an increasing trend showing further increments in bid offers would increase acceptance, i.e., willingness of the rancher to accept incentive payments to practice silvopasture. Because a linear relationship between bid amounts and WTA responses is expected a slight decline in WTA responses at higher bid amounts may result in a relatively lower estimate of the value.

We anticipated that natural attributes of a ranch would positively influence ranchers' silvopasture adoption or willingness to accept the incentive offer. For ranches that are predominantly under improved pasture, changing to silvopasture might cause the rancher to forego more forage benefits. Thus, improved pasture is expected to have a negative influence on the likelihood of silvopasture adoption. Recreational use of ranchland depends on the natural attributes of the land including vegetation. If ranchers are receiving recreation benefits, we hypothesize that they would be more likely to participate in silvopasture practices and that hunting, fishing, and horseback-riding variables will have a positive impact on rancher's adoption decision.

Socioeconomic variables that we expect to potentially influence ranchers' WTA are ranch size, cattle herd size, household income, and respondents' age, education, and affiliation with environmental organizations. We also use Florida Department of Environmental Protection regions (as dummy variables) to account for regional differences in ranchers' adoption of silvopasture.

3.4. Model results

Econometric models were used to analyze the WTA responses and to estimate the mean WTA using DC CVM. Specifically, we estimated a logit model to analyze

Table 1: Logit models of ranchers' willingness to adopt silvopasture practices

Variable and Definition	*Expected Sign*	*Co-efficient*	*Standard Error*
Payment Offer and Opportunity Costs:			
PRICEP, payment per lb of beef produced (U.S. $/lb)	+	1.0030**	0.4593
ACRE, land area under the ranch	±	-4.51E-05	6.10E-05
ACRES, square of the land area	±	8.46E-07	1.13E-06
NORTH, 1 if the ranch is in northern Florida	±	0.5451	0.3776
CENTRAL, 1 if the ranch is in central Florida	+	-0.1508	0.3480
ACCESS, travel distance to the nearest city (miles)	+	0.0110	0.0081
Natural Attributes:			
WNUM, number of wildlife species on the ranch	+	0.0914*	0.0499
CKST, 1 if a creek or stream is on the ranch	+	-0.3447	0.2610
MARSH, 1 if ranch contains any marshy area	+	0.5380*	0.2838
LLPINE, 1 if the current forest cover is longleaf pine	±	0.6109**	0.2725
IMPPAST, 1 if ranch is primarily improved pasture	-	0.0037	0.0040
Recreation Benefits:			
HUNT, 1 if recreational hunting is allowed	+	0.6109**	0.2696
FISH, 1 if recreational fishing is allowed	+	-0.2078	0.2674
HBACK, 1 if horse back riding is allowed	+	-0.5677	0.4690
Socioeconomic Characteristics:			
INC, Annual household income (U.S. $000)	-	-0.0166	0.0114
INCS, Square of the household income	±	0.0934	0.0712
AGE, Age of the respondent (years)	±	-0.0097	0.0079
EDU, Formal education of the respondent (years)	±	0.0452	0.0518
MEMB, 1 if the respondent is a member of any environmental organization	+	-0.3446	0.3812
CONSTANT		-0.9317	1.0429

*Log-L = 221.81; N = 366; Chi Sq. = 45.73**; Correct Predictions = 66.53%*

** Statistically significant at p<0.10, ** Statistically significant at p<0.05.*

rancher's willingness to adopt silvopasture practices. Signs and statistical significance of the estimated coefficients suggest that results are consistent with *a priori* expectations. The variable representing the incentive payment offer (*PRICEP*) is positive and statistically significant suggesting a higher probability of adoption if the payment is higher. This result is consistent with demand theory (Table 1).

3.5. Dependent variable (ranchers' participation responses) recorded as Yes or No to incentive offer.

The results indicate that the existence of natural attributes will increase adoption. The variables representing wildlife habitat (*WNUM*), marshlands (*MARSH*), and longleaf pines (*LLPINE*) have a positive impact on ranchers' adoption of silvopasture. These results have at least two interpretations. First, the presence of natural attributes on ranchlands would diminish the productive use of the land for other agricultural purposes, thereby incurring lower opportunity costs of land use changes, which leads to a greater likelihood of ranchers adopting silvopasture practices. Second, ranchlands with these attributes are better suited for multipurpose use such as pasture and outdoor recreation. As hunting and other outdoor recreational uses by families, friends, and recreation clubs are popular on these lands, we would expect that the presence of natural attributes would encourage ranchers to adopt silvopasture.

In addition, we accounted for the effect of recreational use of ranchlands, such as hunting, fishing, and horseback riding, in our analysis. The coefficient on the variable representing the presence of recreational hunting (*HUNT*) is positive and statistically significant. Although recreational fishing and horseback riding variables, however, indicate negative influence, the coefficient is insignificant.

The positive sign on the *ACCESS* variable suggests that increased distance from an urban center corresponds with an increased likelihood of ranchers adopting a silvopasture, however, the coefficient is statistically insignificant. To assess the influence of ranchers' economic and demographic factors on their adoption decision, we incorporated income, education, age, and their membership in environmental organizations into the model. None of these variables, however, were found to be statistically significant.

We estimated ranchers' mean WTA for their adoption of silvopasture practices. On average, a premium price of $0.19 per pound of beef is required by ranchers to adopt silvopasture practices. [3] This estimate is approximately equivalent to the direct payment of $9.32 per acre per year estimated in Shrestha and Alavalapati (2003) using the same sample. Our estimates of ranchers' WTA for silvopasture practices is comparable with previous studies on farmers' willingness to participate in conservation programs in the U.S. (e.g., Cooper & Keim, 1996; Lant, 1991; Lohr & Park, 1994). Lant (1991), for example, reported that the average annual payment under the Conservation Reserve Program was $48.93 per acre per year. A relatively lower WTA estimate in our study may be due to the complementary nature of cattle and tree farming systems. Unlike Conservation Reserve Programs where farmers face more restrictions, silvopasture requires only modest changes from current ranching practices.

Using the WTA estimates, we calculated the total annual incentive payments required for silvopasture adoption statewide. The United States Department of Agriculture (USDA, 1997) Census of Agriculture shows that Florida has more than 2 million cattle resulting in annual sales of approximately $91.7 million. Current cattle sales data show that annual cash receipts from cattle are $360.5 million (FAS, 2002). Thus, a roughly 20% reduction in cattle output in response to the 20% ranchland set side under tree cover translates into an opportunity cost of $72.0 million. This shows that our estimate of ranchers' WTA for their adoption of silvopasture land use is close to the opportunity cost of silvopasture adoption. However, the total cost of the silvopasture program in Florida would be lower than this estimate since not all ranchers are expected to switch to silvopasture.

4. EXTENSIONS IN CONTINGENT VALUATION

Application of contingent valuation methods (CVM) to estimate the value of environmental goods was much debated in early 1990s. Validity and reliability of the value estimates using hypothetical market scenarios for environmental goods and services were raised as major concerns. However, extensive research and diverse applications of CVM in the past several years have enabled researchers to identify various biases and account for many of those anomalies (Carson, Flores, & Meade, 2001). While some of these concerns and skepticism of CVM persist, these methods remain at the forefront in environmental valuation.

We illustrated an application of the dichotomous choice CVM in estimating ranchers' WTA incentive payments to adopt agroforestry practices. A National Oceanic and Atmospheric Administration (NOAA) Blue Ribbon Panel co-chaired by two Nobel Laureate economists (Kenneth Arrow and Robert Solow) suggested that the dichotomous choice elicitation format is incentive compatible and provides a valid estimate of environmental values (Arrow et al., 1993). However, there are many other elicitation formats in CVM than those presented here including open-ended questions, iterative bidding games, and payment card elicitations as mentioned earlier.

Conjoint experiments offer an alternative approach in stated preference valuations. This approach follows an experimental design that elicits preferences by requiring respondents to make repeated choices. Conjoint, or choice, experiments have been applied to derive environmental values in many recent studies (e.g., Adamowicz, Boxall, Williams, & Louviere, 1998; Holmes & Adamowicz, 2003; Shrestha & Alavalapati, 2002). The most useful feature of the choice experiment design is that values for multiple attributes can be estimated from a single study, such as the value for wildlife habitat preservation, water quality improvement, and carbon sequestration. Shrestha and Alavalapati (2002) used a choice experimental design to estimate the environmental benefits of silvopasture practices in terms of limiting pollution runoff, sequestering atmospheric carbon dioxide in tree biomass, and improving wildlife habitat in the Lake Okeechobee watershed, Florida. While this method is appealing in estimating environmental values, it is relatively complex in terms of research design, data collection, and analysis.

Combining DC CVM with revealed preference methods allows the strengths of the both methods to be utilized in estimating environmental values (Adamowicz, Louviere, & Williams, 1994). The combined stated and revealed preference approach not only enriches the data but also utilizes multiple sources of data that can potentially improve the reliability of value estimates (Louviere, Hensher, & Swait, 2000).

5. DISCUSSION AND CONCLUSIONS

Formulation of conservation incentive programs requires two pieces of information: (1) the value of environmental goods and services of environmentally friendly practices, as perceived by the public, and (2) the value of compensation required for landowners to undertake these practices. Because of the large proportion of non-use values associated with agroforestry, stated preference methods can be useful for estimating environmental benefits of agroforestry practices. We illustrated an application of the dichotomous choice contingent valuation method using a silvopasture case study from Florida to estimate ranchers' minimum willingness to accept incentive payments for silvopasture practices. The WTA measure is theoretically equivalent to the maximum willingness to pay for the same good by consumers if the good was indeed provided. Our results show that Florida ranchers would adopt silvopasture if an additional $0.19 per pound price premium was paid for cattle raised on silvopasture lands, which would cost the public $91.7 million every year.

If society demonstrates strong preferences for such environmental goods and services, the demand for these goods in term of their willingness to pay can also be estimated using contingent valuation or stated preference methods. Shrestha and Alavalapati (2002) applied the choice experiment approach of stated preference method to estimate public willingness to pay for environmental benefits of silvopasture adoption in the Lake Okeechobee Watershed, Florida. Their results indicated that an average household would pay $137.97 per year for five years for the additional environmental benefits associated with ranchers adopting silvopasture practices. In the context of developing countries, valuing environmental services may be more challenging because households in subsistence economies may not be able to fully express their willingness to pay for environmental benefits although they care about environmental services.

Application of WTA elicitation provides an appropriate means to estimate the environmental values of agroforestry land uses. The WTA elicitation essentially approaches the valuation problem from the suppliers' perspective and estimates the marginal value needed to compensate landowners for the supply of environmental goods. This approach has several important applications in agroforestry. First, it has direct policy relevance because policy makers receive information on how much the public must spend on agroforestry to internalize environmental benefits. Necessary institutional arrangements and planning can be made to implement the policy that encourages landowners to adopt agroforestry. Second, in most cases, since landowners have exclusive property rights on their land, asking landowners' WTA is more relevant and direct than the WTP approach. Third, most of the agroforestry benefits are obtained with incremental changes in current land use practices, thus, it is likely that the cost of producing these goods can be much lower than the willingness of the general public to pay for them. Fourth, WTA considerations are familiar to stakeholders involved in agricultural practices in countries, including developing countries, where some level of government support is a common practice. For these reasons, a WTA estimate can be a fairly realistic measure of the environmental benefits of agroforestry.

Because of multiple objectives and the multi-dimensional nature of agroforestry and farmland conservation programs, environmental benefits generated from these programs may vary. If the state were to purchase complete land rights for extended periods, total environmental values could be internalized. Payments for forgoing partial rights or implementing specific conservation practices such as filter strips or riparian buffers, on the other hand, would account for only partial environmental benefits. The economic valuation process and application of valuation methods in these situations can be vastly complicated.

Identifying an appropriate valuation method should be based on the valuation context and environmental goods in question. For example, if the environmental benefits are complementary or weakly-complimentary to the goods and services traded in competitive markets, production function and hedonic price methods can be used to estimate the benefits. It is likely that well maintained agroforestry parcels with sustained productivity will result in higher land prices per acre. In such a case, the observed positive price differentials of agroforestry parcels would be partially attributable to the environmental value of agroforestry land use. It is, however, important to note that each parcel of farmland poses several unique attributes and that these lands are not frequently traded in the market so market information on land transactions and prices may be inaccessible.

As wildlife habitat enhancement is one of the important benefits of agroforestry, recreational use values may be significant. Recreational hunting, wildlife viewing, camping, hiking, biking, and horseback riding are some of the potential recreation uses of agroforestry lands. The value of recreation uses of these lands can be estimated using travel cost methods. Using visitors' travel cost and frequency of visit data, recreation trip demand models can be estimated. Consumer surplus estimated from recreation trip demand models is a valid estimate of the recreation use value of agroforestry. The only caveat to this concept is that recreational uses in private forests or agroforestry lands are still limited. Therefore, the travel cost model may not estimate the representative recreational demand for the environmental benefits of agroforestry.

Another approach to indirectly approximate the benefits of agroforestry would be to estimate the environmental damage caused by conventional agriculture or forestry practices that otherwise would have been avoided by adopting agroforestry. Thus, the estimate of damage such as due to soil erosion or pollution runoff can be attributable to the benefit of agroforestry. This approach, however, takes into account only partial benefits, leaving other important environmental benefits out of the analysis.

Because of continued development of valuation methods and their application in diverse fields, more choices will be available in valuation methods and elicitation techniques for agroforestry valuation. Some of the important fields to watch for these developments are environmental and natural resource economics, health economics and epidemiology, transportation economics, and business and marketing research.

6. NOTES

[1] Pasture improvement includes designing site drainage, growing improved forage grasses such as bahia grass (*Paspalum notatum*), and applying inorganic fertilizers (Boggess et al., 1995). As cattle pastures are developed with improved grasses and accommodate higher stocking density of animals, more feed and fertilizer are imported into the watershed, which causes nutrient enrichment and eutrophication in streams and lake waters.

[2] While both WTA and WTP elicitations are theoretically equivalent measures of environmental benefits, researchers often find that estimates using WTA elicitation are greater than WTP estimates (Mitchell & Carson, 1989). Use of WTA implies that the property rights are held by the respondents, ranchers in this case. Shrestha and Alavalapati (2002) used WTP elicitation to estimate public demand for environmental services of silvopasture practices in the Lake Okeechobee watershed, Florida.

[3] This estimate is slightly higher than a recent estimate obtained using a quadratic form of the incentive payment to account for non-linearity in utility function (Shrestha and Alavalapati, 2003) resulting in an estimated mean WTA of $0.15 with a 95% confidence interval of $0.004 - $0.283.

7. APPENDIX

Ranchers' decision-making process to adopt or not adopt silvopasture practices can be viewed as utility maximizing behavior of households (Cooper & Keim, 1996; Lohr & Park, 1994). Although the variables entering into ranchers' utility function are often unobservable and utility functions are unknown to the researcher, they can be viewed as a function of deterministic and random components (McFadden, 1974) as,

$$V_{ij} = v_{ij} + \varepsilon_{ij} \quad (1)$$

where V_{ij} is the conditional indirect utility of individual rancher i from alternative land use j, v_{ij} is the deterministic component of the model and ε_{ij} is the random component. Selection of silvopasture over conventional ranching implies that the utility of v_{i1} is equal to or greater than that of v_{i0}, where $j = 0, 1$ representing conventional ranching and silvopasture, respectively. Thus,

$$v_{i1}(y + c; x) + \varepsilon_{i1} \geq v_{i0}(y; x) + \varepsilon_{i0} \quad (2)$$

where, y is rancher i's income, c is the incentive payment, and vector x is socioeconomic attributes of the rancher which affect their adoption decision. Variable c can be interpreted as $c^* + \delta$ where c^* is the required incentive payment and δ is the pecuniary cost of conventional ranching minus the pecuniary cost of silvopasture practices (Cooper & Keim, 1996). Thus, c can be considered as a "net" incentive payment. Let,

$$v_{ij}(y; x) = \gamma_j + \alpha y, \quad (3)$$

where $\alpha > 0$, $\gamma = x'\beta$, and β is vector of estimated coefficients. Then, the rancher is willing to accept c if

$$\gamma_1 + \alpha(y + c) + \varepsilon_{i1} \geq \gamma_0 + \alpha y + \varepsilon_{i0} \, . \quad (4)$$

Overall, the researcher can only analyze the probability of ranchers' choice of an alternative over another. The probability of rancher i choosing alternative j, $p(\cdot)$, may be expressed as

$$p_{ij}\{c \geq WTA\} = p\{v_{i1}(y + c; x) + \varepsilon_{i1} \geq v_{i0}(y; x) + \varepsilon_{i0}\}, \quad (5)$$

where WTA is the minimum incentive payment c required by ranchers to change from conventional ranching to silvopasture. If the difference between the two underlying utility functions (v_{i1} and v_{i0}) is positive, the rancher will adopt silvopasture practices upon receiving c. The utility difference model can be expressed as

$$\Delta v = v_{i1} - v_{i0} = \gamma + \alpha c \quad (6)$$

where $\gamma = \gamma_1 - \gamma_0$. Assuming the error terms of the utility functions are independent and identically distributed (iid) and follow a logistic distribution, the choice probabilities can be estimated using a logit specification (Kingsbury & Boggess, 1999; Lohr & Park, 1994; Maddala, 1999). The logistic distribution used to model the probability of adoption may be expressed as

$$p(j = 1 | x, c) = [1 + e^{-\Delta v}]^{-1} = [1 + e^{-(x'\beta + \alpha c)}]^{-1} \quad (7)$$

where the logistic model is specified as the probability of a 'yes' response to silvopasture adoption if the incentive offer is a price premium per pound of beef produced.

In addition to analyzing explanatory variables influencing ranchers' adoption decisions, we can estimate WTA as the welfare measure using the utility difference model. The mean WTA can be estimated using the predicted value of the WTA function estimated at the mean value of the covariates (Cameron, 1988; Shyamsunder & Kramer, 1996). In essence, the net WTA from a logit model is estimated through the transformation of the logit model coefficients by dividing the constant term and all slope coefficients other than the bid variable by the coefficient on the bid variable. With this transformation the coefficients will have ordinary least squares interpretation as

$$\begin{aligned} E(\text{WTA}_i) &= E(X_i\beta + \eta_i) \\ &= \overline{X}\, b + E(\eta_i) \\ &= \overline{X}\, b \end{aligned} \quad (8)$$

where b is the vector of transformed slope coefficients and $\overline{X}$ is the vector of explanatory variable means. Thus, WTA is the sum of the products of transformed coefficients and the means of the respective variables. If the dollar bid amount is the only explanatory variable in the logit model, for example, WTA is the negative ratio of the constant term and the slope coefficient of the bid variable ($-\beta_o/\beta_{bid}$).

8. AUTHORS' NOTE

This research was made possible by the financial support from the USDA (Initiative for Future Agriculture and Food System) and the Florida Agricultural Experiment Station.

9. REFERENCES

Acharya, G. (2000). Approaches to valuing the hidden hydrological services of wetland ecosystems. *Ecological Economics, 35*, 63-74.

Adamowicz, W., Boxall, P., Williams, M. & Louviere, J. (1998). Stated preference approach for measuring passive use values: Choice experiments and contingent valuation. *American Journal of Agricultural Economics, 80*, 64-75.

Adamowicz, W., Louviere, J., Williams, M. (1994). Combining revealed and stated preference methods for valuing environmental amenities. *Journal of Environmental Economics and Management, 26*, 271-292.

Alavalapati, J. R. R., & Nair, P. K. R. (2001). Socioeconomic and institutional perspectives of agroforestry. *World forests, society and environment: Markets and policies* (pp. 71-81). In M. Palo & J. Uusivuori (Eds.). Dordrecht, The Netherlands: Kluwer Academic Publishers.

Arrow, K. R., Solow, R., Portney, P., Leamer, E., Radner, R., & Schuman, H. (1993). Report of the NOAA panel on contingent valuation. *Federal Register, 58*, 4601-4614.

Boggess, C.F., Flaig, E. G., & Fluck, R. C. (1995). Phosphorus budget-basin relationships for Lake Okeechobee tributary basins. *Ecological Engineering, 5*, 143-162.

Cameron, T. A. (1988). A new paradigm for valuing non-market goods using referendum data: Maximum likelihood estimation by censored logit regression. *Journal of Environmental Economics and Management, 15*: 335-379.

Cannell, M. G. R. (1999). Growing trees to sequester carbon in the UK: Answers to some common questions. *Forestry, 72*, 237-247.

Carson, R. T., Flores, N. E., & Meade, N. F. (2001). Contingent valuation: Controversies and evidence. *Environmental and Resource Economics, 19*, 173-210.

Clason, T. R. & Sharrow, S. H. (2000). Silvopasture practices. In H. E. Garrett, W. J. Rietveld, & R. F. Fisher, (Eds.), *North American agroforestry: An integrated science and practice*, (pp. 119-148). Madison, Wisconsin: American Society of Agronomy, Inc.

Cooper, J. C., & Keim, R. W. (1996). Incentive payments to encourage farmer adoption of water quality protection practices. *American Journal of Agricultural Economics, 78*, 54-64.

Environmental Protection Agency (EPA). (1995). *Water quality functions of riparian forest buffer systems in the Chesapeake Bay watershed.* [Technology Transfer Report EPA 903-R.95-004]. Nutrient Subcommittee of the Chesapeake Bay Program, United States Environmental Protection Agency:

Florida Agricultural Statistics Service [FAS], 2002. *Beef cattle and calf inventory by county.* Livestock, dairy, and poultry summary, Florida Agricultural Statistics Service.Florida Department of Agriculture and Consumer Services, Tallahassee, Florida.

Feather, P., Hellerstein, D., & Hansen, L. (1999). *Economic valuation of environmental benefits and the targeting of conservation programs: The case of the CRP.* [Agricultural Economic Report No. 778]. Washington DC: USDA Economic Research Service.

Freeman, A. M., III. (1993). *The measurement of environmental and resource values: Theory and methods.* Washington, D.C.: Resources for the Future.
Garrett, H. E., Rietveld, W. J., & Fisher, R. F. (Eds.). (2000). *North American agroforestry: An integrated science and practice.* Madison, Wisconsin: American Society of Agronomy.
Garrod, G., & Willis, K. G. (1999). *Economic valuation of the environment: Method and case studies.* Northampton, Massachusetts: Edward Elgar.
Holmes, T. P., & Adamowicz, W. L. (2003). Attribute-based methods. In P. Champ, K. Boyle, & T. Brown, (Eds.), *A Primer of Non-market Valuation* (pp. 171-219). Dordrecht, The Netherlands: Kluwer Academic Publishers.
Kingsbury, L., & Boggess, W. (1999). An economic analysis of riparian landowners' willingness to participate in Oregon's Conservation Reserve Enhancement Program. Selected paper for the annual meeting of the Agricultural Economics Association, Ames, Iowa.
Lant, C. L. (1991). Potential of the Conservation Reserve Program to control agricultural surface water pollution. *Environmental Management, 15*, 507-518.
Lohr, L., & Park, T. (1994). Discrete/continuous choices in contingent valuation surveys: Soil conservation decisions in Michigan. *Review of Agricultural Economics, 16*, 1-15.
Louviere, J. J., Hensher, D. A., & Swait, J. D. (2000). *Stated choice methods: Analysis and application.* Cambridge, United Kingdom: Cambridge University Press.
Maddala, G. S, (1999). *Limited dependent and qualitative variables in econometrics.* (Econometric Society Monographs No. 3). Cambridge, United Kingdom: Cambridge University Press.
McFadden, D., 1974. Conditional logit analysis of qualitative choice behavior. *Frontiers in econometrics.* P. Zarembka, ed., pp. 105-142, Academic Press, New York.
Mitchell, R.C., (2000). On designing constructed markets in valuation surveys. *Environmental and Resource Economics, 22*, 297-321.
Mitchell, R. C., & Carson, R. T. (1989). *Using surveys to value pubic goods: The contingent valuation method.* Washington, DC: Resources for the Future.
Morrison, J. L., & Humphrey, S. R. (2001). Conservation value of private lands for Crested Caracaras in Florida. *Conservation Biology, 15*, 675-684.
Nicholson, W. (1998). *Microeconomic theory: Basic principles and extensions* (7th ed.). Orlando, Florida: Dryden Press.
Pattanayak, S. K., & Butry, D. T. (2003). Forest ecosystem services as production inputs. In E. O. Sills & K. L. Abt, (Eds.), *Forests in a market economy*, (pp 361-378). Dordrecht, The Netherlands: Kluwer Academic Publishers.
Pimentel, D., Harvey, C., Resosudarmo, P., Sinclair, K., Kurz, D., McNair, M. et al. (1995). Environmental and economics costs of soil erosion and conservation benefits. *Science, 267*(5201), 1117-1123.
Randall, A. (1987). *Resources economics: An economic approach to natural resource and environmental policy* (2nd ed.). New York: John Wiley & Son.
Salant, P., & Dillman, D. (1994). *How to conduct your own survey.* New York: John Wiley & Son.
Sedjo, R. A. (2001). Forest 'sinks' as a tool for climate-change policymaking: A look at the advantages and challenges. *Resources, 143*, 21-24.
Shrestha, R. K., & Alavalapati, J. R. R. (2002). Valuing environmental benefits of agroforestry using attribute-based stated preference method. Paper presented in the 2nd World Congress of Environmental and Resource Economists, June 24-27, Monterey, California.
Shrestha, R.K., & Alavalapati, J. R. R. (2003). Cattle ranchers' willingness to accept for silvopasture adoption: A contingent valuation approach. Manuscript submitted for publication.
Shyamsunder, P., & Kramer, R. A. (1996). Tropical forest protection: An empirical analysis of the costs borne by local people. *Journal of Environmental Economics and Management, 31*, 129-144.
Sugden, R. (1999). Public goods and contingent valuation. In I. J. Bateman & K. G. Willis (Eds.),*Valuing environmental preferences: Theory and practice of the contingent valuation method in the US, EU, and developing countries*, (pp. 131-151). New York: Oxford University Press.
United States Department of Agriculture (USDA). (1997). *1997 Census of Agriculture: National, state, and county tables.* Washington D.C.: USDA.
Varian, H. R. (1992). *Microeconomic analysis* (3rd ed.). New York: W.W. Norton.
Wade, M. A., Minton, T., & Delargy, L. (2001). Pasture improvement practices utilized by Florida cattlemen. *Florida Scientist, 64*, 4-5.

MAXWELL MUDHARA AND PETER E. HILDEBRAND

ASSESSMENT OF CONSTRAINTS TO THE ADOPTION OF IMPROVED FALLOWS IN ZIMBABWE USING LINEAR PROGRAMMING MODELS

1. INTRODUCTION

For the past three decades, poverty alleviation and food security have been the primary concerns of development practitioners in sub-Saharan Africa (SSA). The inherent poor fertility of most soils in the tropics and subtropics constitutes a major constraint to sustainable crop production by smallholder farmers in the SSA region (Myers, Palm, Cuevas, Gunatilleke, & Brossard, 1994; Smaling, Nandwa, & Janssen, 1997). Low crop productivity is a cause of food insecurity. Use of improved fallows is one of the interventions proposed for improving soil fertility in a sustainable manner, thus combating food insecurity.

Leaving fields fallow to regenerate soil fertility has been a traditional practice in Zimbabwe, as in the rest of SSA. The practice is being abandoned due to the increase in population pressure and the accompanying shortage of land. To boost agricultural production, some governments in SSA embarked on programs to promote the use of chemical fertilizers. In Zimbabwe, widespread promotion of chemical fertilizers was undertaken after independence in 1980. Besides ensuring that fertilizers were readily available, price interventions were also used. Between 1980 and 1991, fertilizer prices under the direct control of the government were set at levels considered affordable to farmers to increase fertilizer demand and use. Free fertilizer was distributed to smallholder farmers in 1980, 1986, 1992, and various years after 1992. Free fertilizer distributions were meant to boost smallholder farmers' demand for fertilizer and cushion them from the inability to purchase fertilizers following droughts. In 2002 and 2003, fertilizer was allocated as input loans since free handouts were insufficient to meet farmers' requirements.

Research on viable agroforestry technologies for the smallholder farmers has intensified in recent years (e.g., Buresh & Cooper 1999; Franzel 1999; Franzel, Phiri, & Kwesiga, 2002; Keil 2001; Kwesiga, Franzel, Place, Phiri, & Simwanza,

J. R. R. Alavalapati & D. E. Mercer, Valuing Agroforestry Systems, 201–218.

1999; Pisanelli, Franzel, De Wolf, Rommelse, & Poole, manuscript submitted; Place, Franzel, De Wolf, Rommelse, Kwesiga, Niang et al., 2002; Swinkels, Franzel, Sheperd, Ohlsson, & Ndufa, 1997). This has mostly been in response to escalating fertilizer prices and declining production levels in the sector. Available literature shows that these technologies are complex and present challenges for assessing their potential adoption by farmers. It is desirable that the possible adoption of the technologies be assessed before dissemination so that resources can be used judiciously for extending the technologies. However, it is particularly challenging to assess possible adoption *ex ante*, or before the technologies are disseminated. Most studies have evaluated the adoption of the technologies after dissemination once farmers have had time to evaluate the technologies and adopt them, i.e. as *ex post* assessments (Franzel 1999; Kwesiga et al. 1999).

Since 1992, the International Centre for Research in Agroforestry (ICRAF) and Department of Research and Specialist Services (DR&SS) conducted research on soil fertility management using improved fallow agroforestry technologies at Domboshawa Training Centre and Makoholi Experiment Station. Technical analysis indicated that the proposed interventions increased maize (*Zea mays*) yields (Dzowela 1992, 1993; Mafongoya & Dzowela 1998, 1999). However, the technologies had not been evaluated in the context of the whole farm in relation to all activities that compete for limited resources at the household level. Rohrbach (1998) observed that technologies need to be more profitable than alternative investment opportunities on the farm as a whole. Since the adoption potential of the technologies was unknown, extension personnel could not identify the target farmers for the complex technologies. Partly for this reason, the technologies had not been extended to smallholder farmers.

This chapter presents the use of a linear programming (LP) model for simulating the livelihoods of smallholder farmers and assessing the potential adoption of improved fallows in the Mangwende Communal Area of Zimbabwe before the technologies were disseminated. In addition, the chapter discusses how the model was used for determining the adoption potential for households of different resource levels. The likely impacts of adopting the technologies on the livelihoods of households under different scenarios are assessed.

2. METHODOLOGY

2.1. Characterization of research site

The survey for this study was conducted in Mangwende Communal Area (CA), lying between 17° 22' and 17° 56' S and between 31° 31' and 32° 09' E, in Mashonaland East Province of Zimbabwe. The CA lies in the northeast of the country, some 90 kilometers (km) from Harare, the capital city of Zimbabwe. Annual rainfall is between 800 and 950 millimeters (mm), the bulk of which falls between the end of October and the end of March. Predominant soils in the area are derived from granite parent materials. Therefore, the soils have low inherent fertility and require high levels of external inputs to improve fertility.

The smallholder farmers of Mangwende CA are typical of other smallholder farmers all over the world. They have limited resource levels, multiple activities and poor access to services such as extension. The farming system is complex, with households relying on various activities to sustain their livelihoods given their limited resources. Fertilizer purchases are decreasing over time, suggesting that farming activities are threatened and food insecurity exists in some years.

Crop production is the major activity for the farm households in Mangwende CA. Maize (*Zea mays*), the dominant subsistence and cash crop, occupies approximately 70 percent of the cultivated area. Other major crops are groundnuts (*Arachis hypogaea*) and sunflower (*Helianthus annuus*). Vegetable gardens are widespread, producing a variety of vegetables for consumption and cash. Maize (*Zea mays*) planting is staggered to ensure that farmers spread out labor requirements and minimize the risk of total crop failure. Cattle are the dominant livestock, and about half of the farmers own cattle as a financial back up and for draft power. Households also depend on members who do not live on the farm for cash remittances and non-cash support. The following define the level of resources on the farm:

- Number of people in the household;
- Age and gender of the head of the household;
- Farm size; and
- Number of cattle owned (and the accompanying access to cattle manure).

The LP model accommodated household resource levels in various ways. The number of people on the farm were determined by age and gender, which were then reflected in the labor constraints and consumption requirements. The area size of the farm was input into the model as a land constraint. Cattle ownership determined the technologies that households could use. For example, households owning cattle could use ox-drawn cultivators for weeding, while non-cattle owners had to hire cattle at a cost and did hand weeding. In addition, only cattle owners could have fields on which manure had been applied.

As resource levels vary across households, technologies were suitable in varying degrees to households with different resources. Nevertheless, some similarities existed across households; for example, maize (*Zea mays*) was the major cash and food crop in the area and was planted by every household. In addition, households practiced dryland farming, i.e., they did not use irrigation.

Interviews revealed that some households experienced food shortages during the year, and therefore, food production was their first priority. While farmers were aware of the secondary benefits of some crops, such as soil fertility enrichment, they only planted crops for food and cash. This indicated that new technologies were needed for satisfying the farmers' food and cash requirements. Any technology should not increase household food insecurity or decrease the cash available for discretionary spending.

2.2. Data collection

Both primary and secondary sources were utilized for collecting data. For primary data, 3 wards were randomly selected from the 28 wards in the CA. A ward is the first administrative unit in the district. Lists of the names of households in each village in the selected wards were compiled from the lists kept by village heads. A household was defined as people who shared the same kitchen. Thirty-five households were selected randomly from each ward, to give a sample size of 105 households. Enumerators recruited from within the research area were specifically trained in the administration of the questionnaires used in the survey. In addition to the questionnaires, farmer focus group discussions were conducted to address general issues applicable to most households. Data obtained from focus group meeting on the labor required for conducting different activities are presented in Appendix A. Technical coefficients for improved fallows were obtained from published results of experiments conducted in the area by the FSRU and by ICRAF-Zimbabwe at Domboshawa Training Centre. Table 1 presents the yields realized for different fallow treatments during experimentation.

2.3. The household LP model

A five-year LP model was constructed in Excel to reflect the livelihood system of the smallholder farmers. The models were solved using the Premium Solver Plus V3.5 for Excel. Details of the theoretical formulation of the model and its construct are presented in Appendix B. The complete LP model is presented in Mudhara (2002). The model was run for each of the 99 sampled households. Household specific parameters from the questionnaire survey and technical coefficients from group interviews were used for each respective household. The objective function in the models was to maximize discretionary household income from farm and non-farming activities, subject to various constraints reflecting household characteristics, such as food security, household composition, draft power ownership, and available

Table 1. Yields of two maize crops used in the model after Sesbania sesban *fallows of different duration at Domboshawa Training Centre, Zimbabwe.*

Fallow duration (years)	*Maize grain yield (t ha^{-1})* [a]			
	Without fertilizer		*With fertilizer*	
	Year 1	*Year 2*	*Year 1*	*Year 2*
1	3.8	3.9	4.0	4.1
2	4.0	5.1	3.9	5.3
3	3.1	5.6	5.3	6.5

[a] *Average maize yield on farmer's fields = 1.60 t ha^{-1} (Standard Deviation = 1.58 t ha^{-1})*
[b] *Fertilizer was applied at 38, 21 and 14 kg ha^{-1} of N, P and K, respectively.*
Source: Dzowela, (1992, 1993); Mafongoya & Dzowela (1999).

arable land. Discretionary household income is the income available to the household after it has satisfied basic requirements cash.

The model was validated in two stages to ensure that it reflected the farmers' livelihood strategies. The first stage of validation, conducted in the research area, involved discussing the results of computer simulations with a sub-sample of participating households. Fertilizer prices were varied to see the model's prediction of how households were likely to respond, and the results were discussed with the farmers. Suggestions from these meetings were incorporated into the model. Farmers agreed that the model simulated their livelihoods well, especially in its ability to identify resources that were limiting in their farming operations.

In the second stage of validation, the data on levels of activities as reported by the farmers during the questionnaire survey were compared to the model results. Model results were obtained from the data generated after running the model for all households. Two variables, the area planted to maize (*Zea mays*) and the area under unmanaged fallow, were compared. The area planted to maize (*Zea mays*) was important for validation because maize (*Zea mays*) was the major staple and cash crop for which the potential for adoption of improved fallow technologies was being evaluated. The new technologies compete with crops for land or replace the land under unmanaged fallow. The difference in the size of unmanaged fallow in the model and reported by farmers during interviews had to be insignificant. Therefore, the unmanaged fallow area was used for validation. After validation of the models, activities for improved fallows were included in the LP models to assess whether the new technologies had a potential to be adopted, all other factors being equal.

To determine the technologies that households would adopt, the LP model was run for each household. The average results for each type of household are reported.

3. RESULTS AND DISCUSSION

Smallholder farmers have limited resources and multiple objectives (Ellis, 1992; Hildebrand, 1986; Netting, 1993), thus making their livelihood systems complex. However, this generalization of their characteristics can be mistaken to imply that they are homogeneous. The survey results reveal the existence of household diversity in various respects including resource levels, composition of the membership of the household and the activities conducted. Resource levels dictate the ability of households to exploit opportunities that different technological innovations offer.

The diversity among households makes it difficult to develop technologies compatible to all the different household types. Models can be used for matching technologies to specific household types so that extension programs can be designed based on farmers' resources and their compatibility with the technologies. Household LP models, sensitive to the diverse household characteristics, can match households and technologies by determining the technology that is both, compatible to the resources of the specific household and satisfies the stated objective function of that household. By pointing out the technologies that are compatible with the farmers' resources and meet the farmers' objectives, such LP models may increase the efficiency and productivity of extension services.

3.1. Agroforestry improved fallow technologies

In agroforestry, improved fallows are technologies where leguminous shrubs or trees, planted and left to grow during the duration of the fallow, are cut down and pruned at the end of the fallow period. The biomass remaining at the end of the fallow period is incorporated into the soil. For better establishment of the *Sesbania sesban* improved fallow, seedlings are planted. This is the recommended practice and is the approach used in the model. This is in contrast to the method where multi-purpose trees are seeded directly into the soil. Agroforestry improved fallow technologies are complex as they have several developmental components and involve multiple interactions among the components. Therefore, to assess the potential adoption of such technologies, farmers' objectives and their resources must be considered. Since all activities conducted on the farm compete for household resources, an evaluation of potential adoption of improved fallow technologies also should consider their contribution to households in comparison to alternative activities (i.e., the opportunity costs). To achieve this, the yields realized and the inputs (and timing of their application) for improved fallow technologies studied in research station experiments were incorporated in validated household LP models. The LP models were validated as truly simulating households of different socio-economic backgrounds. Options of improved fallows of one, two and three-year durations introduced into the household LP model are presented in Figure 1. One-year improved fallows could be planted during the year 1, 2, 3, or 4 of the 5-year model. Two-year fallows could only be planted in year 1 and year 2 of the model. Three-year fallows could be planted in year 1 of the model. Two seasons of maize were planted after each fallow had been removed from the field.

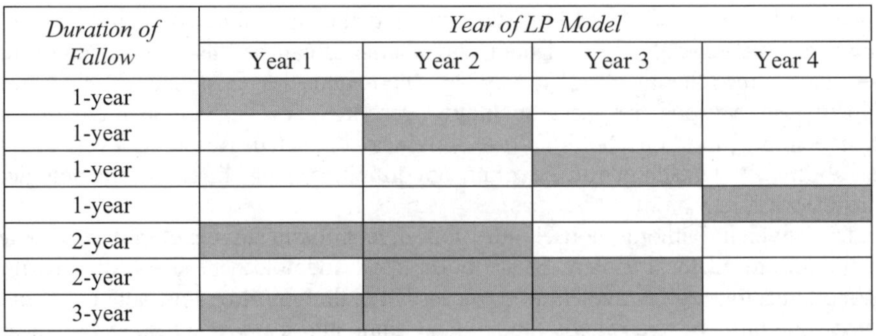

Figure 1. The sequence in which improved fallow options and the following maize crop are introduced into the household LP mode. Shaded = improved fallow; unshaded = cultivation.

Data on labor requirements and yields were recorded during experimentation at Domboshawa Training Centre. Maize (*Zea mays*) yields realized following improved fallows are presented in Table 1. The distribution of the yields that farmers reported was skewed to the left of the mean yield realized on the experiment station (Figure 2). In the figure, point A (2.0 Mg ha^{-1}) is the average maize (*Zea mays*) yield realized by farmers in 2001 while point B (3.50 Mg ha^{-1}) is the average maize (*Zea mays*) yield following pigeon pea and C (4.50 Mg ha^{-1}) the average maize (*Zea mays*) yield following *Sesbania sesban* obtained at the experiment station. In 1994/95, maize (*Zea mays*) yields on plots grown continuously with maize (*Zea mays*) on station were 2.71 Mg ha^{-1} without fertilizer and 3.3 Mg ha^{-1} when 38 kg ha^{-1} of nitrogen was applied. Therefore, yields realized on experimental plots were reduced by 40% to bring them to the same level as the yields that farmers are likely to realize in their own fields (International Maize and Wheat Improvement Center [CIMMYT], 1988).

3.2. Potential adoption of Sesbania sesban *improved fallow*

Results of the LP model indicate that households should adopt *Sesbania sesban* when it is the only improved fallow technology available. When *Sesbania sesban* and pigeon pea are both available at the same time, without allowing pigeon pea seed to be marketed, farmers should realize more cash for discretionary spending by

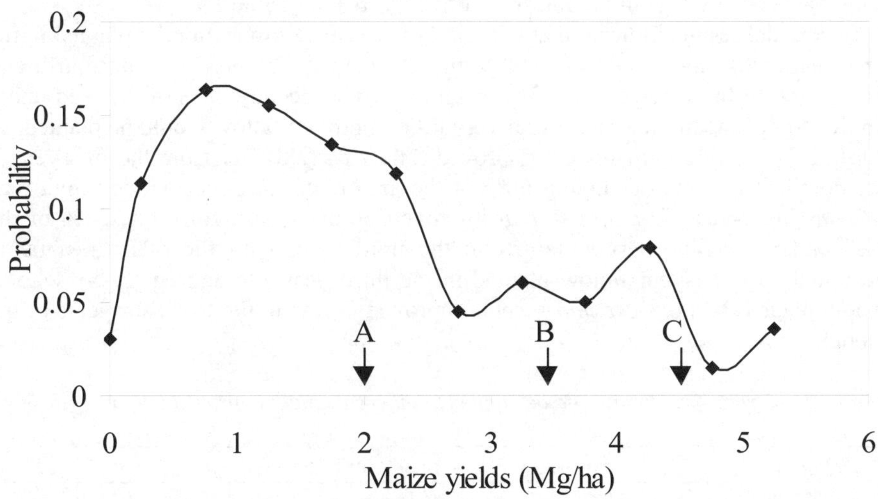

Figure 2. Probability distribution of farmers' maize (Zea mays) yields vs. station yields.

A = 2.0 Mg ha^{-1} or the average maize (Zea mays) yield realized by farmers in 2001.
B = 3.50 Mg ha-1 or the average maize (Zea mays) yield following pigeon pea.
C = 4.50 Mg ha-1 or the average maize (Zea mays) yield following Sesbania sesban.

adopting *Sesbania sesban* improved fallow. At the time of the study, pigeon pea seeds were not marketed in Zimbabwe, making it reasonable to exclude the marketing of pigeon pea seeds.

Over the five years that the model spanned, fallows of different duration could be adopted. Results of running the model on each of the sampled households indicate that most households adopt improved fallows in the first year, with the number of adopters falling with time. In the first year, 81% of the households would adopt 1-year improved fallows, 51% would adopt the 2-year improved fallow and 59% would adopt the 3-year improved fallow. In the second year, the one-year fallow adopted in year 1 should be planted to maize. In this same year, in addition to the 2- and 3–year fallows that should be already planted in the first year, 12% of the households would adopt 1-year improved fallow and 5% would adopt 2-year improved fallow. In the third year, the 2-year fallows planted in the first year and the 1-year fallows planted in the second year are planted to maize. In this third season, 48% of the households would adopt a 1-year fallow while 16% would adopt a 1-year fallow in the fourth year. The average areas that would be planted to each of the improved fallows are presented in Table 2.

In the first year, 1-year *Sesbania sesban* improved fallow would be planted on an average of 0.55 ha, occupying, on average, 63% of the area under improved fallow. The one-year fallow that would be established in the first year would be planted to maize (*Zea mays*) during the second and third years (see Figure 1). In the third year, 0.24 ha of one-year fallow are established. The fallow land would then be planted with maize (*Zea mays*) in the fourth and fifth years (see Table 1).

LP model results indicate that 0.25 ha of a 3-year fallow would be planted in the first year. This land would then be planted to maize (*Zea mays*) in the fourth and fifth years. In the first year, the 3-year fallow would occupy 28% of the land under improved fallow. In the second year very little improved fallow would be planted, so that the 3-year *Sesbania sesban* improved fallow carried over from the first year is the dominant fallow, occupying 68% of the area under *Sesbania sesban* improved fallow. The 3-year *Sesbania sesban* improved fallow would constitute 51% of the area under *Sesbania sesban* fallow in the third year, with the other area being occupied by the 1-year fallow planted in the third year. On aggregate, households would plant 1.2 ha of *Sesbania sesban* improved fallow in the first four years of the model.

Table 2. Average area that would be planted to different length improved fallows over time.

Fallow duration	*Year*			
	1st	*2nd*	*3rd*	*4th*
		Area (ha)		
1-year	0.55	0.04	0.24	0.04
2-year	0.08	0.08	0	0
3-year	0.25	0.25	0.25	0

Figure 3 shows the fertility management over time of maize (*Zea mays*) that should result after the introduction of *Sesbania sesban* improved fallow. It compares the maize (*Zea mays*) area planted on land fertilized using the conventional means (i.e., chemical fertilizers and cattle manure) and maize planted on *Sesbania sesban* improved fallows. On average, from the second to the fifth year, maize (*Zea mays*) on improved fallow would occupy 0.6 ha, or 60% of the area under maize (*Zea mays*), out of an average farm size of 2.6 ha.

The introduction of *Sesbania sesban* improved fallow would result in an increase in income for discretionary spending. Results show that income for discretionary spending for non-owners of draft power would increase significantly (p = 0.009) from US$318, realized before introducing the improved fallow, to US$353, after introducing the fallow. This is a 10% increase in income. The income for discretionary spending for draft power owners increases from US$444 to US$451 after the introduction of *Sesbania sesban* improved fallow. This 2% increase, however, is not statistically significant (p = 0.6) when comparing across sampled households. Therefore, although the income of households without draft power would remain lower than for owners of draft power, introduction of *Sesbania sesban* improved fallow might be more beneficial to non-owners of draft power than owners.

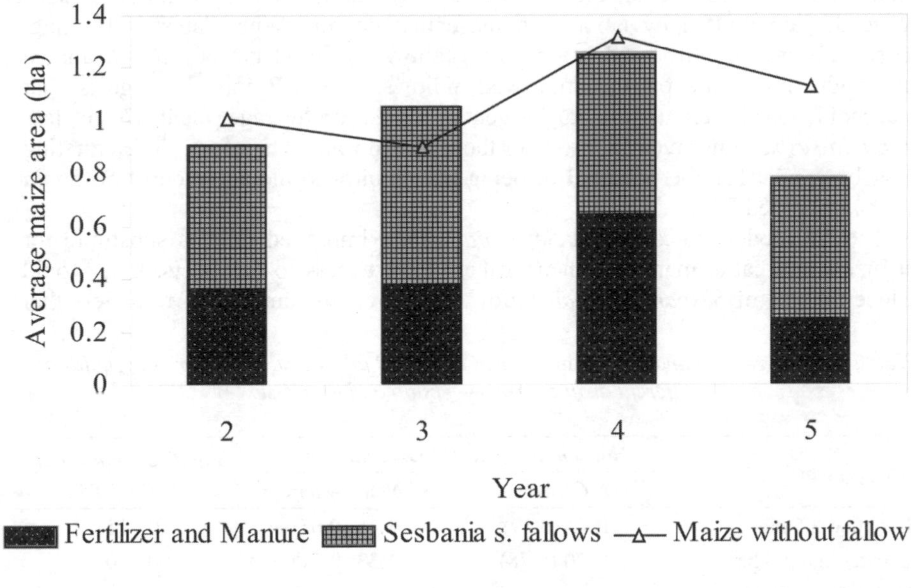

*Figure 3. Maize (*Zea mays*) area that should be under different soil fertility options over time, when* S. sesban *improved fallow is available.*

Model results after introducing the *Sesbania sesban* improved fallow show that income for discretionary spending is significantly different between households with one to four members working fulltime on the farm ($p = 0.0001$). The average annual income for discretionary spending for households with one member working fulltime on the farm, after the adoption of improved fallows, would, be US$327. The income for households with two, three, and four members working fulltime on the farm would be US$364, US$527, and US$582 per annum, respectively. However, only households with one member working full-time on the farm would experience a significant increase ($p = 0.03$) in income for discretionary spending following the introduction of *Sesbania sesban* improved fallow into the system. Therefore, although *Sesbania sesban* improved fallow can be widely adopted, only households with one member working fulltime on the farm are expected to benefit in terms of income for discretionary spending. Demographic survey results show that female-headed households (FHH) have the least number of members working on the farm. However, their income should remain the lowest across different numbers of household members working fulltime on the farm. Households with larger farms should be able to adopt more improved fallows compared to those with smaller farms (Table 3).

Data collected from the household suggest that fifty-three percent of the households in the sample differentiated labor by gender. When differentiating labor by gender, male labor carries out specified tasks, different from those carried out by female labor. The LP model results show that more households differentiating labor by gender are limited by labor than those that do not. When labor is limiting, households are less able to adopt improved fallows, as less labor becomes available for conducting work on the improved fallows. The LP model suggests that households that differentiate labor by gender would, on average, plant 270 m^2 less *Sesbania sesban* improved fallow than those that do not. Therefore, differentiating labor by gender has the potential of being an inhibitor to the adoption of *Sesbania sesban* improved fallow.

The LP model indicates that *Sesbania sesban* improved fallows substitute for fertilizers and cattle manure. When farmers have access to fertilizers, they would reduce the use of *Sesbania sesban* fallow. However, the amount of fertilizers that

Table 3. Average area under Sesbania sesban *improved fallows adoption for households of different draft power ownership and farm size.*

Size of farm	*Household ownership of draft power*		*Significance level of differences*
	Owners	*Non-owners*	
Less that 2.5 ha	0.73 (0.59)[a]	1.15 (0.65)	0.022
Greater that 2.5 ha	1.70 (0.78)	1.33 (0.71)	0.150
Significance level of differences	0.001	0.480	

[a] *Numbers in parentheses are Standard Deviations*

non-owners of draft power have should not affect the area of *Sesbania sesban* improved fallows they plant. Non-owners of draft power do not have access to cattle manure and are more in need of *Sesbania sesban* improved fallow.

At the time of conducting the research, no market outlets for pigeon pea seed existed in the communal areas of Zimbabwe. Therefore, early adopters of pigeon pea fallows would not be able to capitalize on increased returns from selling the pigeon pea grain. However, it was anticipated that, over time, as entrepreneurs realize that quantities viable for marketing were being produced, the marketing outlets might develop. Initially, the government might have to purchase the seed to promote the use of pigeon pea improved fallow. Assessments using the LP model suggest that a price of Z$2,500/Mg (US$45/Mg) for pigeon pea grain would allow farmers to adopt the pigeon pea improved fallow. The price should make farmers adopt an average of 0.2 ha per household of pigeon pea improved fallows. Figure 4 shows the increase in the area planted to pigeon peas as the price of its seed is increased in the LP model. The increase in the area planted to pigeon peas would be accompanied by a fall in the area under *Sesbania sesban* improved fallows.

4. CONCLUSIONS

In the LP model, when both *Sesbania sesban* and pigeon pea improved fallows are optional technologies for farmers, a market for pigeon pea seeds would be needed for farmers to maximize their objective function through adopting pigeon pea improved fallow.

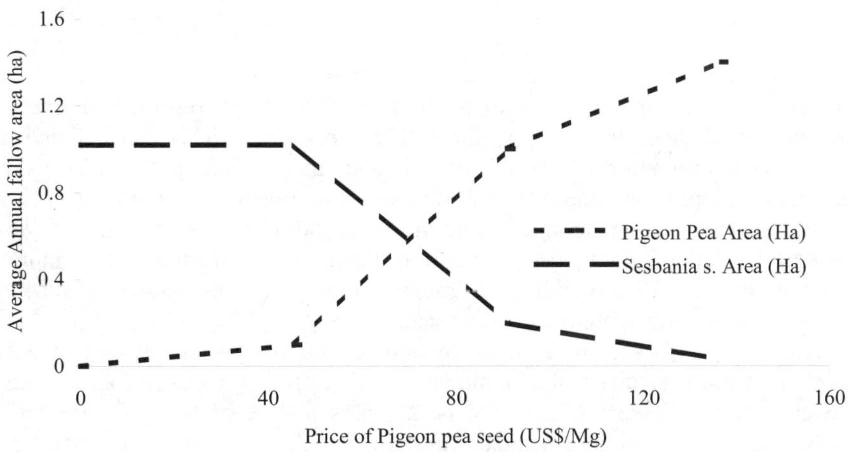

Figure 4. Average area planted to Sesbania sesban and pigeon pea fallows over 5 years at different pigeon pea seed prices.

Several recommendations can be drawn from the study. First, given the economic and social circumstances facing smallholder farm households in Zimbabwe, they stand to benefit from adopting improved fallow technologies. This suggests that there is potential for the adoption of improved fallows. Given the high rate of inflation in Zimbabwe (estimated to be above 600% per annum at the end of 2003) and the lack of chemical fertilizer subsidizes, fertilizer prices are likely to continue rising. LP model findings suggest that the percentage of adopters and the area under fallows may also increase. Therefore, these technologies should be extended to smallholder farmers.

The study shows that the behavior of smallholder farmers can be used to simulate LP models for determining farmers' potential adoption of technologies and for assessing the impact of such technologies on household welfare, measured in various ways such as income for discretionary spending. In addition, the impacts of various policies (e.g. fertilizer prices or availability of markets) can also be estimated through the LP model. In the same light, the model can be employed to determine the effect of change in available household labor or sources of non-farming income, as might result from HIV/AIDS, on the welfare of the household.

Ex ante analyses are useful in project development because they can save time, funds, and effort for development agencies and local peoples. From a scientific point of view, however, *ex ante* analysis should be compared with *ex post* whenever feasible. Granted, in long-term forest studies for timber products, for example, *ex post* comes much too late to be of any use. However, since some of these fallows have three year cycles, it should have been possible to obtain *ex post* data on who really adopted and statistically analyze the comparison. Further research is required on the actual adoption of the improved fallows.

5. LIMITATIONS

Improved fallows should be allowed to complement the resources farmers are already using, rather than competing with the resources. This study shows that, among other factors, draft power (which also means access to cattle manure) determines adoption of improved fallows, yet experimental work has not included manure use into the technologies using improved fallows. The experimental design assumes that farmers do not use chemical fertilizers in combination with manure and improved fallows. The possibility of farmers bringing the three sources of fertility to one plot was not considered in experimental work.

This study is based on a survey conducted in Mangwende Communal Area, which is a representative of communal areas in similar agro-ecological regions. However, Mangwende CA may not be a representative of the drier areas where smallholder farmers are also located. Mangwende CA was used because some of the technologies had been tested in this site, and it had similar agro-ecological characteristics as Domboshawa Training Centre where station experiments were conducted. The performance of these technologies may not be the same in drier agro-ecological regions.

The model results represent the long-term position in the adoption of the technologies, when farmers have full information about technologies and peak adoption has occurred. The model does not determine how long the adoption process takes before peak adoption is attained, which will depend on how the technologies are extended to farmers. In this respect, an efficient extension organization will hasten the adoption process. Nevertheless, the households that might adopt the technologies and the area they adopt, in the long run, were identified in the model.

Being an *ex ante* assessment of the potential adoption of the improved fallows, there are other factors that may impact the adoption of the technologies which were not considered here. For example, the scarcity of fuel wood might encourage farmers to adopt improved fallow as they can harvest firewood from the fallows.

The LP model has often been criticized because of its assumption of linear relationships between variables. This weakness is also acknowledged here. Nevertheless, the weakness was minimized by splitting the production function into several sections, which could be assumed to approach linearity. Thus the shortcoming was addressed to limit the bias of findings of the study.

6. APPENDIX A. LABOR REQUIREMENTS FOR CROP PRODUCTION ACTIVITIES

Crop	*Operation*	*Days per Hectare*
All crops	Land preparation with ox-drawn plough	20
Planting		
Maize	Using ridges[a]	5
	Hand hoe planting/Wire	17
	Hand hoe planting	19
Groundnuts	Making ridges then harrowing	11
	Cover seed with feet	15
	Planting in lines marked with a wire	26
Finger millet (Including rice, Sorghum)	Ridge	1
	Broadcasting	1
	Dropping and covering with foot	1
	Planting on raise beds	10
Sweet potatoes	Planting on ridges	14
	On flat bed	15
Weeding		
	Cultivator	3
	Hand weeding after cultivation	16
Maize	Hand weeding without cultivation	27
	Using plough	5
	Hand weeding after hand hoe planting	64

APPENDIX A, (cont.)

Crop	*Operation*	*Days per Hectare*
Groundnuts	Using cultivator	4
	Hand weeding after cultivation	24
	Hand hoes alone	50
Sunflower	Cultivating	7
	Hand hoe weeding	55
Sweet potatoes	Cultivator making ridges	1
	Hand weeding on Ridges after Cultivator	10
Fertilizer Application		
	Basal fertilizer:	
	Dropping after germination (no holes)	4
Maize	Covering with cultivator	3
	Digging holes and placing fertilizer	3
	Dropping Top dressing per station	2
Harvesting		
Groundnuts	Pulling, Plucking, Transporting, Packing	50
Maize	Cutting, Stacking, De-husking, Transporting	73
	Processing	35
Finger millet	Cutting, Threshing, Packing	50
Sunflower and Soybeans	Cutting, Threshing, Packing	35
Cotton	Picking and Packing	47
	Cutting Stalks	5

[a] *Planting in plough furrows requires an additional person for plowing.*

7. APPENDIX B. A DESCTIPTION OF THE LP MODEL

The objective of the model is to maximize the sum of annual discretionary cash over five years. The year is further divided into quarters. When there is cash surplus in one quarter, that cash is transferred to the next quarter to meet the future expenses and contribute towards the cash objective function. Sources of cash vary in each quarter as the farming activities vary throughout the year. Some non-farming activities, such as poultry production and beer brewing, when undertaken on the farm, can bring in cash each quarter of the year. Other activities, including selling crops, only bring cash in one quarter of the year.

The activities included in the model were the following: maize (*Zea mays*), groundnuts (*Arachis hypogaea*), Bambara nuts (*Vigna subterranea*), finger millet

(*Eleusine coracana*), sunflower (*Helianthus annuus*), cotton (*Gossypium herbaceum*), sweet potatoes (*Ipomoea batatus*), soyabeans (*Glycine max*), horticulture, hiring out labor, vending, poultry production, beer brewing and selling, house construction, making peanut butter, molding bricks for sale, purchase maize (*Zea mays*) to meet own food requirements, and managing cattle manure. All crops, except groundnuts (*Arachis hypogaea*), can be planted early or late. Early planted crops are planted in the first quarter of the year. Late planted crops are planted in the second quarter. Groundnuts (*Arachis hypogaea*) are only planted early, in the first quarter.

The theoretical form of the model is as follows:

Maximize: $Z = \sum_j c_j X_j$

Subject to: $\sum_j a_{ij} X_j \leq b_i$ for all I; and $X_j \geq 0$

Where: Z = Objective function, which maximizes the sum of the gross margins of all activities undertaken in the model.

X_j = j^{th} production activity of the farm, e.g., maize (*Zea mays*), cotton (*Gossypium herbaceum*), and cattle.

c_j = Forecast gross margin of X_j,

a_{ij} = The amount of input i needed to operate activity j on a unit of a resource, such as a hectare of land;

b_i = Available supply of the resource i.

An LP model for one year is organized as a tableau consisting of inputs, activities, constraints, and the objective function. The input section has a list of the inputs needed for all activities. All the inputs listed are not necessarily used by all activities. Activities are represented as columns. They account for the different production techniques through the input-output coefficients. For each activity, the coefficients are linear and constant. In the tableau, the constraints vector, often called the right hand side (RHS), consists of the resources available to the farmer or the constraints to be met.

LP model can handle the multiple activities undertaken on smallholder farms. Consumption requirements are incorporated in the model as constraints, determined by the size of the household, making it ideal for smallholder farmers who have to meet food security requirements. Activities compete for scarce resources in the model, therefore, each activity can only be included in the model at the expense of other activities, that is, there is an opportunity cost incurred in the inclusion of an activity. Figure A-1 shows the linkages between yearly models in the five-year model.

Year 1 model				
Year 1 to 2 transfers	Year 2 model			
	Year 2 to 3 transfers	Year 3 model		
		Year 3 to 4 transfers	Year 4 model	
			Year 4 to 2 transfers	Year 5 model

Figure A-1. Structural form of each year's LP model.

7.1. Model constraints

The constraints are represented by the mathematical representation: $\sum_j a_{ij} X_j \leq b_i$.

In essence, each constraint states that the use of the resources in each of the activities, X*j*, which the household could undertake, cannot exceed the resources that the household possesses.

The land constraint: This constraint ensures that the area under crops in a given year cannot exceed the arable land that the household possesses. Land is allocated to different crops. In addition, manure can be applied to the land and only maize (*Zea mays*) is planted on the land where manure has been applied.

Garden area constraint: The area planted to maize (*Zea mays*) in the garden cannot exceed the size of the garden. Garden maize (*Zea mays*) is planted in the third quarter. Therefore, its cycle spans two years, planted in the third quarter of one year and then harvested in the second quarter of the following year. In the model, the garden area planted to maize (*Zea mays*) in one year is transferred into the next season through the end-of-season transfers. The model distinguishes land with manure from that without. Manured area is planted to maize (*Zea mays*) in two successive seasons.

Labor constraint: Households primarily use family labor. Labor can also be hired in when cash is available for meeting hiring costs. Surplus labor can be hired out to earn cash. Farm activities are conducted at different times of the year, so the opportunity cost of labor also varies within the year. Some households differentiate labor by gender, while others do not. The model was sensitive to the differentiation and timing of activities that households followed.

Cash constraint: This variable is the cash used for farming from the beginning to the end of the season. Cash is obtained from sources such as sales of the previous year's crop, remittances, and non-farming activities. Use of credit has been declining over the years. Approximately 6% and 26% of the sample farmers used credit in 2000 and 1990, respectively. Households start the year or quarter with a stock of cash carried over from the previous year or quarter. Exogenous cash is injected into the system through remittances from employed members staying away from home. At the end of the season, any surplus cash is transferred to the income at the end of the quarter. The household also needs to pay for expenditures incurred during that quarter. For example, households with children going to school would need to buy school uniforms at the beginning of the year and pay school fees at the beginning of each school term.

The model had to be sensitive to technology specific cash requirements. Therefore, expenditure on each crop depends on the stage of development of the crop. For example, early-planted maize (*Zea mays*) is top-dressed at a different time from that of the late-planted crop. Crops under different levels of management also require different financial resources, e.g., maize (*Zea mays*) produced using chemical fertilizer alone requires expenditure for basal fertilizer at planting. This is not required in maize (*Zea mays*) produced using a combination of cattle manure and chemical fertilizer.

Non-farming activities: The limits on the level of non-farm activities that households could engage in were determined from survey data. Without placing limits, the model would have allowed some households to exceed the stipulated limits, which would have been unrealistic since such a scenario reflects an over supply on the markets. Farmers avoid over supply. For beer brewing, for example, they indicated that they took turns to ensure that the person selling in any day was guaranteed buyers.

Consumption requirements: Typically, semi-commercial smallholder farm households store some food for subsistence consumption. The quantity of the staple food crops that each household stored every season depended on the size of the household and was obtained during the survey. Maize (*Zea mays*) requirements for subsistence were expressed as a regression equation to allow variations in household composition during sensitivity analysis to be captured through the consumption regression model. The regression model was linked with the consumption constraint in the model. Maize (*Zea mays*) for subsistence requirements is met from transfers from the proceeding quarter and from working on other people's farms. A minimum level of subsistence requirements has to be met in each quarter.

8. REFERENCES

Buresh, R. J., & Cooper, P. J. M. (Eds.). (1999). The science and practice of short-term improved fallows. *Agroforestry Systems, 47*, 1–356.

Dzowela, B. H. (1992). *Southern Africa Development Coordinating Committee (SADCC–ICRAF) agroforestry project, Zimbabwe: Progress report of 1991-92.* Harare, Zimbabwe: ICRAF.

Dzowela, B. H. (1993). *Southern Africa Development Coordinating Committee (SADCC–ICRAF) agroforestry project, Zimbabwe: Progress report of 1993.* Harare, Zimbabwe: ICRAF.

Ellis, F. (1992). *Agricultural policies in developing countries.* Cambridge; New York: Cambridge University Press.

Franzel, S. (1999). Socioeconomic factors affecting the adoption potential of improved tree fallows in Africa. *Agroforestry Systems 47*, 305–321.

Franzel, S., Phiri, D., & Kwesiga, F. (2002). Assessing the adoption potential of improved fallows in eastern Zambia. In S. Franzel & S.J. Scherr (Eds.), *Trees on the Farm: Assessing the Adoption*

Potential of Agroforestry Practices in Africa, (pp. 37–64). Wallingford, United Kingdom: CAB International.

International Maize and Wheat Improvement Center (CIMMYT). (1988). From agronomic data to farmer recommendations: An economics training manual. Mexico: CIMMYT Economics Program.

Hildebrand, P. E. (1986). *Perspectives on farming systems research and extension.* Boulder, Colorado: Lynne Riener Publishers.

Keil, A. (2001). Improved fallows using leguminous trees in eastern Zambia: Do initial testers adopt the technology? Unpublished masters thesis, Georg-August University, Goettingen, Germany.

Kwesiga, F. R., Franzel, S., Place, F., Phiri, D., & Simwanza, C. P. (1999). *Sesbania sesban* improved fallows in eastern Zambia: their inception, development, and farmer enthusiasm. *Agroforestry Systems, 47*, 49–66.

Mafongoya, P. L., & Dzowela, B. H. (1998). Soil fertility replenishment through agroforestry systems in two contrasting agro-ecological zones in Zimbabwe. In D. Hikwa, B. Maasdorp, A. Mushaka, L. Tapfumaneyi, & J. Nyakabau (Eds.), *Soil Fertility Improvement through Agroforestry in Zimbabwe: Proceedings of the Second National Agroforestry Seminar: 9–10 October 1997, Harare. Zimbabwe,* (pp. 31–42). Harare, Zimbabwe: Scientific Association.

Mafongoya, P. L., & Dzowela, B. H. (1999). Biomass production of tree fallows and their residual effect on maize in Zimbabwe. *Agroforestry Systems, 47*, 139–151.

Myers, R. J. K., Palm, C. A., Cuevas, E., Gunatilleke, I. U. N., Brossard, M. (1994). The synchronisation of nutrient mineralisation and plant nutrient demand. In P. L. Woomer & M. J. Swift (Eds.), *The Biological Management of Tropical Soil Fertility*. New York: John Wiley & Sons.

Mudhara, M. (2002). Assessing the livelihood system diversity of smallholder farm households: Potential of improved fallows in Zimbabwe. Unpublished doctoral dissertation, Food and Resource Economics, University of Florida, Gainesville, Florida.

Netting, R. M. (1993). *Smallholders, householders*. Palo Alto, California: Stanford University Press.

Pisanelli, A., Franzel, S., De Wolf, J., Rommelse, R., & Poole, J. (2002) The adoption of improved tree fallows in western Kenya: Farmer practices, knowledge and perception. Manuscript submitted for publication.

Place, F., Franzel, S., De Wolf, J., Rommelse, R., Kwesiga, F., Niang, A. et al. (2002). Agroforestry for soil fertility replenishment: Evidence on adoption processes in Kenya and Zambia. In C. B. Barrett, F. Place, & A. A. Aboud (Eds.), *Natural Resources Management in African Agriculture: Understanding and Improving Current Practices*, (pp. 155–168). Wallingford, UK: CAB International.

Rohrbach, D. (1998). Developing more practical fertility management recommendations. In S. R. Waddington, H. K. Murwira, J. D. T. Kumwenda, D. Hikwa, & F. Tagwira (Eds.), *Soil Fertility Research for Maize-based Farming Systems in Malawi and Zimbabwe. Proceedings of a Soil Fertility Network Results and Planning Workshop held during 7–11 July 1997 at Africa University, Mutare, Zimbabwe* (pp. 237–244). Harare, Zimbabwe: Soil Fertility Network and CIMMYT–Zimbabwe.

Smaling, E. M. A., Nandwa, S. M., & Janssen, B. H. (1997). Soil fertility is at stake. In R. J. Buresh, P. A. Sanchez, & F. Calhoun (Eds.), *Replenishing Soil Fertility in Africa* (Special Publication 51, pp. 47-61). Madison, Wisconsin: Soil Science Society of America.

Swinkels, R. A., Franzel, S., Shepherd K. D., Ohlsson E. L., & Ndufa, J. K. (1997). The economics of short rotation improved fallows: evidence from areas of high population density in western Kenya. *Agricultural Systems, 55,* 99–121.

PAUL H. THANGATA, JANAKI R. R. ALAVALAPATI AND
PETER E. HILDEBRAND

METAMODELING AGROFORESTRY ADOPTION

Assessing factors influencing adoption of improved fallows in southern Africa using an integrated linear programming and econometric model

1. INTRODUCTION

The number of poor people has increased over the past decades in most parts of Sub-Saharan Africa (SSA). Most of these poor people are also food insecure. The increasing populations and the need for increased food production in most SSA countries have led to land shortages and pressure for continuous arable cultivation. Continuous cultivation with unsustainable agricultural production methods has put pressure on the soil ecology, leading to the depletion of soil organic matter (SOM) and nutrients (Smaling, 1993). The result has been a severe decline in soil fertility (Buresh, Sanchez, & Calhoun, 1997; Sanchez, Shepard, Soule, Buresh, Izac, Mokwunye et al., 1997) causing low food production levels that cannot sustain the growing population. The World Bank (1996) reported that the per capita food production in SSA declined by 1.0% annually between 1980 and 1993.

Agroforestry is viewed as a tool for sustainable agricultural development. It is perceived as an effective means of improving soil fertility and agricultural productivity; increasing fodder and food supplies; protecting watersheds; sequestering carbon; and conserving biodiversity. Given the choice of agroforestry technologies available, research has shown that a plethora of social, cultural, and economic issues including age, education, income of the households, awareness and attitude of the households, and the extent of change agent contact will influence the rate of adoption of agroforestry (Adesina, Mbila, Nkamleu, & Endamana, 2001; Alavalapati, Luckert, & Gill, 1995; Ayuk, 1997; Thangata & Alavalapati, 2003).

J. R. R. Alavalapati & D. E. Mercer, Valuing Agroforestry Systems, 219–236.

One of the agroforestry technologies being researched and promoted in southern Africa is improved fallows. An improved fallow is a system in which short duration trees or herbaceous species are planted in rotation with cultivated crops to improve soil fertility, by substituting for the replenishment benefits of natural fallows. Improved fallows have shown the potential to replenish soil fertility and thereby increase crop yields. They also control soil erosion and provide firewood to smallholder farmers (International Centre for Research in Agroforestry [ICRAF], 1997; Kwesiga, Franzel, Place, Phiri, & Simwanza, 1999).

However, in many smallholder farming systems, farmers' decision making about the choice and management of agroforestry practices is an integral part of the overall strategy for ensuring subsistence and household cash income. Therefore, to some extent, the long-term practice of agroforestry by farmers depends on the existence of multiple benefits from agroforestry products; institutional development; and agricultural and other public policies. Also important is that small-scale farmers often need some form of assistance especially when establishing nurseries for soil fertility replenishing, multipurpose trees like sesbania (*Sesbania sesban* (L) Merr.) and tephrosia (*Tephrosia vogelii* Hook, f) for improved fallows (ICRAF, 1997).

This chapter presents an innovative approach in assessing adoption of agroforestry technologies. Specifically, this study focused on improved fallows of two leguminous shrubs native to Africa, sesbania and tephrosia. Linear programming was used extensively to address this issue. However, this approach does not shed much light on the statistical significance of the results obtained. To overcome this limitation, we need new approaches to data gathering and analysis. Using a case study from Malawi, we present metamodeling, a hybrid model, which integrates linear programming and an econometric analysis to analyze determinants of agroforestry adoption. First, a linear programming model is developed from survey data. Second, results of linear programming are used in specifying an econometric model. Third, this hybrid model, which we call a metamodel, is used to analyze factors determining the adoption of agroforestry practices.

This approach would be appropriate in cases where the number of farmers to be interviewed in a survey is limited or when certain sociocultural factors do not permit the researcher to elicit responses for certain survey questions. For example, in this study we considered two scenarios; one simulating a seed selling incentive, the other without. Details of modeling development and specification, and the model steps are presented in sections 2.5 and 2.6 of this chapter, respectively.

2. THE METAMODELLING APPROACH

2.1. The basic principle in metamodeling

Mathematical models such as linear programming have a long tradition in agricultural economics and have been used by many farm management specialists in farm planning (Brady, 1998). These models can also be used to understand household behavior and assess policy alternatives. However, linear programming

(LP) models are deterministic in nature because they use fixed, non-random values to specify the model. This poses challenges in solving many LP problems as it is difficult for researchers to conclude if estimates derived from LP are statistically significant or not. In such cases some studies have suggested the use of stochastic linear programming (Sengupta, 1970; Sen & Higle, 1999). However, Kleijnen and Sargent (1997) have suggested the use of metamodeling to better understand the behavior of a simulation model.

A metamodel is a model of a model (Kleijnen, 1998; Kruseman, 2000). The use of metamodeling is quite new in some areas; however, regression metamodels from simulation output results are commonly used in farm planning. Most people are more familiar with meta-analysis than metamodeling. Although there are differences between meta-analysis and metamodeling, in terms of their application, we note more similarities. Glass, McGaw, and Smith (1981) described a meta-analysis as a perspective that uses many techniques of measurement and statistical analysis. While meta-analysis is commonly used to quantitatively integrate the findings of multiple, but related, research studies (Glass et al., 1981; Schafer, 2001), a metamodel on the other hand can use the input and output data from other models such as LP mathematical models. Therefore, we see that the main difference is that a meta-analysis is a summary of findings from many empirical studies (Glass et al., 1981), while metamodeling summarizes findings of one replicated study or many. In this paper, however, we prefer the term metamodeling since we use the input and output data from a linear programming model as a base to conduct econometric analysis.

While LPs can produce data for a metamodel, a metamodel can also be used to predict the optimal values of an objective function in LP in response to various levels of the constraints. Metamodels are popular because they are easy to handle and interpret, especially in cases where LP models are used to generate results for base scenarios. A metamodel can also be used to perform sensitivity analyses and therefore has been proposed as a post-model analysis tool (Kruseman, 2000).

Although the relationship of the parameters in the metamodel does not explain the process guiding the decision-making, they provide insights into the behavior of the mathematical models (Johnson, Bauer, Moore, & Grant, 1996) and the apparent relationship between input and output variables (Kruseman, 2000). Metamodels may have different goals; nonetheless, Kleijnen and Sargent (1997) have generalized the goals into three categories:

- Problem Identification: to statistically better understand the problem entity, such as the direction of output results. For example, in the case of regression metamodeling, whether output increases or decreases with an increase in one of the factors.
- Prediction: to forecast behavior, and to assist in the verification and validation of simulation models.
- Optimization: to determine the set of values that gives an optimal solution of the objective function.

Following Johnson et al. (1996), Kleijnen and Sargent (1997), Kleijnen (1998), and Kruseman (2000), this chapter develops a regression metamodel based on the results of an LP model to assess smallholder farm models in Malawi. This is done for two reasons: first, to statistically verify and post validate the results of the LP model, and second, to statistically assess key factors and constraints that impact the adoption of improved fallows.

2.2. The ethnographic linear programming model (ELP)

The LP model used in this study is ethnographic in nature. It is an adaptation of linear programming that has been developed at the University of Florida by Hildebrand, Breuer, Cabrera, & Sullivan (2003) and the Farming and Livelihood Systems Group (FLSG) (Bastidas, 2001; Cabrera, 1999; Kaya, Hildebrand, & Nair, 2000; Litow, 2000; Mudhara, 2002; Thangata, Hildebrand, & Gladwin, 2002). Ethnographic linear programming is a means of quantifying ethnographic data, mostly qualitative, and is thus both descriptive and analytic. ELP simulates the farmers' strategies by choosing from different alternative livelihood activities available to farmers in the farming system and representing different degrees of crop intensity, labor, and land saving techniques available. It takes into account their respective costs, constraints, and advantages.

The ELP models can increase researchers' understanding of the complexity and diversity of smallholder farming systems. ELP is very effective in evaluating various policy instruments, and assists in uncovering farm household decision-making regarding land planning and allocation. This is because it takes into account three important aspects observed by Kuyvenhoven, Heerink, and Ruben (1999):

(i) available resource endowment (land, labor, capital)
(ii) the multiple objectives of the farm household (profit or end year cash, home food consumption, desire for education, etc.)
(iii) the market conditions (prices, access, etc.).

The ELP is represented by the general format:

$$\text{Maximize } z = cx$$
$$\text{subject to}$$
$$Ax \leq b$$
$$x \geq 0$$

where z in this model is the discretionary cash income farmers have at the end of the year after using their constrained resources (represented by the rows of the matrix) to engage in different livelihood activities (represented by the columns of the matrix). C is the row vector of enterprise year-end cash, x is a column vector of enterprise (activity) levels (and all activities are greater than or equal to zero). A is a matrix of technical coefficients, and b is a column vector of farm resource endowments or consumption requirements. The rows of the matrix also represent other goals of the farming households; for example, meeting necessary cash expenses and providing household food security.

The consumption constraints in the model reflect the need for the households to first satisfy household food requirements before marketing any surplus. To specify consumption constraints, minimum food requirements for the household are specified for each crop. The model was implemented in Microsoft Excel® (Microsoft, 2000a) spreadsheet. Microsoft Visual Basic® (Microsoft, 2000b) was used to make calling and solving different households easy and flexible. The premium add-in solver (Frontline Systems, 1999) for Excel was used to handle the large number of variables.

2.3. The data source

Data used to demonstrate this approach are from a survey from Kasungu, Malawi. Household surveys and informal interviews were the main source of primary data for the ELP models (Thangata, 2002). The data included labor availability and use, household composition and food requirements. A total of 40 households were interviewed. Secondary agronomic data such as yields from the first year after fallows were obtained from research stations, on-farm trial and publications (Thangata, 2002; Thangata, Hildebrand, & Gladwin, 2002).

Ethnographic linear programming models were developed for each of the 40 households interviewed. To test the models' prediction ability, and to validate and check areas where the models needed improvements, discussions were held with farmers to see whether the models' preliminary output results adequately depicted what they produce and how they produce it.

2.4. Resource constraints: land, cash and labor

Farm holdings in Kasungu are slightly larger than the national average (Thangata, 2002). In this study, on average the female-headed households have larger land holding sizes than male-headed households.

Since farmers have limited cash available, they often depend on agricultural loans. To be eligible for a loan, a household has to plant tobacco and the interest rate for loans in the 2000/2001 season was 55%. To reflect this, a borrowing activity was included in the model such that, when needed, a household can borrow tobacco inputs and repay with interest at the end of the season. The model also allows for cash to be transferred from one year to the next to be used for purchasing agricultural inputs after satisfying household requirements. This activity allows the models to take only the required amount of credit depending on the farm's cash shortfall. Farmers, especially female headed households (FHH), split the fertilizer meant for tobacco and apply a portion of it on their maize crop. For details see Thangata (2002).

The model separates labor inputs by gender and by month. Each adult male or female provides 25 labor days in a month. Labor supply in any calendar month is the total amount of labor available from the contribution of all household members and hired labor. The model allows for households to hire labor and pay the workers a lump sum payment at the end of the season. Because women are responsible for childcare, the number of infants (under 5 years old) reduces the female labor contribution. For school-going adolescents, food consumption and labor contributed vary according to whether they live at home or school. As the children in the household age, they contribute more labor but also require more food. No new births were considered in the models.

2.5. Model development and specification

At the initial stage of diffusion of improved fallow technologies in the region, the World Agroforestry Centre-Southern African (ICRAF-SA) program and other NGOs were buying seeds from sesbania and tephrosia producers to give to new farmers. The seed selling became the variable of interest in this case. To evaluate whether this additional income from improved-fallow seeds enhances adoption, and to test under what conditions farmers would adopt improved fallow technologies, two scenarios were tested. In scenario 1, farmers do not sell sesbania or tephrosia seeds; in scenario 2, there is a market for the seeds.

First, a model without agroforestry was developed to simulate a situation where farmers did not adopt the agroforestry technology. This model takes into consideration all the activities on the farm as a whole (this study did not consider livestock as most of the farmers interviewed had none). We used this model for validation. Then two scenarios when agroforestry technologies are available are simulated. Second, all cropping activities are evaluated together with improved fallows. Third, the improved fallow model is simulated with an option in which farmers also sell seeds from the improved fallows in addition to soil fertility benefits. In both cases, households were modeled without a seed selling activity and then with a seed selling activity.

In all cases, the model was a ten-year, dynamic linear programming model. The matrix of technical coefficients is identical for all households. It is the resource endowments, including household compositions that vary across households.

2.6. The model steps

Figure 1 presents the steps leading to the metamodeling technique. As reported by Thangata (2002) and Thangata et al. (2002), after the data collection stage and the basic linear programming model (stage 1), a general agroforestry model was developed (stage 2). Two scenarios were run from the agroforestry model, one with a seed-selling incentive (stage 2) and the other without a seed-selling scenario (stage 3). Results from the two scenarios at stage 3 were then used in the econometric regression metamodel (stage 4).

The basic idea in this metamodeling is to use simulation results (output) and input data from the agroforestry adoption model as variables for statistical

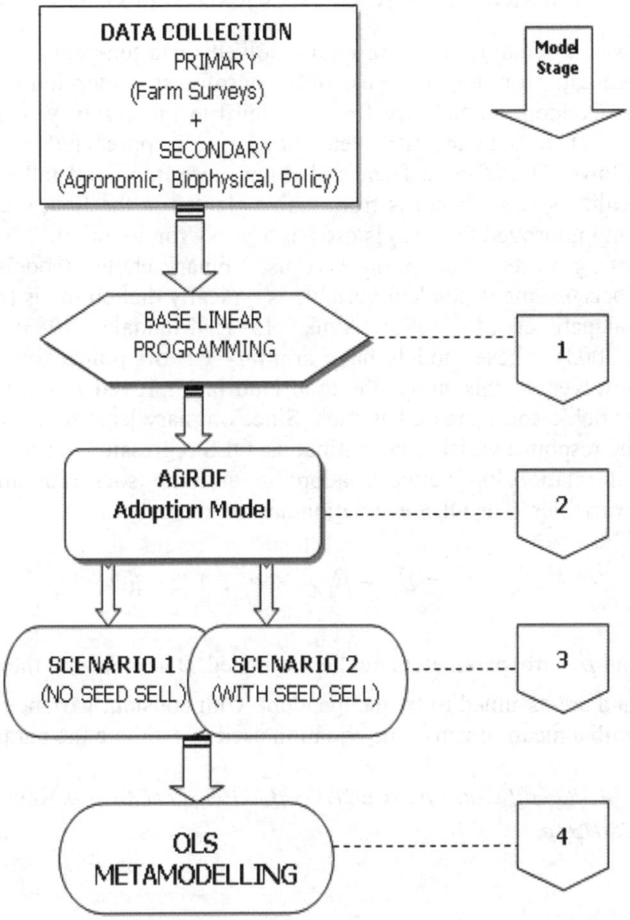

Figure 1. Schematic representation of the methodology and steps for the metamodel.

estimation purposes. In this case the two scenarios in the agroforestry adoption model (stages 2 and 3), are treated as a field experiment with two treatments: one, a group of farmers planting improved fallows where the seeds were bought and two, a group of farmers who planted improved fallows but without a market to sell the seed. The runs for the two scenarios produce independent outputs, and hence different data sets with different model behavior, leading to the regression metamodeling (stage 4). The output from the two agroforestry adoption models becomes the source for the data (rather than household data) used in the econometric model to allow for the exploration of possible future outcomes.

3. MODEL DESCRIPTION AND APPROACH

The model was a dynamic linear program, solved for a ten-year period. However, the study used data from the fifth year of the agroforestry adoption model (stage 2) because there is adoption stability from the third to the eighth year in the models (Thangata, 2002). It is in the fifth year that farmers appreciate the benefits from improved fallows. Therefore, a farmer planting more improved fallows in the fifth year (after realizing some benefits from maize planted in the first, second and third years following improved fallows) is used as a proxy for adoption.

Studies of agroforestry adoption have used binary choice models (tobit, logit, and probit), because the dependent variable is typically dichotomous (Adesina et al., 2001; Alavalapati et al., 1995; Ayuk, 1997; Maddala, 1983; Thangata & Alavalapati, 2003). These models have arbitrary cut off points for the dependent variables. However, in this study, the total land in improved fallows per farm is a continuous variable and is treated as such. Since ordinary least squares (OLS) works well when the response variable is continuous, OLS regression is used to investigate the statistical relationship between adoption and the socioeconomic and other explanatory variables. The OLS estimation has the formula:

$$y_i = \beta_o + \beta_1 \chi_i + e_i \quad i=1,2,\ldots,n \tag{1}$$

where β_o and β_1 are parameters to be estimated. The e_i's are the errors of the prediction, and are assumed to be independent, with constant variance and normally distributed with a mean of zero. The equation used to estimate the parameters is:

$$E(Y_i) = \beta_o + \beta_1 MLABR + \beta_2 FLABR + \beta_3 INCENTIVE + \beta_4 NHH + \beta_5 TTLAND + \beta_6 MZECONS + \beta_7 LOAN + \beta_8 CSHTR + \varepsilon_I \tag{2}$$

The independent variables for the regression analysis were chosen on the basis of prevailing knowledge of factors influencing adoption generally and on the basis of those likely to be particularly pronounced in the study area. The dependent and the explanatory variables used are:

y_i = TTLIF, the dependent variable, land in hectares planted to improved fallows
X_1 = *MLABR*, male labor in a household. (Days/month)
X_2 = *FLABR*, female labor in a household. (Days/month)
X_3 = *INCENTIVE*, a seed selling activity as an incentive, 1, if yes and 0, otherwise.
X_4 = *NHH*, number of people in the household.

X_5 = *TTLAND*, total land cultivated (Hectares).
X_6 = *MZECONS*, total household maize requirements (kg/year)
X_7 = *LOAN*, amount of input loan taken (US$).
X_8 = *CSHTR*, amount of cash transferred (savings) for next season (US$)

Both male, (MLABR), and female (FLABR) labor were hypothesized to be positively correlated with adoption of improved fallows, as was a seed selling activity (INCENTIVE). The variables number of people in the household (NHH), total land cultivated (TTLAND), and total household maize requirements (MZECONS) were anticipated to be positively correlated with adoption. Conversely, the variables the amount of input loan taken (LOAN) and amount of cash transferred (savings) for next season (CSHTR), were expected to be negatively correlated with adoption. The assumption is that a larger loan (at a 55% interest rate) may mean a reduced likelihood of planting more improved fallows because farmers are expected to plant tobacco if they take a loan. In contrast, if farmers keep some cash to purchase their own inputs, they are more likely to purchase chemical fertilizers for tobacco growing which is a more profitable crop. The metadata used in the OLS regressions are presented in the appendix section.

4. THE EMPIRICAL MODEL

In the empirical model the aim is to isolate the factors that determine the adoption of improved fallows estimated with ordinary least squares (OLS).

4.1. Empirical Results and Discussion

Table 1 presents summary statistics for the variables used in the regression. On average, female headed households (FHH) had about 0.8 ha planted to fallows while male headed households (MHH) had on average 0.7 ha. The FHHs used more land (TTLAND) than the MHHs, on average 2.0 ha and 1.8 ha respectively. The MHHs needed more loans for inputs, LOAN, than the FHHs. However, on average, the FHHs saved more cash (CSHTRF) than male households. In addition, MHHs and FHHs both had the same number of male labor days (MLABR) available per month (68 days). The number of female labor days (FLABR) was less in both cases, with 48 days for the MHHs and 57 for the FHHs. The results also show that there is no difference between the two groups (MHH and FHH) in the number of household members (NHH).

The correlation matrix presented in Table 2 shows that multicollinearity was not a concern since none of the variables were strongly correlated.

Figure 2 is a bar graph of the mean (hectares) of total land used, total land planted to improved fallows, and total land planted to maize for the no seed sell and with seed sell options. It is clear that there was a difference in the amount of total

Table 1. Summary statistics of variables used in the OLS regression metamodel.

Variable Name	*Total Sample (n=80)*		*Male headed households (n=64)*		*Female headed households (n=16)*	
	Mean	*Std. Dev*	*Mean*	*Std. Dev*	*Mean*	*Std. Dev*
TTLIF	0.8	0.3	0.7	0.3	0.8	0.5
MLABR	68	34	68	29	68	49
FLABR	50	25	48	25	57	22
TTLLAND	1.8	0.9	1.8	0.8	2.0	1.2
INCENTIVE	0.5	0.5	0.5	0.5	0.5	0.5
NHH	6.1	2.9	6.2	2.8	6.0	3.1
MZECONS	1294	541	1285	493	1328	721
LOAN	12	14	12	15	10	12
CSHTRF	64	72	62	69	70	87

Table 2. Correlation matrix of explanatory variables.

	MLABR	*FLABR*	*TTLLAND*	*INCENTIVE*	*NHH*	*MZECONS*	*LOAN*	*CSHTRF*
MLABR	1.000							
FLABR	0.437	1.000						
TTLLAND	-0.032	-0.121	1.000					
INCENTIVE	0.357	0.178	-0.196	1.000				
NHH	0.046	-0.283	-0.111	0.167	1.000			
MZECONS	-0.581	-0.313	-0.275	-0.068	-0.532	1.000		
LOAN	0.064	-0.006	-0.173	0.443	0.058	-0.046	1.000	
CSHTRF	-0.316	-0.119	-0.533	-0.411	-0.069	0.212	0.332	1.000

land used in the two scenarios of the agroforestry adoption model. However, in order to check the significance of the difference, *t*-tests were performed.

Table 3 summarizes t-tests for the total land used, land in improved fallows and all maize planted in the agroforestry adoption model comparing the without seed selling to the with seed selling option models in the fifth year. Table 3 shows that there was no statistically significant difference between the land planted to improved fallows with and without a seed selling incentive. This is due to the fact that when the households get cash from the seed selling activities, they invest in profitable crops such as tobacco. However, the total land cultivated with and without a seed selling option shows a statistically significant difference. Table 3 also shows a statistically significant difference in the land planted to all maize.

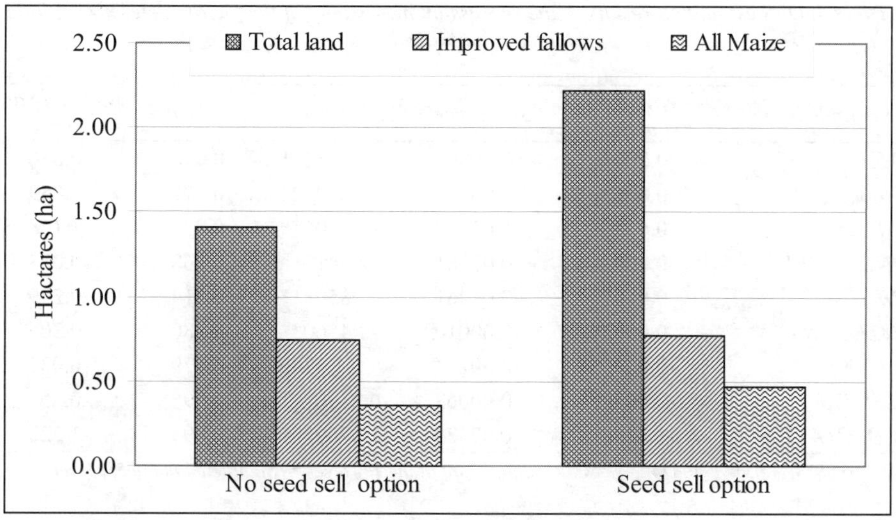

Figure 2. Bar graph of mean total land used, land in improved fallows and all maize planted in the with and without seed selling option agroforestry model.

Table 3. T-tests for simulated improved fallow land, total cultivated for the agroforestry adoption models with seed selling (With) and without seed selling (No) scenarios.

	Total land cultivated		*Improved Fallows*		*All Maize*	
	Seed sell	*No seed sell*	*Seed sell*	*No seed sell*	*Seed sell*	*No seed sell*
Mean	1.41	2.22	0.74	0.77	0.36	0.47
Variance	0.29118	0.88078	0.08658	0.13304	0.01913	0.06618
t-Statistic	-8.8955		-0.6476		-4.6473	
P (two-tail)	0.0000		0.2605		0.0000	

5. ORDINARY LEAST SQUARES (OLS) REGRESSION METAMODEL

Table 4 presents the Ordinary Least Squares (OLS) regression results, where the dependent variable is the amount of land planted to improved fallows, a proxy for adoption. The OLS regression metamodel was estimated with the procedure in Shazam version 8 (Shazam, 1997; White, 1997). The R^2 is statistically significant, and accounts for almost 88% of the variance. The variables male available labor (MLABR) and female available labor (FLABR) are positive but not significant. This could be due to the fact that both households have the same amount of male labor or that adoption of improved fallows is gender neutral.

Table 4. OLS estimation results of the regression metamodel of the factors determining land planted to improved fallows.

Variable Name	*Coefficient Estimate*	*Std. Error*	*Asy. t-ratio*	*Partial Corr*	*Elasticity at means*
MLB	0.0008	0.00099	0.8281	0.098	0.056
FLB	0.0012	0.00119	0.9920	0.117	0.059
TTLLAND	0.6446	0.05078	12.690***	0.833	1.168
INCENTIVE	-0.1658	0.07081	-2.3410**	-0.268	-0.083
NHH	0.0407	0.01461	2.7850***	0.314	0.249
MZECONS	-0.0002	0.00010	-2.4560**	-0.280	-0.305
LOAN	-0.0031	0.00204	-1.5340	-0.179	-0.037
CSHTRF	-0.0039	0.00063	-6.2390***	-0.595	-0.252
CONSTANT	-1.2232	0.07324	-16.700***	-0.893	-1.223

*R^2 = 0.8873; Adj. R^2 = 0.8746; N=80; ***Significant at the 1% level ** Significant at the 5% level.*

As expected, the regression coefficient of total land used (TTLAND) is positive and significant (α = 1%). The INCENTIVE, the scenario to sell or not sell seed, has a negative sign and is significant (α = 5%). The number of people in a household (NHH) is positive and significant at the 1% level, while the amount of maize required for the household, MZECONS, is negative and significant at the 5% level. The variables LOAN, the amount of cash required for inputs, and the amount of cash transferred for use in the following season (CSTRF), are both negative. LOAN is not significant, while CSTRF is significant at the 1% level.

As expected the adoption of improved fallows responded most to total land cultivated (TTLAND). It shows that, holding other variables constant, a unit increase (ha) in cultivated land would positively influence the land planted to improved fallows by 0.6 ha. The negative sign on the INCENTIVE variable may appear surprising at first, but a critical analysis of this variable makes sense. It would appear that as the households get more cash from the incentives, they diversify into other profitable crops and reduce the land in improved fallows. To understand this better, the model was run several times, with and without a seed selling option. The results showed that once a seed selling activity was included in the model, farmers stopped taking fertilizer input loans. However, they grew more tobacco using cash from the seed selling activities. It seems, therefore, plausible to suggest the cash benefit from the seed selling is transferred to more profitable crops. Once that is satisfied, no more improved fallows are planted. This is supported by the t-tests results presented in Table 3, showing no statistically significant difference between the land planted to improved fallows with or without a seed selling incentive.

The variables NHH and MZECONS are related but different. With an increase in the number of people in the household, NHH, the adoption of improved fallows is expected to increase. However, when more maize is required for consumption,

MZECONS, there is a decrease in improved fallow lands. Having more people in the household, NHH, implies that there is enough labor available assuming land is not a constraint. Nonetheless, this does not necessarily mean that the household will require more food for consumption. On the other hand, more maize consumption could be due to hired labor. Since the model has a provision to hire labor when needed as long as it is paid for, and the workers be fed in case of long term hired labor. This would necessitate more food. This is supported by the negative correlation between NHH and MZECONS (Table 2). Since the hired labor would require more food and cash at the end of the season, the hiring household is forced to grow a cash crop such as tobacco so as to be able to pay for the hired labor at the end of the season, resulting in less land planted to improved fallows. Therefore the need for more maize does not necessarily mean that the household will plant more improved fallows. More people in the household might provide enough family labor that does not require payment, and hence the household can plant more improved fallows.

The variables LOAN (the amount of loan for agricultural inputs) and CSTRF (the amount of cash transferred for use in the following season) have the expected signs and are significant, showing that households that take more loans plant fewer improved fallows. This is because households that take more loans are required to pay back the loan with a 55% interest. They, therefore plant more tobacco, the cash crop. In cases where they have some savings (CSTRF), they are able to purchase chemical fertilizers. Hence, they might not want to plant improved fallows as they are focusing on growing enough tobacco to be able to repay the loan. The variables LOAN and CSTRF account for 18% and 60% respectively, of the model's explanatory power.

6. RECOMMENDATIONS AND CONCLUSION

This chapter introduced the use of metamodeling as a post model validation tool for linear programming models. Specifically, an OLS regression metamodel was used to check the results from an agroforestry adoption, linear programming model. The results suggest that the most important variables in the adoption of improved fallows are total cultivated land and number of people in the household. The findings of the regression metamodel support the agroforestry adoption LP model results.

The results from this study have also shown that adoption of agroforestry is gender neutral. This lend support to the earlier finding by Thangata et al. (2002) that in central Malawi, both male and female headed households can adopt the improved fallow technology. The findings have also shown that farmers with access to land are going to adopt improved fallows, with or without the extra incentive of being able to sell improved fallow seeds. Farmers who are able to save cash for future use are less likely to plant more improved fallows. This also applies to those with access to more credit as they tend to grow more tobacco.

The major implication arising from the results of this study is that efforts to encourage adoption of improved fallows should be directed to farmers with enough land. While it is beyond the scope of this paper, the fact that our results have shown that the technology is gender neutral, ICRAF and the Malawi government extension services can use the gender neutrality of this technology as a pathway to addressing gender inequitabilities in rural farming systems.

The approach to modeling presented in this chapter is relatively new in most agroforestry studies and the findings might not be generalizable. An advantage from the approach we have used is that it is possible to identify important variables in adoption studies using small data samples from surveys. Nonetheless, there is a need for future studies to refine this methodology and go further to introduce more than one variable of interest. Also important is for future studies to evaluate the adoption of improved fallow species separately to check farmers' preferences for the two species used in this study (Sesbania and tephrosia). We would however, like to emphasize that studies that intend to use the approach presented in this chapter, the objective of the regression metamodel should be to accurately reproduce the LP simulation results and statistically provide useful insights into farm-level responses.

7. APPENDIX

Output data from linear programming models used in the OLS regression metamodel.

No.	TTLIF	*MLABR*	*FLABR*	INCE-NTIVE	*NHH*	*TTLAND*	*MZECONS*	*LOAN*	*CSHTR*
1	43.3	25	72	1	5	2.2	1250	0.00	57.64
2	72.2	50	83	1	6	2.1	1437.5	0.00	119.01
3	40.7	31	68	1	8	2.5	1375	0.00	93.23
4	47.2	25	8	1	2	0.5	875	14.31	0.00
5	38.0	87	37	1	7	3.0	1625	0.00	123.45
6	46.1	175	83	1	11	6.0	2687.5	0.00	270.32
7	39.7	50	62	1	5	4.0	1000	0.00	163.66
8	70.6	91	81	1	9	2.8	2000	0.00	196.90
9	36.5	112	87	1	10	4.8	2250	0.00	232.41
10	43.6	66	22	1	4	1.9	1187.5	0.00	94.46
11	61.3	150	75	1	9	2.2	2250	0.00	213.90
12	33.3	58	39	1	5	2.2	1250	0.00	143.66
13	32.7	75	25	1	3	1.8	1000	0.00	50.22
14	49.9	83	58	1	7	2.6	1625	0.00	151.17
15	41.1	39	116	1	10	2.3	1687.5	0.00	93.37
16	43.9	25	34	1	3	2.2	937.5	0.00	47.92
17	29.7	45	54	1	7	2.6	1250	0.00	99.18
18	38.9	68	31	1	5	1.7	1125	0.00	90.12
19	30.8	62	25	1	3	1.8	625	0.00	59.50

APPENDIX (cont.)

No.	TTLIF	*MLABR*	*FLABR*	INCE-NTIVE	*NHH*	*TTLAND*	*MZECONS*	*LOAN*	*CSHTR*
20	69.5	97	88	1	12	2.8	1812.5	0.00	224.45
21	55.5	47	25	1	5	1.3	1000	0.00	53.70
22	85.5	50	52	1	3	2.1	750	0.00	124.01
23	69.2	100	76	1	7	1.8	1500	0.00	171.54
24	67.7	83	58	1	7	2.1	1625	0.00	184.63
25	33.4	95	39	1	8	2.5	1375	0.00	83.36
26	31.9	50	25	1	3	1.5	750	0.00	54.10
27	28.6	25	50	1	3	1.6	750	0.00	52.58
28	76.5	50	33	1	3	1.3	687.5	0.00	106.75
29	77.4	50	37	1	4	1.550	750	0.00	94.66
30	26.8	83	47	1	6	2.700	1125	0.00	103.74
31	42.8	50	24	1	3	2.116	562.5	0.00	103.13
32	39.3	62	22	1	5	1.300	937.5	0.00	78.05
33	65.4	50	25	1	3	1.300	750	0.00	57.12
34	56.6	56	31	1	5	1.500	1000	0.00	70.63
35	31.6	58	33	1	5	2.100	1125	0.00	68.04
36	32.9	133	83	1	14	4.700	2875	0.00	244.21
37	37.3	108	64	1	10	4.650	1625	0.00	204.45
38	32.6	56	43	1	7	2.300	1125	0.00	100.68
39	62.6	39	31	1	5	1.100	937.5	0.00	66.40
40	45.2	54	41	1	8	1.267	1250	0.00	111.90
41	50.8	25	72	0	5	1.603	1250	14.34	0.00
42	39.2	50	83	0	6	1.659	1437.5	16.13	0.00
43	15.5	31	68	0	8	0.645	1375	21.48	0.00
44	51.3	25	8	0	2	1.881	875	11.19	0.00
45	43.8	87	37	0	7	3.239	1625	18.90	0.00
46	48.0	175	83	0	11	1.169	2687.5	0.00	73.11
47	47.2	50	62	0	5	2.267	1000	12.87	0.00
48	52.9	91	81	0	9	2.354	2000	25.09	0.00
49	49.9	112	87	0	10	1.329	2250	52.77	0.00
50	49.6	66	22	0	4	2.230	1187.5	13.69	0.00
51	54.7	150	75	0	9	1.433	2250	22.29	0.00
52	50.4	58	39	0	5	1.112	1250	35.66	0.00
53	52.3	75	25	0	3	1.709	1000	37.79	0.00
54	52.4	83	58	0	7	1.828	1625	40.20	0.00
55	46.5	39	116	0	10	1.137	1687.5	17.15	0.00

APPENDIX (cont.)

No.	TTLIF	*MLABR*	*FLABR*	INCE-NTIVE	*NHH*	*TTLAND*	*MZECONS*	*LOAN*	*CSHTR*
56	51.6	25	34	0	3	1.363	937.5	13.33	0.00
57	47.6	45	54	0	7	1.379	1250	13.36	0.00
58	41.8	68	31	0	5	0.957	1125	15.43	0.00
59	34.1	62	25	0	3	2.850	625	12.82	0.00
60	47.8	97	88	0	12	1.173	1812.5	0.00	90.51
61	43.7	47	25	0	5	0.991	1000	38.96	0.00
62	32.5	50	52	0	3	1.800	750	12.38	0.00
63	53.6	100	76	0	7	1.875	1500	0.00	54.76
64	35.1	83	58	0	7	1.837	1625	17.61	0.00
65	45.9	95	39	0	8	0.944	1375	0.00	55.95
66	48.3	50	25	0	3	0.897	750	11.08	0.00
67	46.7	25	50	0	3	0.927	750	9.78	0.00
68	44.1	50	33	0	3	1.054	687.5	10.76	0.00
69	19.8	50	37	0	4	2.019	750	12.89	0.00
70	37.8	83	47	0	6	1.058	1125	0.00	86.67
71	37.5	50	24	0	3	1.246	562.5	15.28	0.00
72	48.3	62	22	0	5	0.897	937.5	17.49	0.00
73	44.9	50	25	0	3	1.247	750	9.78	0.00
74	40.7	56	31	0	5	1.613	1000	14.68	0.00
75	44.0	58	33	0	5	3.514	1125	21.81	0.00
76	47.8	133	83	0	14	1.905	2875	0.00	89.09
77	49.3	108	64	0	10	1.266	1625	20.72	0.00
78	47.7	56	43	0	7	1.100	1125	13.50	0.00
79	53.2	39	31	0	5	1.474	937.5	11.77	0.00
80	22.4	54	41	0	8	1.733	1250	13.88	0.00

TTLIF= the dependent variable, land in hectares planted to improved fallows
MLABR=male labor in a household. (Days/month)
FLABR= female labor in a household (Days/month)
INCENTIVE= a seed selling activity
NHH= number of people in the household
TTLAND= total land cultivated (Hectares)
MZECONS= total household maize requirements (kg/year)
LOAN=amount of input loan taken (US$)
CSHTR= amount of cash transferred (savings) for next season (US$)

8. REFERENCES

Adesina, A. A., Mbila, D., Nkamleu, G. B., & Endamana, D. (2001). Econometric analysis of the determinants of adoption of alley farming by farmers in the forest zone of southwest Cameroon. *Agriculture, Ecosystems and Environment, 80,* 255–265.

Alavalapati, J. R. R., Luckert, M. K., & Gill, D. S. (1995). Adoption of agroforestry practices: a case study from Andhra Pradesh, India. *Agroforestry Systems, 32,* 1–14.

Ayuk, E. T. (1997). Adoption of agroforestry technology: the case of live hedges in central plateau of Burkina Faso. *Agricultural Systems, 54,* 189–206.

Bastidas, E. (2001). Assessing potential response to changes in the livelihood system of diverse, limited-resource farm households in Carchi, Ecuador: Modeling livelihood strategies using participatory methods and linear programming. Unpublished doctoral dissertation, University of Florida, Gainesville, Florida.

Brady, M. V. (1998). Experimental integration of a root zone leakage model (SOIL-N) with a farm level economic model. TP 3- Integration of Mathematical models (Swedish Water Management Research Programme [VASTRA] Working Paper 5) Department of Economics, Swedish University of Agricultural Sciences, Uppsala, Sweden.

Buresh, R. J., Sanchez, P. A., & Calhoun, F. G. (Eds.). (1997). Replenishing Soil Fertility in Africa SSSA. (Special Publication No.51): Madison, Wisconsin: Soil Science Society of America.

Cabrera, V. (1999). Farm problems, solutions, and extension programs for small farmers in Cañete, Lima, Peru. Unpublished masters thesis, University of Florida, Gainesville, Florida.

Frontline Systems. (1999). Premium Solver for Excel for Windows. Incline Village, Nevada: Author.

Glass, G. V., McGaw, B., & Smith, M. L. (1981). Meta analysis in social research. London: Sage Publications.

Hildebrand, P. E., Breuer, N. E., Cabrera, V. E., Sullivan, A. J. (2003). Modeling diverse livelihood strategies in rural livelihood systems using ethnographic linear programming. (Staff paper SP 03-5). Food and Resource Economics Department, Institute of Food and Agriculture Sciences, University of Florida, Gainesville, Florida.

International Center for Research in Agroforestry (ICRAF). (1997). Annual Report 1996. Nairobi, Kenya: ICRAF.

Johnson, K. E., Bauer, K. W., Jr., Moore, J. T., & Grant, M. (1996). Metamodeling techniques in multidimensional optimality analysis for linear programming. *Mathematics and Computer Modeling, 23*, 45-60.

Kaya, B., Hildebrand, P. E., & Nair, P. K. R. (2000). Modeling changes in farming systems with the adoption of improved fallows in southern Mali. *Agricultural Systems, 66*, 51-68.

Kleijnen, J. P. C. (1998). Experimental design for the sensitivity analysis, optimization and validation of simulation models. In J. Banks (Ed.), *Handbook of Simulation*. New York: John Wiley and Sons.

Kleijnen, J. P. C., & Sargent, R. G. (1997). A methodology for the fitting and validation of metamodels in simulation. (Center discussion paper No. 97116). Katholieke University Brabant, Tilburg, The Netherlands. Retrieved April, 2002 from http://greywww.kub.nl:2080/greyfiles/center/1997/doc/116.pdf..

Kruseman, G. 2000. Bio-economic household modeling for agricultural intensification. (Mansholt Studies 20). Mansholt Graduate School, Wageningen, The Netherlands.

Kuyvenhoven, A., Heerink, N., & Ruben, R. (1999). Economic policy in support of soil fertility: which interventions after structural adjustment? In E. M. A. Smaling, O. Oenema, & L. O. Fresco (Eds.). *Nutrient Disequilibria in Agroecosystems: Concepts and Case Studies*, (pp.119-140). Wallingford, United Kingdom: CAB International.

Kwesiga, F., Franzel, S., Place, F., Phiri, D., & Simwanza, C. P. (1999). *Sesbania sesban* improved fallows in Eastern Zambia: Their inception, development and farmer enthusiasm. *Agroforestry Systems, 47*, 49-66.

Litow, P. A. (2000). Food security and household livelihood strategies in the Maya Biosphere Reserve: The importance of milpa in the community of Uaxactún, Petén, Guatemala. Unpublished master's thesis, Agricultural Education and Communication, Institute of Food and Agriculture Science, University of Florida, Gainesville, Florida.

Maddala, G. S. (1983). Limited dependent and quantitative variables in econometrics. (Econometric Society Monograms 3). Cambridge, United Kingdom: Cambridge University Press.

Microsoft. (2000a). Microsoft Excel version 9.0 for Windows [Computer software]. Seattle, Washington: Microsoft Corporation.

Microsoft. (2000b). Microsoft Visual Basic 9.0 for Windows [Computer software]. Seattle, Washington: Microsoft Corporation.

Mudhara, M. 2002. Assessing the livelihood system diversity of smallholder farm households: Potential of improved fallows in Zimbabwe. Unpublished doctoral dissertation, Food and Resource Economics, Institute of Food and Agriculture Science, University of Florida, Gainesville, Florida.

Sanchez, P. A, Shepherd, K. D., Soule, M. J., Buresh, R. J., Izac, A.-M. N., Mokwunye, A. U. et al. (1997). Soil fertility replenishment in Africa: An investment in natural resource capital. In R. J. Buresh, P. A. Sanchez, & F. G. Calhoun (Eds.). Replenishing Soil Fertility in Africa. (Soil Science Society of America [SSSA] Special Publication No.51, pp. 1-46). Madison, Wisconsin: SSSA.

Schafer, William D. (2001). Replication: a design principle for field research [Electronic version]. *Practical Assessment, Research & Evaluation, 7*(15).

Sen, S, Higle, J. (1999). An introductory tutorial on stochastic linear programming models. *Interfaces, 29*(2), 33-61.

Sengupta, J. K. (1970). Stochastic linear programming with chance constraints. *International Economic Review. 11*(1): 101-116.

Shazam (1997). Shazam user guide. Retrieved February 2002, http://shazam.econ.ubc.ca/intro/logit1.htm.

Smaling, E. M. A. (1993). Soil nutrient depletion in sub-Saharan Africa In H. van Reuler, & W. H. Prins (Eds.), *The Role of Plant Nutrients for Sustainable Crop Production in Sub-Saharan Africa* (pp. 53-67). Leidschendam, The Netherlands: Dutch Association of Fertilizer Producers.

Thangata, P. H. (2002). The potential for agroforestry adoption and carbon sequestration in smallholder agroecosystems of Malawi: An ethnographic linear programming approach. Unpublished doctoral dissertation, Natural Resources and Environment, University of Florida, Gainesville, Florida.

Thangata, P. H., & Alavalapati, J. R. R. (2003). Agroforestry adoption in southern Malawi: The case of mixed intercropping of *Gliricidia sepium* and maize. *Agricultural Systems, 78,* 57–71.

Thangata, P. H, Hildebrand, P. E., & Gladwin, C. H. (2002). Modeling agroforestry adoption and household decision making. *African Studies Quarterly* [Electronic version], *6*, (1/2).

White, K. J. (1997). Shazam user's reference manual version 8.0. New York: McGraw-Hill.

World Bank. (1996). African Development Indicators 1996. Washington, D.C.: World Bank.

EVAN MERCER AND ANN SNOOK

ANALYZING *EX-ANTE* AGROFORESTRY ADOPTION DECISIONS WITH ATTRIBUTE-BASED CHOICE EXPERIMENTS

1. INTRODUCTION

Although many cases of successful agroforestry extension efforts exist (for examples, see Chapter 2), all too often attempts to promote agroforestry have resulted in low adoption rates, with farmers reluctant to adopt new or improved agroforestry systems or abandoning agroforestry shortly after establishment. As a result, the recent increase in research on the adoption of agroforestry innovations has been motivated largely by the perceived gaps between advances in agroforestry science and extension (Mercer, in press). The theoretical and empirical literature on adoption of agroforestry innovations has been reviewed by Pattanayak, Mercer, Sills, and Yang (2003) and Mercer (in press). Significant progress has been made, especially in using binary choice regression models for *ex-post* analyses to examine how past adoption decisions are correlated with variables describing farmers, their farms, demographics and socio-economic conditions. These *ex-post* analyses have been useful for increasing our understanding of who adopts first, identifying communities and households to target as potential early adopters, and developing policies to promote agroforestry. However, the *ex-post*, binary choice regression studies have contributed little to the problem of designing agroforestry systems that appeal to potential adopters because they are not able to examine how farmer preferences vary for different combinations of characteristics of agroforestry alternatives.

Although a variety of reasons contribute to low adoption rates, they often result from inadequate assessments of farmers' preferences, priorities, and constraints prior to designing new agroforestry systems (Current, Lutz, & Scherr, 1995; Mercer & Miller, 1998). Therefore, rigorous *ex-ante* analyses that are able to provide predictive understanding of farm households' land-use decisions and the relative importance of the characteristics of land-use systems demanded by farmers should provide valuable information to project planners and agroforestry system designers.

J. R. R. Alavalapati & D. E. Mercer, Valuing Agroforestry Systems, 237–256.

Although recent progress has been made in *ex-ante* adoption analysis using a farming systems approach (Current et al., 1995; Barrett, Place, & Aboudk, 2002; Franzel & Scherr 2002), systematic, quantitative *ex-ante* assessments of adoption are relatively rare, partly because, as Franzel & Scherr (2002) point out, some scientists believe they have been too "soft" or "subjective."

In this chapter, we describe a quantitative, econometric based method for *ex-ante* analysis of the adoption potential of new agroforestry systems and provide an example of its application in southeastern Mexico. The method we apply, generally referred to as "conjoint analysis," originally developed by market researchers, is a survey-based technique that focuses attention on the trade-offs people make between the attributes of alternative goods and services. The basic requirement for any conjoint-based analysis is that the products or services tested are treated as sets of distinct attributes (or features) with a limited set of variations (or levels) for each attribute (feature). Eliciting individuals' stated preferences between goods and services with different attribute combinations allows the analyst to evaluate the importance of different attributes of the good or service, compare alternative versions of the good or service on each of the important attributes, and estimate the probability of purchase (adoption) of different attribute combinations (Louviere, 1988, 1994).

The most common application of conjoint analysis has been assisting firms in the design of new, multi-attribute products; a problem with many similarities to designing multi-attribute land use systems like agroforestry. Conjoint allows one to determine the combination of attributes of products (land-use systems) that consumers (farmers) are most likely to purchase (adopt). In analyzing agroforestry adoption potential, respondents evaluate alternative land-use systems and make trade offs among various features of the land use systems, selecting combinations of attributes (features) as better than others. Therefore, conjoint analysis can be used to assess the economic and non-economic criteria farmers use to manage their lands, how farmers value different attributes of land use systems, how these values affect adoption and subsequent management behavior, and determine the characteristics of agroforestry systems most likely to be adopted.

Although market researchers have used conjoint for new product design since the 1970s (Green & Wind, 1975), natural resource economists have only recently begun to apply conjoint analysis for valuing environmental goods (Adamowicz, Louviere, & Williams, 1994; Holmes, Zinkhan, Alger, & Mercer, 1998; Opaluch, Swallow, Weaver, Wessells, & Wichelns, 1993) and analyzing land-use decision making (Baidu-Forson, Waliyar, & Ntare, 1997; Zinkhan, Holmes, & Mercer 1997). Attribute-based choice experiments (ACE), a subset of conjoint analysis, were developed about 20 years ago in response to economists' concerns with the theoretical limitations of the typical conjoint analysis ranking and rating studies (Bennett & Blamey, 2001b; Louviere, 2001). Traditional ranking/rating models impose a variety of theoretical and practical problems for economists, including: i) problems comparing ranking or rating data across respondents, ii) respondents may have problems ranking large numbers of alternatives, iii) rating or ranking alternatives are not typical problems faced by consumers, and iv) traditional conjoint analyses are based on statistical and mathematical considerations rather than economic or behavioral theory (Bennett & Blamey, 2001b; Louviere, 2001).

ACE addresses problems with traditional conjoint analysis by asking respondents to choose between alternatives rather than rank or rate the alternatives. As a result, the pattern of choices allows one to model the probability of choosing a particular alternative in terms of the attributes used to describe that alternative (Bennett & Blamey, 2001b). ACE models are also consistent with the sound, well tested, and long-standing theory of random utility. Holmes and Adamowicz (2003) provide a thorough explication of the application of random utility theory to attribute-based choice experiments (see the Appendix for details). ACE models assume that the attributes convey utility to the respondent and that the level of utility the respondent associates with an alternative determines the probability that he/she will choose that alternative. By regressing the stated choices on attribute levels, a wealth of information can be gleaned regarding preferences for the individual attributes, and the probability of choosing programs with any combination of attributes can be predicted (Bennett & Blamey, 2001b).

In this chapter, we present a case study applying attribute-based choice experiments to the problem of designing new agroforestry systems. First we describe the study site, the Calakmul Biosphere Reserve in southeastern Mexico, and the methods we used to design the attribute-based choice experiment and analyze the data. Then we present results of the experiment and discuss how they could be used for improving agroforestry system and project design. This chapter is not intended to provide the reader with all the tools needed to immediately undertake an attribute-based choice experiment. Rather, we hope to provide an example of how these techniques can be applied to the difficult problem of *ex-ante* analysis of farmer demand for new land use systems like agroforestry. The large and growing literature on the intricacies of implementing ACE in natural resource settings can be accessed through recent reviews by Bennett and Blamey (2001a) and Holmes and Adamowicz (2003).

2. CASE STUDY SITE

The objective of this case study project was to develop information to improve the adoption potential of agroforestry projects in southeast Mexico. Research was conducted in the buffer zone of the 723,000 hectare (1.7 million acre) Calakmul Biosphere Reserve in southeastern Campeche, Mexico (Figure 1) which was created in 1989 to protect the last great frontier for Mexicans in search of farmland. Following the improvement of roads in the area in the 1970s, immigration to the Calakmul area increased sharply with poor people looking for land to cultivate.

With a population of about 15,000, Calakmul consists of the core bioreserve area where settlement is prohibited, a buffer zone of 72 communities called *ejidos* and a few privately owned properties (Bosque Modelo, 1997). *Ejidos* vary in size from

Figure 1. Map of Calakmul Biosphere Reserve, Campeche, Mexico.

100 to 50,000 hectares and from 10 – 150 members with each member family having equal rights to the use of their *ejido's* communal forest and agricultural lands. The agricultural allotments vary between *ejidos,* ranging from 25 to 50 hectares, while communal forest areas vary from 250 – 25,000 hectares per *ejido*.

The flat terrain of Calakmul is punctuated by low hills with elevations ranging from 205 to 270 meters above sea level. Rainfall is unpredictable both in distribution throughout the year and in amount, typically ranging between 600 and 1600 millimeters per year, with a dry season occurring between February and May. Tropical semi-deciduous forest is the primary natural vegetation with a mosaic of high graded old forest and large areas of secondary forests of staggered aged stands. The most abundant tree species are chiclé *(Manilkara zapota*) and breadnut or ramon *(Brosimum alicastrum*) valued respectively for their latex and leaves for fodder. The most important commercial timber species are mahogany *(Swietenia macrophylla*) and Spanish cedar (*Cedrela odorata)*.

Agriculture is dominated by a slash and burn system known locally as *milpa*. Fields are cleared by cutting and burning the forest or forest fallow and then planted primarily with corn (the primary subsistence crop) often in association with beans and squash. Fields are typically cropped for 2 or 3 years and then left to fallow for anywhere from 3 – 15 years. With a typical household having 3-5 hectares of *milpa* in production each year, corn yields are highly variable from year to year and from field to field (250 kg/ha – 2.0 t/ha) (Snook, 1996).

Limitations to production are unpredictable and insufficient rainfall, shallow soils, lack of money to invest in improved production techniques, seasonal labor shortages, lack of technical expertise in agronomy and forestry, and poor access to markets. As a result, there is currently little hope for most Calakmul residents to move beyond subsistence agriculture.

Forests, however, provide an additional, important source of cash crops such as honey, timber, chiclé latex, and construction materials for home use and the local market. Until 1991, there was virtually no effort to plant timber trees outside of the forest areas in the active forestry *ejidos*[1]. In the early 1990s, however, an intercommunity organization, CRASX, (Regional Agrosilvocultural and Services Council of Xpujil) initiated the Calakmul Agroforestry Project to develop and disseminate agroforestry technologies in Calakmul to restore the agricultural and forest resource base while improving farm production and conserving forest cover. Between 1991-96, the Project provided 225 timber trees and 110 fruit trees to each farmer who agreed to plant the trees in association with agricultural crops in one hectare agroforestry plots (Snook & Zapata, 1998). Approximately 700 hectares were established with the goal of providing short, medium, and long term production starting with annual crops, followed by fruits, and finally timber[2]. Between 1995-97, another tree planting project concentrated only on native trees without the fruit tree component. The project provided, free of charge, 21 native tree species which were to be planted in individual or community managed plots, often in association with crops[3].

In 1997, the International Centre for Agroforestry Research, (ICRAF) initiated two studies to examine agroforestry adoption in Southeastern Mexico. The first used traditional *ex-post* analysis based on binary choice regressions of revealed preferences to examine the characteristics of past agroforestry adopters (Mercer, Snook, Haggar, & Sosa, in press). The revealed preference analysis found that households most likely to have previously planted trees on their farms were the more educated, more experienced, relatively wealthier households that immigrated from nearby states in the Yucatan peninsula, and who had cleared larger amounts of their forestland. The second approach, reported here, applied attribute-based choice experiments to examine how farmers value different attributes of agroforestry systems and which combinations of attributes are most likely to be adopted. The goal was to provide information to assist in the design of new agroforestry systems and projects that would be more attractive to farmers. In the ACE study, farmers were presented with a series of agroforestry systems with varying attribute combinations and asked to choose the system they preferred. They were also allowed to choose to reject all new systems and continue with their current land-use system. Details on methods, data and results are presented next.

3. METHODS

The main steps in implementing a choice-based experiment are (Bennett & Adamowicz, 2001):

1. *Survey and Experimental Design,* which consists of: i) characterizing the decision problem to be analyzed; ii) selecting and defining attributes and values for each attribute level; iii) constructing the choice tasks, alternatives or profiles to be presented to the respondent; iv) developing and pre-testing the questionnaire; and v) establishing sample frames and sizes (usually determined by the trade off between accuracy and data collection costs).
2. *Data Collection*: a variety of methods for data collection have been utilized for ACE ranging from pencil and paper direct interviews, to telephone and mail surveys, to computer aided surveys. Type of data collection method is determined by costs and appropriateness for the population being analyzed.
3. *Model Estimation*: typically multinomial logit (MNL) models are estimated with maximum likelihood procedures, although the particular issues being examined and nature of the data will determine the most appropriate estimation method.
4. *Policy/Decision Support Analysis*: analysts are usually interested in determining the relative values of different attributes of the products or land use systems being analyzed and the most desired combination of attributes (i.e., trade-off analysis in choosing between combinations with different attributes) to be used for system design and decision support tools and developing policies to promote desired systems.

3.1. Survey and experimental design

Designing a choice based survey instrument is typically a lengthy process of information gathering through key informant interviews, community meetings, focus groups, and extensive field testing. Casey, Mercer, and Snook (1999) provide a detailed discussion of the survey instrument design process used in this project. A series of focus groups with farmers, agricultural and forestry technicians and extension personnel, and local ICRAF professionals were used to develop the five attributes (each with 3-6 levels) used to describe the hypothetical agroforestry systems for the choice experiment. Following the initial round of focus groups, 17 potential attributes of agroforestry were identified. These were narrowed down to the 5 considered to be most useful to ICRAF in designing potential agroforestry systems in the area. The attributes and levels are presented in Table 1[4].

Table 1: Attributes and levels for Attribute-based Choice Experiment

Attribute	*Level 1*	*Level 2*	*Level 3*	*Level 4*	*Level 5*	*Level 6*
Extra Labor* (days year^{-1})	5 days	10 days	20 days	30	40	50
Extent of Technical Assistance	1 year	3 years	5 years	n/a	n/a	n/a
Products from System	Timber only	Timber & crops	Timber, fruit, & crops	n/a	n/a	n/a
Source of tree seedlings	Gather for free in Forest	Work in local nursery	Pay for seedlings; delivered to farm	n/a	n/a	n/a
Impact on Forest Conservation	No impact on future forest environment	Forest environment better in future for your children	Forest environment worse in future for your children	n/a	n/a	n/a

Labor variable redefined for regression as Low < 10 days year^{-1}; Medium = 20 to 30 days year^{-1}; High > 30 days year^{-1}

The final five attributes used in the analysis are:

- Number of additional days of labor per year required to implement and maintain the new system (six levels from 5 to 50 additional days of labor per year),
- Number of years that technical assistance will be available to adopters (three levels: 1, 3, or 5 years of technical assistance),
- Types of outputs produced by new system (three levels: i) timber; ii) timber plus crops; and iii) timber plus crops and fruit trees);
- Availability of tree seedlings and how obtained (three levels: gather from forest; work for seedlings in nursery; or pay for seedlings delivered to farm),
- Impacts of the system on the forest environment (three levels: no impact on forests; forest environment better in the future; forest environment worse in the future).

This 3^4 x 6^1 experimental design results in 486 possible agroforestry systems for the respondents to choose between. To produce a more tractable experiment, an orthogonal fractional factorial design (Addleman, 1962; Holmes & Adamowicz, 2003) was used to generate a subset of 64 agroforestry systems that covered the range of variability between all possible combinations. An eight level blocking factor was used to split the 64 plans into eight random blocks so that each final survey contained 8 sets of paired agroforestry systems. Hence, farmers were presented with a series of 8 separate, trichotomous choice experiments each with a pair of alternative agroforestry systems and the status quo (or opt-out) option. The status quo option is provided because the farmers might not prefer to adopt either of the alternative systems.

3.2. Data Collection

Data were collected in 1998 via in-person interviews based on a stratified random sample of farmers in the *ejido*s occupying the buffer zone of the Calakmul Biosphere Reserve. The sample was stratified by *ejido* size with the final sample consisting of 176 farmers in 15 separate *ejido*s. The ACE questions, however, were only asked of those farmers who stated that they would be interested in planting trees on their farms in the future, resulting in a final sample of 142 farmers for the ACE analysis.

Following the collection of socio-economic and household specific data, the interviewer briefly explained the attributes and levels. Then, two hypothetical agroforestry systems were presented to the respondent who was asked to pick the most preferred system (or to choose neither system). The systems were depicted with line drawings combined with written statements provided next to each picture. The combination of the line drawings and verbal descriptions provided by the interviewers ensured that non-literate respondents fully understood the choices. After studying each alternative for a few minutes, respondents were asked to pick their preferred system or "none of the above." This was repeated 8 times for the 8 choice experiments presented to each farmer.

3.3. Data Coding and Model Estimation

The choice problem outlined above required each respondent to choose one of the two agroforestry alternatives to adopt or to decide not to adopt either of the given alternatives. The respondent then repeated this process for 8 different choice sets. Therefore, each respondent provided one response for each choice set which was recorded along with the attribute levels for the two agroforestry choices and the status quo option (as well as the socio-economic data for the individual making the choice). For each respondent there are 8 x 3 = 24 data points. When the status quo attribute levels are known, coding attribute levels for the status quo (none of the above) option is usually handled like the other choice alternatives (Holmes & Adamowicz, 2003). In our case, since no information was available on the attribute levels for the status quo option, zeroes were used to code all the attribute levels for the status quo alternatives. Since we based our analysis on the multinomial logit model (MNL), this approach normalizes utility relative to the status quo option.

Therefore, for these trichotomous choice sets, three lines of data were coded for each choice set with each line representing the dependent variable (i.e., "1" if chosen and "0" if not), the attribute levels for that alternative and the socio-economic characteristics of the respondent. An alternative specific constant (ASC) for the status quo option was created taking on the value of "1" if that line of data described the status quo alternative and "0" otherwise. ASCs account for variability in choice not explained by the attributes or socio-economic variables. ASCs are especially important when an opt-out option is provided, since the attributes of the opt-out alternative are usually not known or non-existent (Holmes & Adamowicz, 2003).

When coding the qualitative or categorical attribute levels, effects codes rather than dummy codes are preferred since the attribute level for the omitted category would be collinear with the regression intercept and no information about the preferences for the omitted level could be obtained (Holmes & Adamowicz, 2003). Effects codes overcome this problem by using "1", "-1", and "0" to code the variables for the attribute levels rather than just "1" and "0" for typical dummy variables. For an attribute with three levels, one level is chosen as the base and two effects codes variables are created for the other two levels. Using the technical assistance variable in our case as an example, one year of technical assistance was chosen as the base level and two variables were coded in the data set (three years (assist3) and five years (assist5) of technical assistance). Whenever the attribute level for the choice alternative was the base (i.e., one year of technical assistance) assist3 and assist5 were coded as "-1". When three years of assistance was the level included in the choice, assist3 was coded as "1" and assist5 was coded as "0". Likewise when five years was the attribute level (assist5 was coded as "1" and assist3 as "0"). Effects codes were used for all attribute variables in this experiment. Using effects codes allows one to easily compute the parameter value for the omitted attribute by simply summing the coefficients of the other two levels of that variable (Holmes & Adamowicz, 2003).

Based on random utility theory, respondents' choices for each choice set were modeled as a sequence of three equations, each of which described the probability of choosing that alternative. The appendix provides an overview of random utility theory and its application with multi-nomial logit models for estimating the preference parameters from choice experiments. The conditional indirect utility, V, can be specified for each alternative as a linear function of the attributes (Bennett & Adamowicz, 2001). Assuming that errors are independently and identically distributed (IID) and follow a type 1 extreme value (Gumbel) distribution, a conditional or multinomial logit regression model was used to estimate the trichotomous choice responses with the levels of the attributes of the systems used as explanatory variables. All variables that remained the same across the respondent's choices (such as income or farm size) drop out of the model (Holmes & Adamowicz 2003; McFadden, 1974).

Assuming no interactions effects, each choice set of 5 attributes and 3 levels (the base level for each attribute is not included in the regression) is described with three linear in parameters models:

Alternative 1: $V_1 = \beta_1 A_1 + \beta_2 A_2 + \beta_3 A_3 + \ldots\ldots \beta_{10} A_{10}$
Alternative 2: $V_2 = \beta_1 A_1 + \beta_2 A_2 + \beta_3 A_3 + \ldots\ldots \beta_{10} A_{10}$
Status quo: $V2 = ASC + \beta_1 A_1 + \beta_2 A_2 + \beta_3 A_3 + \ldots\ldots \beta_{10} A_{10}$
where: β_i = coefficient *i* for attribute A_i
ASC = alternative specific constant for the opt-out alternative.

Since $\frac{\partial V_i}{\partial A} = \beta_i$, the regression coefficients, β_i, can be interpreted as the marginal utility of the attributes, $A_{.i}$.

The STATA (1999) maximum likelihood routine was used to estimate the resulting multinomial logit regression model (MNL) (see Appendix). Using the MNL model, socio-economic characteristics can only be introduced as interactions with either the attributes or the alternative specific constant, MNL models predict the relative attractiveness of each alternative and characteristics that don't vary across alternatives cannot be estimated. More complex models are possible, such as latent class models and random parameter models, that allow incorporation of socio-economic heterogeneity and interaction effects are possible but require experimental design and analysis that are beyond the scope of this chapter. Holmes and Adamowicz (2003) provide a detailed overview and analysis of those models.

4. RESULTS

Descriptive statistics for the entire sample are provided in Table 2. The average farmer had immigrated to Calakmul 11 years prior to the survey, was 38 years old with 4 children, and produced an average annual income of US$1,457. Farmers immigrated to Calakmul from more than 10 other states in Mexico. The education level of the farmers was very low; sixty percent had never finished primary school; only 28.98 percent had finished primary school, and only 9.8 percent had finished secondary school. The average farmer received 49 hectares of land on joining the *ejido*, 39.7 hectares of which was originally under primary forest cover and 8 hectares under secondary fallow. The average farmer had harvested 9.9 hectares of forests (about 1 hectare per year) since joining the *ejido* and, at the time of the interview, had 28 hectares under forest cover, 19 hectares under fallow, and 4.8 hectares in *milpa*.

The results from the maximum likelihood estimation of the multinomial logit model of the trichotomous choice responses of agroforestry system preference are shown in Table 3. Parameters estimates for the omitted (base case) attribute levels were computed as the sum of -1 times the parameter values for the included levels of each attribute. Although the Pseudo R^2 was 0.085, the Chi^2 value of 204.24 and significance (at the .05 or .10 level) of all but three attribute levels suggests a reasonable fit for the model. The regression coefficients can be interpreted as marginal utility values showing the rate at which the respondent's utility increases (or decreases) given a change in the attribute levels. The coefficient on the status

Table 2. Descriptive statistics for entire sample of farmers (n = 176).

Variable	*Mean*	*Standard Deviation*	*Range*
Age of farmer (years)	38	13.7	16-74
Number of children at home	4	2.9	0-12
Income (US$/year)	$1,457	$1,573	$53-8,154
Length of residence in Calakmul (years)	11	6.3	0.3-36
Distance to fields from house (km)	2.81	2.2	0-10
Total land allotted to farmer (hectares)	49	25	0-120
Area in *milpa* (hectares)	4.8	4.1	0-36
Amount of land planted with trees (hectares)	1.63	2.54	0-16
Current amount of forest in farmer's land allotment (hectares)	28	24	0-95
Original amount of forest in farmers land allotment (hectares)	39.7	26.8	0-120
Amount of forest harvested (hectares)	9.9	11.05	0-50
Current amount of fallow forest (hectares)	19	11.7	0-60
Original amount of fallow forest (hectares)	8	11	0-45

quo Alternative Specific Constant (ASC), then shows the marginal utility of the status quo relative to the agroforestry alternatives. Since the ASC is significant (5% level) and a relatively large negative value, farmers appear to strongly prefer the agroforestry alternatives to maintaining the status quo[5].

Figure 2 shows how marginal utility changes with the different levels for each attribute. As expected, the greater length of time that technical assistance is provided the higher the utility and probability of adoption. This may suggest that farmers view the systems as complicated, difficult, and/or risky to adopt without adequate assistance. Additionally, farmers may also perceive additional benefits (not necessarily related to agroforestry) from having access to technical assistance (e.g., participation in other development projects and access to general agricultural advice).

Farmers preferred agroforestry systems that produced both timber and crops over strictly forestry systems that only produced timber, reflecting their preferences for sustainable production of both wood and food products. Interestingly, the least preferred product mix was timber, crops, and fruit trees. This may be due to problems with fruit tree based agroforestry systems that were promoted beginning in 1991. The fruit trees required large amounts of weeding, and many farmers were unable to sell the fruit produced due to transportation problems and an already glutted market for oranges. Gathering seedlings for free from the forest was preferred to working in local nurseries or paying for seedlings.

Table 3. Maximum likelihood estimates of conditional logit analysis of impact of attributes on farmers' preferences for new agroforestry systems (n=142 farmers).

Variable	*Coefficient (preference weight)*	*Standard Error*	*Z-value*
Alternative Specific Constant (Status Quo)	-0.471	0.081	-5.8 [a]
Technical Assistance			
Low Level (One Year)	-0.241	------	------
Medium Level (Three Years)	0.024	0.062	0.39
High Level (5 Years)	0.217	0.075	2.87[a]
Additional Labor Required			
Low Level (5-10 days per annum)	-0.071	------	------
Medium Level (20-30 days per annum)	-0.060	0.099	-0.60
High Level (40-50 days per annum)	0.131	0.071	1.84[b]
Product Mix			
Low Level (only Timber)	-0.061	------	-----
Medium Level (Timber + crops)	0.252	0.064	3.95[a]
High Level (Timber + crops +fruit trees)	-0.191	0.062	-3.06 [a]
Source of Tree Seedlings			
Low Level (gather for free from forest)	0.296	------	------
Medium Level (work for seedlings in nursery)	-0.077	0.073	-1.06
High Level (purchase delivered seedlings)	-0.219	0.065	-3.37 [a]
Impact on Forest Environment			
Low Level (Worse in future)	-0.334	0.067	-4.96 [a]
Medium Level (No Impact)	-0.054	------	------
High Level (Better in future)	0.388	0.064	6.04[a]

Likelihood Ratio $chi^2(11) = 204.24$; Prob > chi^2 (11) = 0.000; Psuedo $R^2 = 0.085$
[a]Significant at the 5 percent level; [b]Significant at the 10 percent level

The environmental impact attribute also performed as expected, with respondents strongly preferring systems that improve future forests to those with no impact. Systems that result in a degraded future forest environment produced strongly negative reactions. Given that the average respondent had cleared almost one fourth of their forestland, this result may suggest that farmers are beginning to recognize the negative impacts of increasing deforestation and desire sustainable land-use systems that will reduce the rates and impacts of deforestation.

At first glance, the labor attributes appear to be counter intuitive with the low and medium labor levels producing negative utility values while the highest additional labor level is strongly positive and significant. However, this likely is due

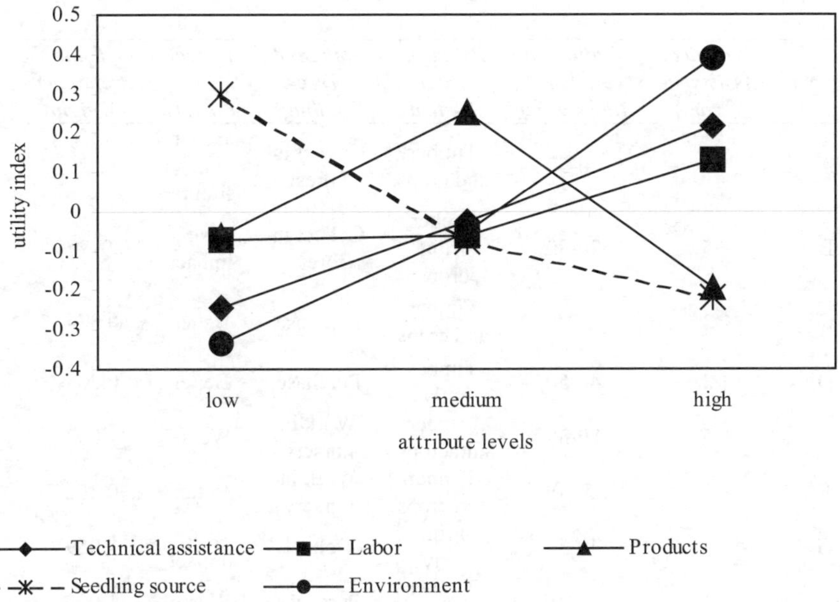

Figure 2. Marginal utilities of agroforestry system attributes.

to the farmers correlating the amount of labor input with the productivity level of the system and assuming that systems that require only a few extra days of labor a year (as in the low and medium additional labor cases) would not produce enough to be worth bothering with.

The relative impacts of the attributes on the farmers' preferences for agroforestry systems are depicted in Figure 3. Relative attribute impact was calculated by constructing a ratio where the numerator is the difference of the maximum and minimum coefficients (i.e., utility) for the levels of that attribute; the denominator of the ratio is the sum of the values in the numerator for all attributes. Surprisingly, the condition of the future forest environment had the greatest impact (31% out of 100%) on the farmers' preferences indicating the strength of farmers' concerns for future generations in their current decision-making. Nearly equal in importance were the source of seedlings (22%) (which may reflect past problems tree planting programs had with the timing of seedling delivery to farmers), technical assistance (20%), and the product mix (19%). The amount of additional labor required for the system (9%) was the least important factor indicating perhaps the existence of

Table 4. Relative agroforestry systems/projects desirability with different attribute combinations.

System	*Technical Assistance (years)*	*Additional Labor (days/year)*	*Products from System*	*Source of Tree Seedlings*	*Impact on Forests*	*Total Preference Weight*	*Rank*
A	5	40-50	Timber and crops	Collect in Forest	Better in Future	1.28	1
B	5	20-30	Timber, fruit, crops	Collect in forest	No impact	0.0127	5
C	3	5-10	Timber and crops	Purchase	Better	0.374	2
D	3	40-50	Timber only	Purchase	Better	0.263	3
E	5	20-30	Timber and crops	Work in nursery	Worse	-.002	6
F	1	5-10	Timber and crops	Work in nursery	Better	0.262	4
G	1	40-50	Fruit only	Collect	No impact	-.059	7
H	1	5-10	Timber only	Purchase	Worse	-1.056	8

excess labor at various times during the year. Labor's relatively low importance reflects the fact that the opportunity costs of labor are low during much of the year, and therefore, labor is not seen as a particularly important constraint.

ACE results also allow one to determine which combinations of the attributes of the agroforestry system and/or project would be most desirable to the farmers. This is accomplished by simply summing the parameter estimates for alternative combinations of attributes (as in Table 4) to determine the total preference weight for that system by the average respondent. The system/project with the highest total preference weight is the most preferred. For the current study the most preferred system is system A with a total preference weight of 1.28. System A would provide 5 years of technical assistance, 40-50 days of additional labor, timber plus annual crops, seedlings gathered from the forest and a better forest environment in the future. The least preferred system is system H (-1.056 total preference weight), which would provide only 1 year of technical assistance, require 5-10 days of additional labor annually, produce only timber, require seedlings to be purchased, and result in a poorer forest environment in the future. Systems B-G, which were designated by members of the ICRAF project staff, range from being perceived negatively overall (E and G) to intermediate but positive preference scores for systems B, C, D, and F.

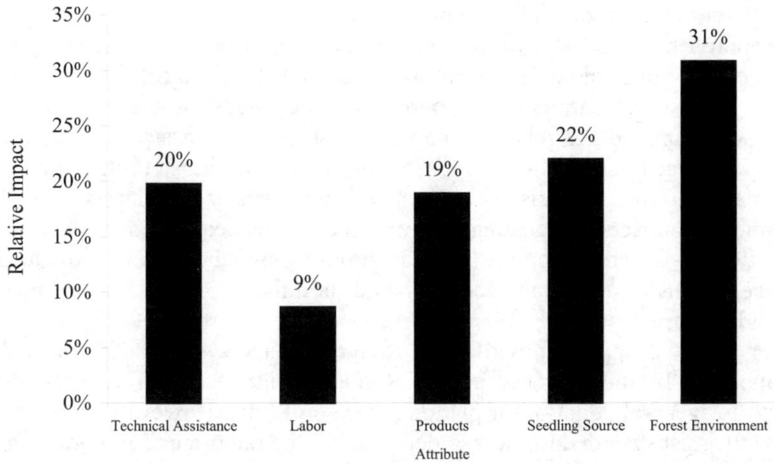

Figure 3. Relative impact of attributes on farmer preferences for new agroforestry system.

5. SUMMARY AND CONCLUSIONS

Achieving the full potential of agroforestry to contribute to sustainable land use requires improving adoption rates. No matter how elegantly designed, efficient, productive or ecologically sustainable, if the system is not adopted by a significant proportion of the target population or communities, impacts are likely to be minor. Realization that adoption rates were lagging behind the science of agroforestry has led many to call for increased research in farmer decision-making with respect to adoption of new agroforestry innovations (Adesina & Chianu, 2002; Alavalapati, Luckert, & Gill, 1995; Bannister & Nair, 2003; Lapar & Pandey, 1999; Nair, 1996; Sanchez, 1995; Thacher, Lee, & Schellas, 1997). In response, an explosion of research in agroforestry adoption began in the mid-1990's, most of which examined *ex-post* adoption decisions (Mercer, in press). Although significant progress has been made in analyzing agroforestry adoption potential prior to implementation of agroforestry extension projects (Current et al., 1995; Barrett et al., 2002; Franzel & Scherr, 2002), development of rigorous, statistically based methods for analyzing potential demand is needed. In this chapter, we present one alternative, attribute-based choice experiments (ACE).

ACE studies derive from the long history of applying conjoint analysis by market researchers to the problem of developing new, multi-attribute goods and services that will be demanded or adopted by consumers, a similar problem to designing new multi-attribute agroforestry systems and projects that will be adopted by farmers. ACE improves on traditional ranking and rating conjoint by being firmly grounded in economic and behavioral theory and by examining respondent preferences in a context that is familiar (i.e., choosing to buy different products or to adopt different farm production systems).

The approach is illustrated with a simple case study from southeast Mexico as an example of how one might approach applying ACE to an agroforestry context of primarily subsistence farmers. Responding to the needs of the research client, ICRAF, we examined the relative importance of technical assistance, labor input, products produced, source of tree seedlings and the impact on future forest environments. These subsistence farmers put heavy emphasis on future environmental impacts, suggesting a strong motive to bequeath a better world to their children. Other important considerations were the amount of technical assistance, means of acquiring seedlings, and the mix of products between timber, crops, and fruit trees.

Other potential applications of ACE to agroforestry system and project design are numerous. In the current case, ICRAF was interested in broadly defined attributes for general adoption of planting trees on farms. However, ACE has large potential to assist system designers in determining the importance and preference for such attributes as: the arrangement of alleys and crops in alley cropping; determining alternative tree and crop species mixes; the relative importance of environmental protection (e.g., erosion control) versus income generation; the distribution of income over time; and the impact of relative risk on adoption decisions. In addition, carefully designed ACE studies could provide quantitative analysis of the potential impact and relative importance of various policy incentives such as: provision of credits; markets, inputs such as seeds or tree seedlings; technical assistance and education; and risk reducing policies like price supports.

6. APPENDIX: APPLYING RANDOM UTILITY AND MULTINOMIAL LOGIT MODELS TO ACE ESTIMATION

The theoretical foundation for the empirical models used to analyze attribute-based choice experiments (ACE) is based on the theory of random utility maximization (RUM). The following explication of applying random utility theory to choice experiments is derived from Holmes and Adamowicz (2003). The basic assumption of RUM is that the true but unobservable utility of a good or service j is composed of both systematic (v) and random components (ε) as in equation 1:

$$U_j = v(x_j, p_j; \beta) + \varepsilon_j \tag{1}$$

where x_j is a vector of attributes of j; p_j is the cost of j; β is the vector of parameters; and ε_j is the random error term with a mean of zero. Whereas respondents know with certainty their choice behaviors, the researcher's knowledge is stochastic since it is based only on the behavior observed during the choice experiment. This uncertainty is modeled with the error term, ε_j. Assuming utility is linear in parameters, the estimation equation for (1) is:

$$U_j = \sum_{k=1}^{k} \beta_k x_{jk} + \beta_p p_j + \varepsilon_j \tag{2}$$

Marginal utilities from equation 2 are the derivatives of U with respect to the βs and the marginal rate of substitution between any two parameters is the ratio of the respective βs. The estimated coefficients for the various attributes. If a price variable *(p)* is included, the marginal rate of substitution between any attribute and the price variable (β_k / β_p) can be interpreted as the marginal value or implicit price of that attribute.

From equation 1, the probability that a respondent will choose alternative i from choice set C is expressed as:

$$P(i \mid C) = P(U_i > U_j) = P(v_i + \varepsilon_i > v_j + \varepsilon_j) \forall j \in C \tag{3}$$

Assuming that errors are independently and identically distributed (IID) and follow a type 1 extreme value (Gumbel) distribution), equation 3 can be re-arranged to show that :

$$P(i \mid C) = P(v_i - v_j > \varepsilon_i - \varepsilon_j) \forall j \in C \tag{4}$$

Equation 4 shows that all variables that are constant across alternatives (for example, individual respondent characteristics such as income, household size, education, etc.) drop out of the model. Assuming that preferences are EV1 distributed (via the unobserved variables) and choices are independent from irrelevant alternatives (IIA assumption), then the multinomial (or conditional) logit model can be applied and the probability of choosing alternative i with scale parameter μ is:

$$P(i \mid C) = \frac{\exp(\mu v_i)}{\sum_{j \in C} \exp(\mu v_j)} \tag{5}$$

If utility is assumed to be additively separable and $\mu = 1$, the probability of choosing alternative i from choice set, C, is:

$$P(i \mid C) = \frac{\exp(\sum_{k=1}^{j} \beta_k x_{ik} + \beta_p p_j)}{\sum_{j \in C} \exp(\beta_k x_{ik} + \beta_p p_j)} \tag{6}$$

Assuming a sample size of N defining y_{in} as (1) if the respondent chooses alternative i and (0) otherwise, and the MNL likelihood function can be expressed as:

$$L = \prod_{n=1}^{N} \prod_{i \in C} P_n(i)^{y_{in}} \tag{7}$$

To estimate the MNL model and determine the values of the βs one substitutes (6) into (7) and maximizes the resulting log likelihood function (equation 8):

$$\ln L = \sum_{n=1}^{N} \sum_{j \in C} y_{in} (\sum_{k=1}^{I} \beta_k x_{ikn} + \beta_p p_{in} - \ln \sum_{j \in C} (\sum_{k=1}^{I} \beta_k x_{xkn} + \beta_p p_{jn}) \tag{8}$$

7. NOTES

[1] By law, active forestry *ejidos* are required to replace trees harvested for timber by enrichment plantings of mahogany and cedar under the forest canopy. Tree survival and growth, however, in the enrichment plantings have been very low (Sosa, 1997).

[2] Personal communication: Acopa, Miguel, Head of the Calakmul Agroforestry Project 1991- 1996.

[3] Personal communication: Mex, G., Coordinator of Bosque Modelo para Calakmul and Uc, C. ICRAF-Mexco.

[4]Production levels were initially considered as one of the most important attributes. During pre-testing, however, it became apparent that respondents were focusing solely on that attribute in their choices. So, it was dropped from the final design to enable analysis of other important attributes, realizing that achieving maximum production and income generation was the number one priority for all systems

[5] This result is not unexpected since only farmers indicating an interest in agroforestry were given the ACE task.

8. REFERENCES

Adamowicz, W., Louviere, J., & Williams, M. (1994). Combining revealed and stated preference methods for valuing environmental amenities. *Journal of Environmental Economics and Management, 26*, 271-292.

Adesina, A. A., & Chianu, J. (2002). Determinants of farmers' adoption and adaptation of alley farming technology in Nigeria. *Agroforestry Systems, 55*, 99-112.

Addleman, S. (1962). Orthogonal main-effect plans for asymmetrical factorial experiments. *Technometrics, 4*(1), 21-46.

Alavalapati, J., Luckert, M., & Gill, D. (1995). Adoption of agroforestry practices: a case study from Andhra Pradesh, India. *Agroforestry Systems, 32,* 1-14.

Bannister, M. E., & Nair, P. K. R. (2003). Agroforestry adoption in Haiti: the importance of household and farm characteristics. *Agroforestry Systems, 57*, 149-157.

Barrett, C. B., Place F., & Aboudk, A. (2002). *Natural Resource Management in African Agriculture: Understanding and Improving Current Practices.* Wallingford, United Kingdom: CABI Publishing.

Baidu-Forson, J., Waliyar, F., & Ntare, B. R. (1997). Farmer preferences for socio-economic and technical interventions in groundnut production system in Niger: Conjoint and ordered probit analyses. *Agricultural Systems, 54*(4) 463-476.

Bennett, J. & Adamowicz, V. (2001). Some fundamentals of environmental choice modeling. In J. Bennett& R. Blamey (Eds.), *The Choice Modeling Approach to Environmental Valuation* (pp. 37-69). Cheltenham, United Kingdom: Edward Elgar.

Bennett, J., & Blamey, R. (Eds.). (2001a). *The Choice Modeling Approach to Environmental Valuation*. Cheltenham, United Kingdom: Edward Elgar.

Bennett, J., & Blamey, R. (2001b). Introduction. In J. Bennett & R. Blamey (Eds.), *The Choice Modeling Approach to Environmental Valuation* (pp. 1-10). Cheltenham, United Kingdom: Edward Elgar.

Bosque Modelo. (1997). Visión Estratégica 1997-2000. Bosque Modelo para Calakmul, Ecología Productiva.

Casey, J., Mercer, E., & Snook, A. (1999). Evaluating farmer preferences for agroforestry systems: survey instrument design. In *Proceedings of the 1998 Southern Forest Economics Workshop* (pp 153-158). Williamsburg, VA.

Current, D., Lutz, E., & Scherr, S. (1995). Costs, Benefits and Farmer Adoption of Agroforestry: Project Experience in Central America and the Caribbean. World Bank Environment Paper Number 14, pp 1-27. Washington, DC: The World Bank.

Franzel, S., & Scherr, S. J. (2002). *Trees on the Farm: Assessing the Adoption Potential of Agroforestry Practices in Africa*. Wallingford, United Kingdom: CAB International.

Green, P.E., & Wind, Y. (1975). New way to measure consumers' judgments. *Harvard Business Review, 53,* 107-111.

Holmes, T., Zinkhan, C., Alger, K., & Mercer, E. (1998). The effect of response time on conjoint analysis: estimates of rain forest protection values. *Journal of Forest Economics, 4*(1), 7-28.

Holmes, T.P., & Adamowicz, W.L. (2003). Attribute-based methods. In P. Champ, K. Boyle, & T. Brown (Eds.) *A Primer on Non-market Valuation* (pp 171-219). Dordrecht, The Netherlands: Kluwer Academic Publishers.

Lapar, M. L., & Pandey, S. (1999). Adoption of soil conservation: the case of the Philippines uplands. *Agricultural Economics, 21*, 241-256.

Louviere, J. J. (1988). Conjoint analysis modeling of stated preferences. *Journal of Transport Economics and Policy, 22*(1), 93-119.

Louviere, J. J. (1994). Conjoint Analysis. In R. Bagozzi (Ed.), *Handbook of Marketing Research* (pp. 223-259). Oxford: Blackwell Publishers.

Louviere, J. J. (2001). Choice Experiments: An overview of concepts and issues. In J. Bennett & R. Blamey (Eds.), *The Choice Modeling Approach to Environmental Valuation* (pp 13-36). Cheltenham, United Kingdom: Edward Elgar.

Mercer, E. (in press). Adoption of agroforestry innovations in the tropics: A review. *Agroforestry Systems*.

Mercer, E., & Miller, R. (1998). Socio-economic research in agroforestry: progress, prospects, priorities. *Agroforestry Systems, 38,* 177-193.

Mercer, E., Snook, A., Haggar, J., & Sosa, M. (in press). Agroforestry adoption in the Calakmul Biosphere Reserve, Campeche, Mexico. *Journal of Crop Production*.

McFadden, D. (1974). Conditional logit analysis of qualitative choice behavior. In P. Zarembka (Ed.) *Frontiers in Econometrics* (pp. 105-142). New York: Academic Press.

Nair, P. K. R. (1996). Agroforestry directions and trends. In P. McDonald & J. Lassoie (Eds.), *The Literature of Forestry and Agroforestry* (pp. 74-95). Ithaca, New York: Cornell University Press.

Opaluch, J. J., Swallow, S. K., Weaver, T., Wessells, C. W., & Wichelns, D. (1993). Evaluating impacts from noxious facilities: Including public preferences in current siting mechanisms. *Journal of Environmental Economics and Management, 24*, 41-59.

Pattanayak, S., Mercer, D. E., Sills, E., & Yang, J. (2003). Taking stock of agroforestry adoption studies. *Agroforestry Systems, 57*,173-186.

Sanchez, P. A. (1995). Science in Agroforestry. *Agroforestry Systems,* 30, 5-55.

Snook, A. (1996). *Annual Report*. International Centre for Research in Agroforestry-Mexico. Chetumal, Quintana Roo, Mexico.

Snook, A., & Zapata, G. (1998). Tree Cultivation in Calakmul, Mexico: Alternatives for reforestation. *Agroforestry Today, 10*(1), 15-18.

Sosa, L. (1997). Evaluación de Plantaciones de Cedro (*Cedrela odorata* L.) y Caoba (*Swietenia macrophylla* King) en Sistemas Agroforestales en la Area de Bosque Modelo , Calakmul, Campeche.

Unpublished masters thesis. División de Ciencias Forestales, Universidad Autónoma Chapingo, Mexico.
STATA. (1999). Stata Reference Manual: Release 6. Volume 1 A-G. College Station, Texas: Stata Press.
Thacher, T., Lee, D., & Schellas, J. (1997). Farmer Participation in reforestation incentive programs in Costa Rica. *Agroforestry Systems, 35*, 269-289.
Zinkhan, F. C., Holmes, T. P., & Mercer, D. E. (1997). Conjoint Analysis: a preference-based approach for the accounting of multiple benefits in southern forest management. *Southern Journal of Applied Forestry, 21*(4), 1-7.

9. AUTHOR'S NOTE

Funding for this research was provided by the USDA Foreign Agricultural Service, Research and Scientific Exchanges Division, the USDA Forest Service Southern Research Station, and the International Centre for Research in Agroforestry. Special thanks are extended to Jim Casey, Carlos Uc, Mauricio Sosa and the farmers of Calakamul for all their support in the data collection process.

RUNSHENG YIN

VALUING THE IMPACTS OF AGROFORESTRY IN NORTHERN CHINA

1. INTRODUCTION

Based on the recent experience of northern China, this chapter attempts to quantify the impact of agroforestry on agricultural productivity at the regional level, and investigate the spatial and temporal relationships between trees and annual crops at the plot level. The chapter will offer empirical answers to the questions of under what circumstances agroforestry can flourish and how it will benefit farm households. It is believed that these analyses will be of broad international interest.

Although China has a rich tradition in agroforestry extending over thousands of years, its current experience is a matter of the past twenty-five years (Yin & Hyde, 2000; Zhu, Cai, Wang, & Yiang, 1991). In the northern plains – a major agricultural region covering Hebei, Henan, Shandong, Beijing, Tianjin, and parts of Anhui, Jiangsu, Shanxi, and Shaanxi, few trees, let alone forests, survived the destruction of wars and exploitation that preceded the founding of the People's Republic in 1949 (Yin, 1994). Average grain production was approximately 800 kilograms/hectare/year (kg/ha/yr), and many areas of farmland were abandoned due to desertification, salinization, course changes in the Yellow River, and other environmental stresses (Henan Statistics Bureau, 1986; Shandong Statistics Bureau, 1988). Living conditions were miserable. Even in the 1990s, 243 million people lived on a land base of 464,000 square kilometers, of which only 26 million hectares was arable farmland – or about 0.08 ha per capita (Zhong, Xian, & Li, 1991). Furthermore, the arable land base was declining at an annual rate of 0.8 percent while the region's population was growing at a rate of 1.5 percent annually (China Agriculture Yearbooks, 1985-1992).

The government directed attention to land rehabilitation in the 1950s. Its initial focus was on engineering and biological measures, including afforestation projects for windbreaks, preventing sand erosion, and controlling salinization. The very limited new resources were quickly depleted, however, after the collectivization and industrialization movements initiated in 1958. Collectivization raised doubts about ownership security for the forest resource, and backyard furnaces for steel

J. R. R. Alavalapati & D. E. Mercer, Valuing Agroforestry Systems, 257–277.

production consumed most of the remaining wood for fuel (China Forestry Yearbook, 1986-1988).

The economic adjustments of the early 1960s promoted afforestation activities like shelterbelts and intercropping for the purposes of protecting the environment and improving agricultural productivity. Unfortunately, the Cultural Revolution and the political struggles that dominated Chinese society for the next decade restricted any real opportunity for environmental conservation and economic growth. Both agricultural and forest production stagnated. Per capita annual income in the communes was only 63 Yuan in 1976 (about US $25), more than one third of rural households were in debt, and many of them lived with insufficient fuel, clothing, and housing.

Following the Cultural Revolution, economic development became the policy priority. Beginning in 1978, the land tenure reform transferred land use rights from the collectives to individual households, which is commonly called the household responsibility system (HRS). Rural households were also given the right to sell their production in excess of assigned quotas at market prices, although quotas were still delivered at planned, often low, prices. In addition to improving incentives and thus productivity, these measures served as means for government officials to reduce the transactions costs and risks of centralized management. The reforms spread rapidly. They received the official sanction of the central government in September 1980, and by 1984 the HRS had expanded to 70% of all rural communities. As a result, agriculture has expanded at an unprecedented pace ever since. Grain yields in the northern plains surpassed 6,000 kg/ha/yr by the late 1980s, and this region is now among the most productive agricultural regions in the world (China Agricultural Yearbook, 1988).

Notably, China's agriculture has witnessed phenomenal technological adjustments in the past several decades, including the expansion of irrigation systems, the application of chemical fertilizers and pesticides, the introduction of improved seeds, and the use of plastic sheeting to concentrate heat and moisture. These technologies have their limitations, however, and several problems have emerged, including a decline in the upper layers of groundwater, a decrease in soil organic content, increasing crop vulnerability to pests and diseases, and crop susceptibility to desiccating early summer winds. Maintaining a healthy environment conducive to steady increases in productivity appears to be beyond the reach of modern technology alone.

Growing trees can provide farm households with timber, fuel, and other products. Properly distributed trees can also be a crucial complementary technology for sustaining agricultural production. Trees reduce wind velocity and modify solar radiation, thereby regulating air and soil temperatures and increasing field moisture. Trees also improve soil nutrition and control erosion. Undeniably, they do compete with annual crops for light and water, and there are limits to the density of tree cover before crop production begins to decline. Nevertheless, China's northern plains were largely deforested by 1970 and the overall potential for trees to contribute to agricultural production was probably positive.

In fact, China's agricultural regions in general, and the northern plains in particular, witnessed an upsurge in agroforestry tree plantings following the introduction of rural reforms in 1978. Forest cover in the northern plains increased from about five percent in 1977 to 13 percent before the turn of the century, thereby greatly easing local shortages of fuelwood and small construction timber, and substantially improving the environment for agriculture (State Forestry Administration, 2000; Zhong et al., 1991). Table 1 shows that by 1988 the region possessed 14.66 million hectares of forest shelterbelts, 4.5 million hectares of farmland intercropped with trees, and 1.7 million hectares of small woodlots, along with more than 5.7 billion trees planted around the "four sides" – villages, houses, roads, and waterways (Ministry of Forestry, 1988).

The impacts of agroforestry are reflected, and thus can be examined, at different levels in different ways. At the regional level, one of the major questions is whether and how the expansion of agroforestry has affected agricultural production. At the plot level, a more relevant question is how trees and crops interact over time and space. Obviously, these questions are closely related; only by clearly answering both of them can the full impact of agroforestry be appreciated. Hence, the objectives of this chapter are to: (1) estimate the effect of agroforestry on agricultural productivity, and (2) assess the spatial and temporal relationships between trees and annual crops. It should be noted that, while these analyses carry great policy significance, they have rarely been done empirically. Given today's worldwide adoption of agroforestry as a means of sustainable rural development, such analyses are in urgent need.

Table 1: Forest resources in selected areas of northern China (1988).

Province	*Area (1,000 ha)*			*Trees along four sides*[c]	*Standing timber volume*[c]	*% Total land area*
	Shelter-belts[1]	*Inter-cropping*[a]	*Wood-lots*[b]			
Henan	2,200	1,987	223.3	1.13	34.24	12.1
Shandong	3,800	1,335	400.0	1.28	35.05	7.4
Jiangsu	2,000	300	482.0	1.20	25.00	8.2
Anhui	1,434	40	276.7	1.15	19.14	13.9
Hebei	3,286	47	208.7	0.88	29.00	9.7
Shanxi	976	700	108.0	0.04	17.00	15.0
Shaanxi	967	90	300.0	0.05	7.96	7.2
Total	14,663	4,499	1,698.7	5.73	167.39	

Source: Afforestation Bureau (Ministry of Forestry, 1988)

[a]. *Intercropping refers to agricultural and tree crops grown intermixed in the same field. The area reported is the intercropped area under tree cover. Woodlots, of course, refer to small, forested areas for fuelwood, construction wood, and perhaps other forest products.*

[b]. *Measured in billions of trees. "Four sides" trees are trees planted around villages and farmhouses, and along roads and waterways.*

[c]. *Measured in cubic meters.*

Below, the paper contains two main parts before the summary section. The first part is devoted to quantifying the effect of agroforestry on agricultural productivity, and the second part is used to examine the spatial and temporal relationships between trees and annual crops in a typical intercropping system.

2. THE EFFECT OF AGROFORESTRY ON AGRICULTURAL PRODUCTIVITY

This part, which tests the hypothesis that agroforestry can contribute to the growth of agricultural productivity, is organized as follows. First, the analytic model is described. Next, the data and variables are discussed. Then, the empirical results are presented.

2.1. Analytical approach

The effects of newly planted trees on the agricultural environment can be evaluated by examining an agricultural production function with terms describing the important policy reforms and the level of agroforestry activity as production shifters. The interpretation is that agricultural production is a function of various standard agricultural inputs: land, labor, and capital of various sorts. Policies like improved tenure provided the incentive to use all inputs more efficiently, and thus production increased. Improved tenure also induced farmers to plant trees in order to improve their agricultural environment. The environmental improvement affected the overall conditions in which agricultural activity operated and, thereby, also acted to improve agricultural productivity.

The above hypothesis is accepted if the estimated parameters on the production shifters are significantly positive. Then, it can be inferred that improved tenure did provide the anticipated incentive for productivity growth, and, as it did, it allowed a longer-term management perspective. Meanwhile, farmers also responded by planting trees to provide environmental services that complemented agricultural productivity.

The basic agricultural production function takes the form

$$\begin{aligned} ln\, Y_{it} &= \alpha + \beta_1\, ln\, A_{it} + \beta_2\, ln\, L_{it} + \beta_3\, ln\, K_{it} + \beta_4\, ln\, F_{it} \\ &\quad + \gamma_1 HP_{it} + \gamma_2 CM_{(it-1)} + \gamma_3 ES_{it} + \gamma_4 T + \sum_{i=2}^{5} \delta_i D_i + \varepsilon_{it} \end{aligned} \tag{1}$$

where *Y* is aggregate production; the log terms on the right hand side are the inputs to agricultural production: *A* is land, *L* is labor, *K* is capital, and *F* is fertilizer; α is a constant, the βs, γs, and δ are parameters for estimation; the subscripts *i* and *t* refer to regional distinctions within the sample population and the data year, respectively; and ε is an independently distributed error term with zero mean.

The second line of the equation describes the production shifters. Two reforms may have had tremendous effects on China's agricultural performance – the tenure shift from collectives to the HRS and market reform. The proportion of farmland converted to HRS (*HP*) measures the tenure reform. The ratio of crop to manufactured input prices (*CM*) is an indication of the degree of market reform that gradually made agriculture a relatively more attractive economic activity. *ES* is the measure of the environmental services provided by agroforestry activities.

HP, CM, and *ES* are exogenous to the production function since it is anticipated that they improved the entire agricultural environment, including the conditions under which all other agricultural inputs (land, labor, equipment, seed, and fertilizer) were used. This formulation is equivalent to identifying agroforestry as a technological innovation from the perspective of land management (Hayami & Ruttan, 1985).

Finally, the *D*s are cross-sectional dummies (one for each of the sample units). They eliminate the effects of localized variations in factors like soil quality and the extent of irrigation. A time trend *T* captures the effects of other unidentified periodic production shifters. China's generally unreliable input price data prevent the use of a pure cost function, and the limited number of observations prohibited the use of a more flexible functional form. These limitations are not a problem, however. Robust Cobb-Douglas results should be sufficient to test the HRS and agroforestry propositions.

2.1.1. Data and variables

The data used in this analysis are from five representative prefectures in Shandong, the largest and most diverse province in the northern plains, for the thirteen-year period from 1978 to 1990 – a period long enough to capture most of the effects of the changes in rural reforms. Thus, the pooled dataset contains 65 observations. The value of aggregate agricultural output is the sum of the annual physical products of grain, cotton, and oilseed, which accounted for over ninety percent of total regional agricultural production, multiplied by their official prices in 1980. This means that the 1980 prices are the common denominators for aggregation. Official prices were the only received prices in 1978, and official procurement and market prices were still not greatly varied by 1990, the final period for the analysis.

The measures of agricultural inputs are total sown area for *A*, total farm labor force for *L*, and agricultural capital *K*, and chemical fertilizer *F*. Sown area is a better measure than cultivated area because sown area captures the effect of multiple cropping. The farm labor is the total rural labor force minus those laborers engaged in the non-agricultural sectors. Agricultural capital is measured with the total horsepower (hp) of farm machinery, plus draft animals counted at 0.7 hp per animal. This may be an overestimate because many farm machines were employed in other activities. As an alternative, draft animals alone as the measure of capital will also be examined. The measure of chemical fertilizer is simply the gross weight of fertilizers consumed. Irrigation is another important capital input, but independent data were unavailable. Irrigation machinery, however, is part of the farm machinery that is captured in the measure of capital. So, the coefficient for capital also reflects the impact of irrigation.

Tree cover (in hectares) is a reasonable proxy for the environmental services provided by trees so long as the trees are well established and well distributed in the immediate areas of agricultural production itself. This proxy and the analysis might be refined further by introducing separate agroforestry shifters: a) for trees intercropped with agricultural crops, and b) for trees in shelterbelts. In each case the proxy would be the ratio of area in the respective agroforestry category to total farm area.

Shandong's Statistics Bureau assisted in compiling the agricultural data; likewise, Shandong's Forest Bureau provided the data for intercropped and shelterbelt areas from the forest surveys of 1977, 1982, and 1988. The forest inventory data for intermediate years are interpolations. Table 2 summarizes the full dataset in index form (for the entire province), with 1980 being the base year for variables other than *HP* and *CM*.

Five mechanisms can be identified, through which trees contribute to the agricultural environment: i) reducing wind velocity and wind erosion; ii) controlling sheet and rill erosion; iii) mediating solar radiation and regulating soil and air temperatures; iv) increasing field moisture; and v) improving soil nutrients. It is expected that farmers in the northern plains were well aware of these potential benefits, where declining upper layers of groundwater, desiccating winds and

Table 2. Indices of all variables, 1978-1990.

Year	*Output*	*Land*	*Labor*	*Capital*	*Fertilizer*	*Tenure reform*	*Market reform*	*Agro-forestry*
1978	87.18	101.34	91.88	84.02	71.58	0.00	79.62	17.68
1979	90.12	100.60	96.47	93.46	81.56	0.01	82.05	19.74
1980	100.00	100.00	100.00	100.00	100.00	0.12	100.00	22.78
1981	120.82	97.89	102.80	109.87	119.73	0.40	99.81	25.75
1982	148.31	97.39	103.32	121.73	142.72	0.67	105.61	28.96
1983	175.57	99.84	103.96	145.43	167.60	0.83	104.91	32.25
1984	199.92	102.84	105.08	158.53	171.10	0.95	105.21	35.01
1985	179.99	105.16	104.29	177.81	179.30	0.99	102.29	38.27
1986	180.72	107.25	104.40	196.65	177.04	0.99	96.36	42.10
1987	195.76	106.32	105.38	216.05	176.92	0.99	103.84	46.14
1988	182.83	106.60	107.61	240.89	194.30	0.99	106.99	46.88
1989	192.31	105.72	109.44	245.22	201.98	0.99	114.50	47.29
1990	195.33	99.15	104.16	248.75	215.55	0.99	153.54	47.06

Agroforestry is the index for combined shelterbelt and intercropped land shares. The agroforestry index excludes the smaller woodlots and four-sides trees, which are not planted for their effect on agricultural productivity.

decreasing soil organic matter were particular problems (Zhu et al., 1991). Intercropped trees are not planted densely enough to form strong wind barriers, but intercropping is particularly important for its contributions to soil temperatures and field moisture, and to soil organic matter. Separating the agroforestry shifter into two independent variables will make it possible to gauge the specific production gains from either shelterbelts or intercropping, and the wind, organic matter, temperature, and moisture effects associated with each.

The objectives of woodlots and "four sides" trees are altogether different and the area they cover is much smaller than the area covered by shelterbelts and intercropped products (Table 1). Woodlots (which are small forest plantations) are primarily sources of construction timber and fuelwood although they may help control wind erosion, and some woodlots are designed to protect riverbanks and control floods as well. "Four sides" trees are planted to improve the ambient environment in areas of heavy human activity, such as the vicinity of home sites. Therefore, the functions of woodlots and "four sides" trees are not so clearly complementary with agriculture. The local forestry agencies recognize the differences and compile data on woodlots and "four sides" trees in a different form from the data for shelterbelts and intercropping. For these reasons, woodlots and "four sides" trees were excluded from the analysis.

One further consideration is the lag between the initial establishment of tree seedlings and any effect of forest cover on the environment. This lag may be short for both intercropped trees and shelterbelts because the environment was so severely depleted, and because farmers in the northern plains plant fast-growing trees. Therefore, it is anticipated that the impacts of young trees probably accumulate rapidly, but the full impacts of various trees may increase well beyond the thirteen-year period covered by the dataset.

2.2. Empirical results

Table 3 reports the regression results. They were corrected for heteroscedasticity due to different prefecture sizes by dividing all input and output variables by the area of cultivated land in the prefecture in 1980. Otherwise, the two regressions are fixed effects, or time demeaned, models (Wooldridge, 2000). While autocorrelation in the error terms was detected, correcting for it produced little improvement. After testing for various possible lags between the initial observations of trees on the land and their first effects on agricultural productivity, one-year lags proved the best fit.

The first two columns of results in Table 3 report the aggregate production function with the two different definitions of capital – farm machines and draft animals combined, and draft animals only. The third column eliminates the physical inputs and focuses on the production shifters as a means of confirming the agroforestry findings from the production functions. In general, the statistical fit, as indicated by the R^2, is satisfying for all three regressions, the coefficients on the agroforestry variables are positive, and the coefficient on intercropping is always

Table 3. Coefficient estimates for agricultural production.

Variable	Full Capital		Draft Animals as Only Capital		Focus on Production Shift Variables	
	P-value	*(t-stat)*	*P-value*	*(t-stat)*	*P-value*	*(t-stat)*
Intercept	0.361	(0.660)	*2.618	(3.829)	*2.456	(13.333)
Agroforestry (*ES*)						
Shelterbelt	0.005	(0.767)	0.001	(0.206)	0.010	(1.204)
Intercropping	*0.007	(2.519)	*0.006	(2.041)	*0.007	(2.033)
Tenure reform (*HP*)	*0.361	(4.225)	*0.551	(6.368)	*0.663	(7.909)
Market reform (*CM*)	0.0002	(0.336)	-0.0001	(-0.277)	0.0001	(0.083)
Capital (*K*)	*0.536	(4.588)	*0.346	(5.642)		
Fertilizer (*F*)	*0.435	(4.212)	*0.311	(2.966)		
Land (*A*)	-0.382	(-1.502)	-0.043	(-0.195)		
Labor (*L*)	-0.064	(-0.189)	-0.172	(-0.549)		
Time trend (*T*)	*-0.065	(-4.525)	*-0.035	(-3.405)		
Prefecture dummies (*D*)	*-0.250	(-2.983)	*-0.510	(-6.223)	*-0.467	(-7.595)
	-0.046	(-0.277)	-0.290	(-1.884)	-0.048	(-0.397)
	*0.214	(2.110)	-0.082	(-0.979)	-0.172	(-1.624)
	-0.118	(-1.048)	*-0.508	(-4.724)	*-0.625	(-5.279)
Degrees of freedom		51		51		55
Adj. R^2		0.958		0.964		0.914

** Indicates significance at the 0.01 level.*

significant. The coefficients on most of the remaining variables also meet expectations, and many of them are significant.

The coefficient on tenure reform (0.34~0.66) is large and significant, while the coefficient on market reform (0.006~0.007) is small and insignificant. These results are consistent with the findings of McMillan, Whalley, and Zhu (1989) and Wen (1993) for China as a whole. They provide specific confirmation for Shandong and the northern plains that conversion to HRS is the primary source of growth in agricultural productivity. The much smaller market reform effect is not surprising. The data in Table 2 show that in most years the changes in this index were too small to make a significant contribution to agricultural revenues.

The coefficients on the two measures of capital (0.35 and 0.54, see Table 3) are not significantly different from each other, and the choice between them does not alter either the agroforestry results or the overall statistical reliability of the regressions. This probably confirms the observation that farmers with more machinery also tend to possess more draft animals, an observation that would be reasonable if draft animals were the major source of field traction and equipment like small tractors was used for other purposes. The coefficient on fertilizer (0.31~0.44) is also positive and significant.

The coefficients on land (-0.04~-0.38) and labor (-0.06~-0.17) are negative but insignificant in Table 3. As revealed in Table 2, there was relatively little variation in the data for the land variable. In fact, land area declined slightly while output expanded rapidly. Therefore, it is not surprising that land is statistically uncorrelated with output. Labor is a different story. Population in the northern plains has been growing at a 1.5 percent annual rate – even while the agricultural land base has been declining. Table 2 indicates that the farm labor increased in Shandong. Therefore, the labor coefficient is less surprising. It is less surprising yet, considering that the agricultural labor statistics are calculated as the employment residual after all other employment has been identified, rather than an actual estimate of agricultural labor – such a residual would be an unreliable predictor.

The sum of the coefficients on productive factors (A, K, L, and F) is less than one, which suggests that the production process is characterized by decreasing returns to scale. It implies that productivity increases have become more and more difficult to obtain over time. This should raise concerns about the outlook for regional agriculture and heighten the importance of exogenous factors like policy shifts (such as *HP* and *CM*) and environmental improvements (such as *ES* from planting trees) that can improve productivity. Since the northern plains is a major region of China's national grain production and a leader in the overall expansion in agricultural productivity, these findings also raise questions about the sustainability of national agricultural development without further improvements in both policy reforms and the agricultural environment.

Finally, the time trend (-0.035~-0.065) is negative and significant (see Table 3). This result differs from that of Lin's (1992), who found a positive trend for all of Chinese agriculture. The current finding may indicate that the most important regional technological changes were embedded as quality improvements in capital and fertilizer. This is consistent with Stone's (1988) depiction for China – other than the increased use of fertilizer and improved seeds, farming technological changes were slow. It could also be due to the reduction in government investment in the northern plains' agricultural infrastructure, including poor maintenance, inadequate expansion of irrigation and other public production facilities, and limited access to extension services (Feder, Lau, Lin, & Luo, 1992). The irrigation infrastructure is greater in the northern plains than in other regions of China, and the decline in its budget is notable.[1]

The specification in the third column of Table 3 adds conviction to the findings from the first two regressions. It eliminates the agricultural inputs and the trend variable in order to focus on the statistical robustness of the production shifters. Tenure continues to be the dominant policy variable. The intercropping coefficient is constant at 0.006-0.007 and significant, regardless of the data or model specification, whereas the shelterbelt coefficient is positive but insignificant and probably smaller. These results suggest that the capacity of intercropped trees to regulate soil and air temperatures and moisture and improve soil nutrients was more important to Shandong farmers than the more limited capacity of young shelterbelts to control wind velocity and erosion. Of course, as the trees mature beyond the 13-year period of the data coverage, the windbreak effects of shelterbelts will increase, and any conclusion about the relative importance of shelterbelts and intercropping may change.

To summarize, tenure reform achieved by the policy change was the dominant source of agricultural productivity growth. Farmers planted trees with confidence that they would recover the benefits of these investments, and agroforestry in general, intercropping in particular, had a rapid and positive impact on agricultural production in Shandong. This conclusion may be extended to the entire northern plains.

Decomposing the growth in agricultural output into its factor shares can shed additional light on the contributions of HRS and agroforestry. The procedure is to a) multiply the coefficients of the production shifting variables by 100, since they are in percentage form, b) calculate the average change in each variable for the time period in question, and c) multiply the results from the first two steps for each productive factor. Table 2 shows that the tenure reform associated with HRS was essentially complete by 1984. Thus, this decomposition might separate the whole period into two sub-periods, 1978-1984 and 1985-1990. The factor shares for all variables in the second regression are reported in Table 4. For each variable, the contribution to agricultural growth is provided in both absolute and percentage terms. The latter is presented in parentheses.

The conversion to HRS explains over half of the total growth in the first period (59.3 percent in Table 4). This is comparable to Lin's (1992) estimate of 47 percent of agricultural growth nationwide due to HRS. McMillan et al. (1989) and Fan (1990) make similar observations. Aside from the change in tenure, the major contributors to agricultural growth were capital, fertilizer, and the environmental services arising from intercropping. Agroforestry, including both shelterbelts and intercropping, contributed moderately – its absolute share was 5.0 (1.1+3.9) during 1978-1984, but increased to 5.4 (0.5+4.9) during 1985-1990. The percentage share of agroforestry was only 5.4 percent for the earlier period, when HRS had its greatest impact and many agroforests were younger; however, it increased to 19.4 percent for the second period as the trees matured. This second period result supports the observations of Zhong et al. (1991) and Zhu et al. (1991) of 10-15 percent increases in crop yields due to agroforestry.

Nonetheless, these estimates may have been conservative for the northern plains. This is because most of the contribution of agroforestry came from intercropping which, by itself, accounted for 4.24 percent of the growth in agricultural productivity between 1978 and 1984, and 17.72 percent between 1985 and 1990 (the small effect of shelterbelts was excluded). The larger effects of shelterbelts should become apparent later as more shelterbelts would have been established and mature after 1990. In any case, given that most of the contribution of agroforestry to agricultural growth was due to intercropping, it is intriguing and insightful to further

Table 4: Accounting for agricultural productivity growth

		1978-1984		1985-1990	
	Regression coefficient	*Change in variable*	*Contribution to growth (%)*	*Change in variable*	*Contribution to growth (%)*
Inputs					
Capital (K)	0.346	74.51	25.78 (28.05)	86.69	29.99 (107.66)
Fertilizer (F)	0.311	99.52	30.95 (33.67)	30.88	9.60 (34.47)
Land (A)	-0.043	1.50	-0.065 (-0.07)	2.88	-0.12 (-0.44)
Labor (L)	-0.172	13.20	-2.27 (-2.47)	4.36	-0.75 (-2.69)
Shifters					
Tenure reform (*HP*)	0.551	0.99	54.55 (59.34)	0.00	0.00 (0.00)
Market reform (*CM*)	-0.0001	25.59	-0.20 (-0.28)	9.29	0.09 (0.32)
Trend (*T*)	-0.035	7.00	-0.25 (-0.27)	5.00	-17.50 (-62.81)
Agroforestry (ES)					
Shelterbelts	0.001	10.84	1.08(1.18)	4.54	0.45 (1.63)
Intercropping	0.006	6.49	3.89(4.24)	8.23	4.94 (17.72)
Residual			2.77 (2.99)		1.16 (4.15)
Total growth			91.92 (100.00)		27.86 (100.00)

investigate the spatial and temporal relationships of intercropping to fully understand its impact. This analysis is presented in the following section.

3. SPATIAL AND TEMPORAL RELATIONSHIPS OF INTERCROPPING

This section further examines the spatial and temporal relationships between intercropped trees and annual crops. First, the materials and methods are described, the data are discussed, and then, the results are presented.

3.1. Materials and methods

Among the intercropped tree species in the northern plains, paulownia (*Paulownia elongata*) was the most popular prior to the 1990s, accounting for about two-thirds of the total area of intercropping (Ministry of Forestry, 1988). Paulownia has a number of favorable characteristics for intercropping. First, it is one of the fastest growing tree species. In a paulownia intercropping (PI) system, an eleven-year-old tree has an average diameter of 38 cm at breast height and a height of 12 m, yielding 0.5 m^3 of timber (He, 1990). Second, even though its light-loving feature requires being planted less densely to maintain a stable crop output throughout its rotation, paulownia's low leaf area index makes it easier to be intercropped. Third, most of its roots are distributed under the tillage layer, which reduces water competition with annual crops (Zhu et al., 1991). Fourth, the late leaf opening up gives wheat, a major summer crop, enough time for flowering and grain filling. Finally, paulownia can reduce the negative effect of hot dry winds on wheat yields. These and other physical, ecological, and biological effects of intercropped paulownia trees on crop growth and yield are brought about through their modifications of the microclimate and soil properties of farm fields.

Because of the extensive practice of multiple cropping (i.e., growing more than one crop on the same land in a year), the PI models are usually expressed as "paulownia + crop 1 + crop 2." Given that irrigated wheat is the only crop in the region that can survive the harsh winter (sown in early October and harvested in late May or early June of the following year), it is universally used as the first crop in all PI models. The second crop can be corn, cotton, soybeans, sweat potato, or peanuts. Because of the associated high yields and returns, the most common models are "paulownia + wheat + beans" (PWB), "paulownia + wheat + corn" (PWC), and "paulownia + wheat + cotton" (PWT). Other than intercropping, there may be interplanting between crop 1 and crop 2. For instance, cotton, which requires a longer growing season, is often planted in the inter-rows of wheat prior to its harvest.

To allow for convenience in farming, the planting densities of paulownia trees have generally been determined by altering only row spacing, from 6 to 50 meters (m). The distance of trees within a row is fixed at 5 m. Planting spacings include 5x6 m (333 trees/ha), 5x10 m (200 trees/ha), 5x20 m (100 trees/ha), 5x30 m (67 trees/ha), 5x40 m (50 trees/ha), and 5x50 m (40 trees/ha). Among all the PI areas, more than 70 percent fall in the range from 5x20 m to 5x50 m. Experiments have shown that across all the spacings, crops have little effect on the growth process of paulownia trees, but paulownia has a significant impact on crop growth processes and yields (Zhu et al. 1991). So, the primary concern has been directed at the variations in crop yield and quality, with the total timber volume of a given area simply derived by multiplying the volume of a single tree by the number of trees planted (He, 1990). To shorten the rotation and maintain a high survival rate, 2-year-old paulownia seedlings are planted. Rotation lengths are around 10 years.

In this section, the physical and financial performance of PI systems will first be measured using the PWB model as an example. A statistical test of the spatial and temporal effects of PI models will then be conducted. The financial evaluation is based on 1983 price and cost data for the region. Cropping costs include expenses for seeds, labor, fertilizer, manure, irrigation, and chemicals. Chemicals cover pesticides and insecticides. Fertilizer includes 60 percent nitrogen, 30 percent phosphorus, and 10 percent potassium. The cost of intercropped trees is determined by multiplying the number of trees per hectare by $0.64/tree. Finally, a five percent discount rate is used.

To estimate the spatial effect statistically, it is hypothesized that at a given time, different densities of paulownia trees result in different crop outputs. To estimate the temporal effect, it is hypothesized that the same density affects crop yields differently over time. These two hypotheses can be tested with a model in which crop output is defined as a function of spacing and time. While the former is a categorical proxy variable, the latter is a trend variable. Given the fact that spacing and time affect crop outputs jointly, an interactive term is included in the model, which is thus specified as an interactive quadratic function of time and spacing:

$$y = \alpha_0 + \alpha_1 x_1 + \alpha_2 x_2 + \alpha_3 x_3 + \alpha_4 x_4 + \alpha_5 x_5 + \tau \qquad (2)$$

where y = crop yields, x_1 = spacing proxy (1 = 5x6 m, 2 = 5x10 m, ..., 6 = 5x50 m, 7 = control), x_2 = square of spacing proxy, x_3 = time in years (1, 2, ..., 12), x_4 = square of time, x_5 = interacting term of spacing proxy and time, $\alpha_1, \ldots, \alpha_5$ = parameters to be estimated, τ = error terms which are independently distributed with zero mean.

3.2. Data

The Chinese Academy of Forestry (CAF) provided the data for this study. CAF established a paulownia intercropping experimental station in Danshan, Anhui in 1983. A randomized block design was adopted to allocate the five treatments – tree spacings (5x6 m, 5x10 m, 5x20 m, 5x40 m, 5x50 m) with three replications. The size of each block was 91,200 m^2 – 760 m (EW) x 120 m (SN). Neighboring blocks were set 150 m apart. Each plot occupied 0.2 ha. Control plots (no trees) were in the open fields outside the intercropping area.[2]

CAF also conducted a series of demonstrations of paulownia intercropping in Luyi and Minquan of Henan, and Danshan of Anhui. These trials, ranging from 16 to 20 hectares in each of the three counties, covered the aforementioned tree spacings of the PI models. Most of the experiments and demonstrations were successfully carried out until the early 1990s, and they generated a tremendous amount of ecological, biological, and economic observations. This dataset, covering 12 years, was derived from the experimental and trial records. Included in the data are yields for three major models with six tree spacings as well as one control. Table 5 summarizes the annual crop yields for the PWB model; Table 6 reports timber yields of the intercropped paulownia trees under different spacings, and the branch and leaf residues that a tree can produce in a rotation of up to 12 years; and Table 7 presents the associated cost and price data.

The dataset has some drawbacks in its generation. First, while the input data in different intercropping plots were recorded at Danshan experimental station, it was not fully compiled. As a result, an average measure of crop input was assumed for a specific PI model based on the assumption that natural conditions and farming intensities were similar in the three counties (He, 1990). Second, each data point on the output side was an average based on the 10-15 original observations from different sites. These treatments removed much of the randomness in the data. Third, the failure to document all of the annual crop yields caused three yield values in the

Table 5. Wheat and bean yields in paulownia intercropping systems.

Year in rotation	*Yield (kg/ha); Tree spacing (m)*						
	5×6	5×10	5×20	5×30	5×40	5×50	Control
Wheat							
1	2,883	2,924	2,945	2,918	2,975	2,979	2,883
2	3,684	3,717	3,704	3,725	3,750	3,744	3,675
3	3,440	3,740	3,896	3,995	1,926	3,965	3,824
4	3,243	3,761	4,088	4,263	4,100	4,184	3,973
5	3,047	3,614	4,274	4,385	4,473	4,475	4,181
6	2,621	3,465	3,816	3,966	4,089	4,175	3,836
7	2,432	3,152	3,588	3,891	3,894	3,942	3,651
8	2,067	2,799	3,470	3,705	3,857	3,891	3,719
9	1,914	2,624	3,312	3,534	3,809	3,902	3,935
10	1,808	2,381	3,042	3,323	3,770	3,846	3,926
11	1,632	2,084	2,981	3,197	3,738	3,863	3,897
12	1,393	1,625	2,658	2,930	3,149	3,723	3,909
Bean							
1	1,259	1,254	1,253	1,223	1,224	1,232	1,220
2	1,317	1,337	1,355	1,367	1,368	1,361	1,292
3	1,047	1,268	1,343	1,374	1,374	1,365	1,301
4	813	1,197	1,331	1,380	1,403	1,368	1,308
5	479	903	1,298	1,349	1,430	1,448	1,353
6	225	657	1,047	1,314	1,416	1,445	1,358
7	65	566	1,134	1,310	1,403	1,442	1,353
8	0	549	1,004	1,277	1,394	1,413	1,337
9	0	384	978	1,272	1,362	1,385	1,344
10	0	258	956	1,254	1,359	1,401	1,359
11	0	131	927	1,244	1,370	1,394	1,374
12	0	0	887	1,211	1,352	1,365	1,389

Source: Chinese Academy of Forestry; cotton and corn yield data from Yin and He (1997).

PWC model and two in the PWB model to be interpolated. Obviously, the task of fully observing and compiling all the input and output data would have been formidable. However, these steps of data manipulation virtually eliminate the opportunity to perform sophisticated statistical analysis. Nevertheless, it is still possible to conduct analyses with the averaged input and output data. Further, it is indeed rare to find such a valuable dataset elsewhere.

3.3. Empirical results

Table 5 reveals that as the density of paulownia trees increases, crop yields decrease; and the longer the tree rotation, the more pronounced are crop yield declines. However, yields of the first crop (wheat) witnessed a pattern of decline different from those of the second (soy beans). During the earlier years of the rotation, wheat yields with less densely intercropped trees could significantly outperform that of the control. Even in later years, the yield reduction was relatively less severe. For instance, wheat yields with the 5x50 m and 5x40 m tree spacings rose by 5.4 and 4.4 percent, respectively, in the first eight years of the intercropping. This plot-level evidence of positive ecological effects of perennial trees on annual crops further lends confidence to the earlier result that agroforestry, especially intercropping, has contributed to agricultural productivity growth. With the highest paulownia density (333 trees/ha in 5x6 m spacing), wheat yields declined to about 1/3 of that produced in the control field at the end of the rotation.

In contrast, the yields of soybeans in the early years of intercropping rose less than those of wheat with lower tree densities (no more than 1.7 percent in the first eight years). Later, the crop yield experienced dramatic reductions. After six years, for example, paulownia trees in 5x6 m spacing virtually eliminated the production of soybeans. For densities of fewer than 100 trees/ha (5x20 m), the yields were reduced by half at the end of the rotation. Obviously, when the shading effect of trees becomes too severe to allow the second crop to grow, farmers would stop cropping to avoid losses (negative net returns). With a certain degree of reduction in crop yields, however, farmers were able to gain timber and tree residuals from intercropped trees. As shown in Table 6, in twelve years the PI system could produce a timber volume ranging from 237 m^3/ha in 5x6 m spacing to 28.5 m^3/ha in 5x50 m spacing. Additionally, tree branches would be available for fuel, and leaves for either soil nutrients or fodder.

Figure 1 depicts the annual net returns of wheat and soybeans in the PWB model, which increased monotonously with decreasing tree densities. Also, time made the impact of density on the net returns more pronounced. Notice that, because part of the decline in the annual net revenues is attributable to discounting, a comparison of financial performance is better made with reference to the control series. On the other hand, Figure 2 suggests that the net returns from paulownia timber are positively correlated with tree densities. With a density of more than 100 trees/ha (5x6 m spacing), the net returns increased sharply. Also, Figure 2 indicates that the net returns under a specific tree density can be altered by the decision of when to cut the trees – in nine or ten years.

If both perennial trees and annual crops are accounted for in the financial assessment, it was found that, because timber revenues dominated those from wheat and soybeans, the higher the tree density the greater was the total net return.[3] This implies that in the PWB model, the direct economic benefits from timber would

Table 6. Paulownia timber volume in intercropping.

Year in rotation	Timber for different tree spacings (m³/ha)						Residues (kg/tree)	
	5×6	5×10	5×20	5×30	5×40	5×50	Branches	Leaves
1	7.0	4.2	2.1	1.4	1.0	0.8	0.3	0.0
2	15.4	9.2	4.6	3.1	2.3	1.8	1.8	0.5
3	26.9	16.2	8.1	5.4	4.0	3.2	5.3	2.2
4	41.6	25.0	12.5	8.4	6.3	5.0	11.8	4.2
5	59.4	35.7	17.8	11.9	8.9	7.1	21.8	7.1
6	80.2	48.2	24.1	16.1	12.0	9.6	36.1	10.8
7	104.1	62.5	31.3	20.9	15.6	12.5	55.2	15.4
8	130.9	78.6	39.3	26.3	19.7	15.7	79.7	21.0
9	160.8	96.6	48.3	32.4	24.1	19.3	110.3	27.6
10	193.6	116.3	58.2	39.0	29.1	23.3	147.4	35.2
11	229.4	137.8	68.9	46.2	34.5	27.6	191.7	43.9
12	237.1	142.4	71.2	47.7	35.6	28.5	243.7	53.7

Source: Chinese Academy of Forestry.

Table 7. Cost and price data for wheat, bean, and paulownia (1983).

Cost ($/ha)							Price ($/kg, $/m3)		
Seed	Fertilizer	Water	Wage	Chemical	Other	Total	Wheat	Bean	Timber
79.75	143.55	47.85	133.98	12.76	31.90	449.80	0.213	0.298	127.6

Source: Chinese Academy of Forestry.

indeed be much larger than those indirect benefits derived from the increase of crop yields. Also, as the farming regime shifted from annual crops only to intercropping with the lowest paulownia density (40 trees/ha in 5x50 m spacing), the total net returns increased substantially. Thus, households could not only meet their demands for various outputs from intercropping, but also capture the greatly increased economics benefits. Now, it is clear why farmers had a strong motivation to adopt intercropping.

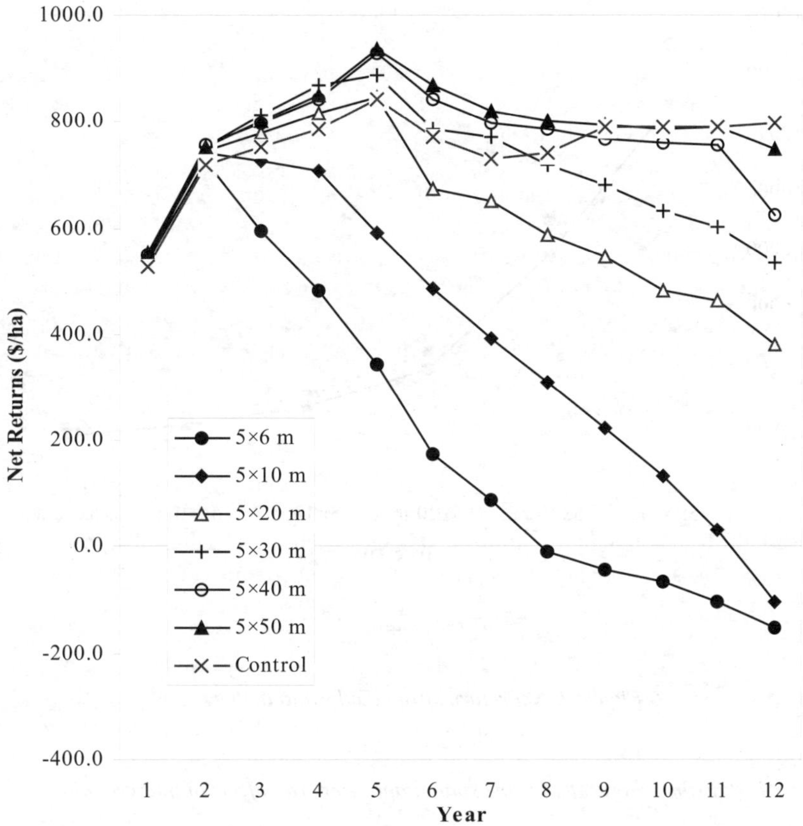

Figure 1. Net returns of wheat and soybeans under various intercropping tree densities.

Table 8 presents the estimated statistical results for the spatial and temporal effects of intercropping. Again, the models fitted well as indicated by the high R^2 values. All the parameters are significant in the wheat equation, while the coefficient of the second-order time term (x_4) is insignificant in the other three equations. This implies that yields of the second crops are linear over time if plotted. To better capture yield dynamics, the equations for corn, bean, and cotton were re-estimated with all the variables inverted, which led to the acceptance of the quadratic terms. As such, both the spatial and temporal effects of intercropped paulownia on crop yields became significant. Since second-order terms are involved in each equation, however, it is difficult to configure these effects. One way to overcome this difficulty is to visualize yield changes by plotting the fit yield curves, which has

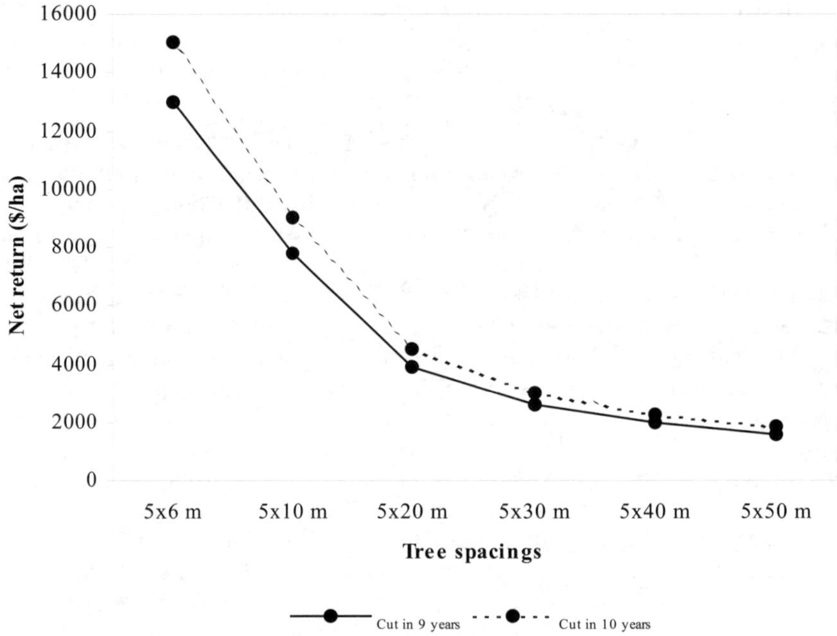

Figure 2. Net returns from paulownia timber.

Table 8. Estimated results for the spatial and temporal effects of intercropping.

Dependent Variable		*Independent variables*						
		Intercept	x_1	x_2	x_3	x_4	x_5	R^2
Wheat	Coefficient	2378	515.5	-67.7	119	-25.3	37.9	0.86
	T value	13.0	7.2	-7.9	3.0	-9.1	9.2	
Corn	Coefficient	5049	-9807	4047	4192	-5042	6647	0.87
	T value	21.1	-10.0	4.9	4.7	-6.3	8.9	
Soybeans	Coefficient	1662	-3138	1177	1270	-1694	2375	0.86
	T value	19.4	-8.9	4.0	3.9	-5.9	8.9	
Cotton	Coefficient	654	-1913	860	3329	-2744	1160	0.89
	T value	10.2	-7.3	3.9	13.8	-12.8	5.8	

In the wheat yield equation, x_1, x_2, ... x_5 represent tree density proxy, square of density proxy, time, square of time, and product of time and density proxy. In other equations, these variables represent inverses of tree density proxy, square of density proxy, time, square of time, and product of time and density proxy. The degree of freedom is 78.

been done in Yin and He (1997). Overall, it can be said that growing trees in farm fields will increase or decrease crop yields, depending on the tree density and rotation length.

It should be emphasized that changes in crop yields should not be the only criterion to judge intercropping. Additional considerations are the benefits derived from timber, which may come at the expense of a certain amount of crop yields. Nevertheless, farmers did not widely adopt high tree density PI regimes, let alone tree farming, due to the existing constraints of crop production responsibilities and food self-sufficiency. Households were responsible for providing a specified amount of grain or cotton annually to the community, in addition to guaranteeing its own food consumption. The PI models with high tree densities, though economically appealing, were detrimental to agricultural production in most of the years. This explains why a majority of the observed paulownia intercropping systems maintained a low tree density.

4. SUMMARY AND CONCLUSIONS

This study has validated the proposition that land tenure reform, by converting collective farming to the HRS, was the major contributor to expanding agricultural productivity in the 1980s. The HRS contributed to tree planting and the expansion of forestry as well, as long-term responsibility contracts for land use gave farmers the confidence that they would obtain the returns in agricultural productivity emanating from their own long-term conservation investments. These investments were one source of the increase in forest cover. It is against this background of incentive structure improvement that agroforestry in northern China has seen its tremendous success over the last 25 years.

There are reasons to believe that the beneficial effects of trees and forests would have only expanded further in the 1990s. First, trees planted in the early 1980s were now approaching maturity and their contribution as windbreaks was reaching its maximum. Other trees planted later were still growing and their contributions were increasing. Farmers' experience in tree and forest management was also improving. Undoubtedly this means broader coverage, better spacing, and greater enhancement for agricultural productivity. Also, the HRS contracts were renewed in the late 1990s and, with that, much of the remaining uncertainty about the rural institutional arrangements dissipated. As a result, the climate for long-term investments in agricultural productivity was enhanced once more. This would have improved the investment climate for some of the longer-term production activities, like planting trees and managing forests, being for commercial or conservation purposes.

The evidence provides convincing support to the proposition: tenure counts – it gives households rights to the fruits of their investments. However, it must be stressed that tenure counts in a stable policy environment, which allows households to plan and invest with confidence. Tenure and the relative stability of the policy environment were central features of China's forestry reforms, and they are the key predictors of the impact of these reforms. It can be anticipated that these two features continued to affect forest practices well beyond the period for which data were available, because the long growing period for forestry means that anything that affected harvest and reforestation decisions in the early 1980s would be reflected in the numbers and volumes of 20-year old trees in the 2000s. It would therefore be interesting to carry out a future follow-up study.

Related to the above observation, it should be made clear that China's agroforestry experience is dynamic. For instance, the PI systems were common in the 1970s and 1980s, when paulownia timber was widely used for house construction and furniture making as well as for export. However, the paulownia wood market was saturated in the 1990s, while the demands for fruits, nuts, crafts, panels, medicinal herbs, and other outputs became strong. As such, many types of cash trees and other timber species have replaced paulownia in intercropping. Market conditions are changing, and farm households' responses to them are evolving.

The evidence also supports the proposition that some of China's increase in forest cover in the late 1980s and 1990s was designed to obtain market and non-market environmental benefits. Households planted trees not only to produce timber and other products for their own consumption and sales, but also to improve their agricultural environment. And they rapidly obtained benefits, even from young trees. Tree planting explained at least 10 percent of the 5,200 kg/ha increase in agricultural productivity in the northern plains.

While China's experience is generally not well understood by the outside world, it has great international implications. From the technical perspective, agroforestry – mixing farming, forestry, and animal husbandry over space and time – is becoming more and more important as population continues to expand while farmland continues to decline in many developing countries. From an economic perspective, however, development along this direction is possible only if the institutional settings that farmers face are conducive and stable – land tenure and market accessibility are such that they feel sufficiently confident with the investment climate for some of the longer-term production activities such as forestry. In this regard, China has rich experiences and lessons to share with other countries.

5. NOTES

[1] Therefore, the trend variable may capture some of the positive effect of technical change as well as negative effects of declining infrastructure, with the latter being the dominant part of the variable. Antle (1983) showed that a measure of infrastructure could be an important explanatory factor in agricultural productivity.

[2] For more details on the experimental design, see Zhu et al. (1991).

[3] Of course, this finding has a lot to do with the relative price weights of wheat, soybeans and timber. During the period of the study, the prices for timber in the northern plains were determined in the open market, whereas the government regulated the prices for wheat and soybeans. Therefore, these results could have been biased toward timber because government prices for wheat and soybeans were relatively low.

6. REFERENCES

Antle, J. 1983. Infrastructure and aggregate agricultural productivity: International averages. *Economic Development and Cultural Change*, *31*(3), 609-619.
China Agricultural Yearbooks. 1985-1992. Beijing: China Agricultural Press.
China Forestry Yearbook. 1986-1988. Beijing: China Forestry Press.
Fan, S.G. 1991. Effects of institutional reform and technological change on production growth in Chinese agriculture. *American Journal of Agricultural Economics*, *73*(2), 266-275.
Feder, G., Lau, L. J. , Lin, Y. J., & Luo, X. P. 1992. The determinants of farm investment and residential construction in post-reform China. *Economic Development and Cultural Change, 41*(1), 1-26.
Hayami, Y., & Ruttan, V. W. 1985. Agricultural Development: An International Perspective. Baltimore, MA: Johns Hopkins University Press.
He, Q. 1990. An economic evaluation of paulownia intercropping. *Journal of Forest Economics*, *4,* 11-30.
Henan Statistics Bureau. 1986. Henan Yearbook. Zhenzhou, China.
Lin, Y.J. 1992. Rural reforms and agricultural growth in China. *American Economic Review, 82*(1), 34-51.
McMillan, J., Whalley, J., &. Zhu, L.J. 1989. The impact of China's economic reforms on agricultural productivity growth. *Journal of Political Economy 97*, 781-807.
Ministry of Forestry. 1988. Resource Bulletin. Beijing, China.
Shandong Statistics Bureau. 1988. Shandong Yearbook. Jinan, China.
State Forestry Administration. 2000. Forestry Development Report, Beijing, China.
Stone, B. 1988. Development in agricultural technology. *China Quarterly 116,* 762-822.
Wen, J. G. 1993. Total factor productivity growth in China's farming sector: 1952-1989. *Economic Development and Cultural Change 41,* 1-41.
Wooldridge, J. M. 2000. Introductory Econometrics: A Modern Approach. Cincinnati, OH : South-Western College.
Yin, R.S. 1994. China's rural forestry since 1949. *Journal of World Forest Resource Management 7,* 73-100.
Yin, R.S., & He, Q. 1997. The spatial and temporal effects of paulownia intercropping: the case of northern China. *Agroforestry Systems 37,* 91-109.
Yin, R.S., & Hyde, W.F. 2000. The impact of agroforestry on agricultural productivity: the case of Northern China. *Agroforestry Systems 50*(2), 179-194.
Zhong, M. G., Xian, C., & Li, Y. M.. 1991. A study of forestry development in the major agricultural areas. In W.T. Yong (Ed.), Studies on China's Forestry Development. Beijing: China's Forestry Press.
Zhu, Z. H., Cai, M. T., Wang, S. J., & Yiang, Y.Y. (Eds.). 1991. Agroforestry Systems in China. Chinese Academy of Forestry, People's Republic of China, and International Development Research Center, Canada.

7. AUTHOR'S NOTE

Prior and partial versions of the material in this chapter appear in Yin and Hyde (2000) and Yin and He (1997).

SHASHI KANT AND EMMA LEHRER

A FRAMEWORK FOR INSTITUTIONAL ANALYSIS OF AGROFORESTRY SYSTEMS

1. INTRODUCTION

A radical change in the analysis of "agroforestry" systems is necessary to determine if these systems meet the sustainability standards. Generally, biological or ecological sustainability has dominated the discussions of sustainable land-use systems, and scientists have argued for the development and adoption of land-use systems that mimic natural ecosystems (Jackson & Bender, 1984). Many of these discussions have overlooked the fact that social and economic sustainability are as critical as ecological sustainability. Sustainable land-use systems should not only mimic natural ecosystems but should also be well integrated within local social and economic systems. The main cause behind this neglect stems from a narrow view of a "system", prevalent in the science stream including the science of agroforestry. For example, Nair (1989), citing references of Arnold and de Wit (1976), and de Wit and Goudrian (1974), defines a system as "a group of physical components, i.e., an assemblage of objects, connected or related in such a manner as to form and/or act as an entire unit." As per this definition, physical components are essential parts of a system[1], but social and economic systems may not have physical components. For example, a country's legal system will have different components or legal instruments – a constitution, federal acts, provincial acts, and municipal acts – connected to each other, but none of them is a physical component. Hence, for the analysis of agroforestry systems, an understanding of a system has to be extended from physical components to physical as well as non-physical components and from physical connectedness to physical and non-physical (connections through people's actions in market as well as non-market situations, social norms and sanctions) connectedness.

In the natural resource and environment literature, specifically in ecological economics, interactions, and connectedness between biophysical, social and economic systems are well recognized (Norgaard, 1981). Various scholars have argued that the usefulness of any resource is determined by the available technology and the surrounding institutional structure (Bromley, 1991). Where a technology is a

J. R. R. Alavalapati & D. E. Mercer, Valuing Agroforestry Systems, 279–302.

way in which physical and human capital inputs are converted to outputs and where an institutional structure consists of a set of rules, compliance procedures, and moral and ethical norms that constrain the behavior of individuals in the interests of maximizing the wealth or utility of principals, or of those holding rights to resource use (North, 1981). An institutional analysis includes an understanding of the main features of resource users and stakeholders, as well as external agents influencing the behavior of the resource users, institutions, contextual factors that influence the effect of institutions on the behavior of principals, resulting actions and interactions among agents, and outcomes. In this characterization of a resource system, a technology captures the biophysical sub-system and an institutional structure captures the socio-economic sub-system. The dominance of a technological perspective of a resource system, or bio-physical sub-system, has been common to almost all fields, including agriculture, forestry, and agroforestry, during the industrialization era. In the last two decades, the institutional side has gained an increasing prominence in forestry and agriculture sectors. Nevertheless, the agroforestry sector has remained relatively uninfluenced by these developments.

Various scholars working in the agroforestry sector have recognized the importance of social and economics aspects. For example, Lundgren (1989) discussed institutional aspects of agroforestry research, but he was using the term institutions as synonymous to organizations. Scherr and Hazell (1994) identified the economic importance of resources, the willingness to invest in long-term, economic incentives, and institutional support as necessary elements to support the adoption of new technologies associated with natural resource management. Some scholars have identified important institutional issues (Rogers, 1995; Pannell, 1999), such as, insecure or inequitable land tenure (Feder & Onchan, 1987; Riddell, 1987), social stigmas associated with the technology (Fujisaka, 1994), non-favorable taxation systems and distortions in pricing system (Bhati, Klijn, Curtotti, Dean, & Stephen, 1992), inability to enforce property rights and institutional inertia (Pannell, 1999). Similarly, Sanchez (1999) discussed the importance of research to support land tenure policy in southeast Asia, but his focus was on science-based agroforestry to produce economically and socially sound results. None of these discussions, however, has treated socioeconomic elements as a sub-system of agroforestry systems, and therefore these discussions are somewhat limited in their ability to capture the connectedness of the elements. The characterization of a resource by two dimensions - technology and institutions – offers a unique opportunity to assess social and economic aspects from an institutional perspective and all the biophysical aspects of technology.

Many social scientists, working with natural resources, such as Oakerson (1986) Ostrom (1990), and Edwards and Steins (1998) have suggested institutional analysis frameworks for common pool resources. Unfortunately, no such framework has been developed specifically for agroforestry systems. In this Chapter, we attempt to fill that significant gap.

To begin we provide an overview of agroforestry and agroforestry institutions, which helps in understanding the context for developing an institutional framework, and identify the main features of the required framework. This is followed by an overview of existing frameworks and a discussion of the proposed framework for agroforestry systems. Finally we conclude with some observations about the uses and further developments of the framework.

2. AGROFORESTRY (AF), INSTITUTIONS, AND INSTITUTIONAL ANALYSIS

Irrespective of the definitions and specifics of AF systems, sub-systems, and practices, the practice of growing tree species and agricultural crops in intimate combination has existed for time immemorial. In the early 1970s, some development agencies such as the World Bank and the Food and Agriculture Organization (FAO) re-oriented their forest policies to include many elements of agroforestry in order to assist the ordinary farmer to increase food production and the traditional forest sector to produce and harvest wood (King, 1989). People adopt agroforestry systems to meet their diverse needs (Current, Lutz, & Scherr, 1995). For example, the main purpose of trees in agroforestry is to provide fuelwood in Guatemala (Urrea, 1995), animal fodder and bedding in Nepal (Gilmour &, Nurse 1995), and increased revenue from tree crops in India, and increased crop production in Kenya (Andreatta, 1998). However, agroforestry systems also contribute, directly or indirectly, to numerous ecological, social and economic outcomes. Ecological outcomes include increased soil retention, increased soil fertility through nutrient cycling; suppressed growth of unwanted weeds; and reduced outbreaks of pests and diseases (Senanayake & Jack, 1998; Wojtkowski, 1998). Economic and social outcomes include increased production of agricultural and tree crops; a more evenly distributed workload; a buffer against unexpected fluxes in market prices, and social cohesion and empowerment.

Agroforestry, therefore, has emerged as a holistic land-use system, in which trees and crops should not be considered independently. Institutional analysis frameworks focused on one sector – either forestry or agriculture - cannot capture the complexities of agroforestry systems. The lack of an institutional analysis framework focused on agroforestry can be attributed to multiple factors. First, agriculture and forestry have emerged independently over the past few centuries for reasons that include the effects of Europe's agricultural and industrial revolutions; neo-classical economic theories of specialization and trade; and the more recently green revolution and tropical deforestation (Lundgren, 1987). The technological developments of the industrial revolution, for example, produced specialty tools designed for harvesting crops or cutting down trees. These developments allowed landowners to specialize in the large-scale production of certain crops, which where then traded instead of being locally consumed. Just as farming and forestry technologies began to specialize, so did the institutions related to these fields, which developed independently into two different schools of thought – one for forestry and one for agriculture. The emphasis on specialization restricted the optimal integration of institutions, and development of an institutional analysis framework sensitive to the integrated nature of agroforestry.

Second, during colonization, European laws and values were imposed throughout the world upon indigenous land-use systems that were intimately linked with local cultural and ecological values. The results of the imposed laws provide a striking example of the lack of local and cultural considerations in many countries. English property law, for example, states that whoever owns the soil, is also the owner to that which is above and below it, where ownership is characterized as the exclusion of others having simultaneous rights to the land (Okoth-Ogendo, 1985). In Africa, however, traditional land tenure institutions separate the land from the objects affixed to it, the latter being generally regarded as the property of whose labor it is (Okoth-Ogendo, 1985). In this case, rights to land and rights to trees or crops planted on the land are seen independently from each other, where the rights to a tree are granted to the individual who plants it, regardless of the land on which it is growing (Kidd & Pimentel, 1992). Similarly, in many colonized countries, trees have played a fundamental role in determining title to land, a concept that was not recognized by European institutions. In areas of Brazil and Indonesia, for example, land traditionally had to be cleared of trees for an individual to establish tenure (Kidd & Pimentel, 1992), while, in many other countries, title to land can actually be acquired by planting trees (Otsuka, Suyanto, Sonobe, & Tomich, 2001). Thus, in order to protect land title, some land tenures strictly prohibit the planting of trees by others. Not only do land and tree tenure definitions vary among cultures, so does the very definition of a "tree". Often, socio-cultural norms determine what is considered a tree regardless of biological definitions. In some African communities, for instance, perennial plants with woody trunks are not recognized as trees if they are used by children (Kidd & Pimentel, 1992).

Similarly, there exist many situations where recently developed institutions, such as national statutory laws and policies, conflict directly with traditional conventions. New institutions, even those developed for agroforestry, may actually be counter-productive if they do not consider traditional land use. In some cases, steps to encourage agroforestry have actually eroded pre-existing agroforestry practices. For example when national statutory laws fail to recognize traditional land rights and prohibit customary use of and access to land (Kidd & Pimentel, 1992; Otsuka et al., 2001). This suggests the need for: i) institutions for agroforestry, which are sensitive to local cultural, social, and ecological conditions, and complimentary to existing formal and informal institutions, and ii) an institutional analysis framework which incorporates institutional diversity across cultural, social, and ecological settings and macro, micro formal, and informal institutions.

Third, the institutional emphasis has been on the inputs (front-end) of agroforestry systems, and institutions related to the process and the outputs of AF system have been neglected. For example, there is a lot of discussion about land tenures (land is one of the inputs), but it is hard to find any discussion of institutions concerning land and tree management, or marketing of outputs. Cases of AF from India and Costa Rica reveal the consequences of the lack of attention given to projects throughout their lifecycle. In both countries, fast-growing tree species were selected to reduce the gap in demand and supply of wood products. The landowners responded on the basis of the current-level of supply and market prices of selected wood products without considering the long period required for production of wood products in which market supply and price conditions may change. In both cases, the combined effect of other landowners acting in the same way caused a significant increase in the supply of these products and ultimately led to market saturation (Current, 1995; Dwivedi, 1992). The resulting market for these goods was so poor, that cases were reported of farmers uprooting their trees after having invested years in their cultivation (Dwivedi, 1992). These two examples reveal the lack of adequate institutions governing agroforestry outputs, and demonstrate the futility of investing resources upfront in the absence of appropriate institutions for production/management processes and outputs. Similarly, because land is only one of the inputs of AF system along with capital, labor, technology, and seedlings, the success of AF systems will depend upon the existence of appropriate institutions for all of these inputs. Hence, an institutional analysis framework should incorporate institutions related to all three stages of the AF system – inputs, production process, and outputs – and their interactions.

In summary, an institutional analysis framework for agroforestry should incorporate institutions for inputs, production process, and outputs; macro-level as well as micro-level and formal as well as informal institutions, and be responsive to social, cultural, and ecological diversity. In addition, the framework should be specific to agroforestry situations, appropriate to evaluate institutions holistically to ensure that they are compatible and complete, and sensitive to the dynamic nature of institutions. In the next section we propose a framework with these desired features.

3. AN INSTITUTIONAL ANALYSIS FRAMEWORK

Institutions, as social constructs, constrain the behavior of individuals, and are meaningless in the absence of subjects – human agents. Hence, a simple categorization of institutions or an institutional analysis without considering the main features of agents; contextual factors, other than institutions, that influence the effect of institutions on the behavior of agents; resulting actions and interactions among agents; and outcomes – cannot provide any meaningful inputs to policy makers and resource managers. The first framework of institutional analysis for natural resource management, specifically for common pool resources, was proposed by Oakerson[2] in 1986. The framework is comprised of four elements: i) physical and technical attributes of a resource; ii) decision-making arrangements; iii) patterns of interaction; and iv) outcomes.

The physical and technical attributes examine sub-tractability, excludability and divisibility aspects of the resource. These attributes are considered 'hard' constraints that can affect outcomes either directly or indirectly by shaping patterns of interaction.

Decision-making arrangements refer to the institutions governing use of the commons, such as who makes decisions, what those decisions will be, whether certain individuals are permitted access, and in what capacity. These rules are divided into three categories: 1) operational rules that determine daily decisions; 2) collective choice rules, which are used by resource users, officials and authorities to determine operational rules; and 3) constitutional level rules[3], which determine the creation, enforcement and modification of collective choice rules. These institutions set the rules for how the commons should be managed. However, rules, exclusively, are inadequate to ensure a particular pattern of behavior unless they reflect the values of the society in which they are to be applied. Therefore, unlike physical and technical attributes, these soft-constraints can only affect the outcome indirectly through their influence on patterns of behavior.

The *patterns of interactions* represent the collective choices and decisions made by individuals in response to physical attributes and institutions. People respond to actions both individually and as part of a group, according to how different strategies impact them collectively and individually. In other words, the costs and benefits associated with each alternative action will determine the decision. The collective choices made by individual strategies emerge as patterns of interactions and, ultimately, determine outcomes.

Outcomes are the physical result of the interactions among the elements of the framework. These end products can then be subject to evaluation, which can determine, for instance, whether they have been produced efficiently or equitably.

Oakerson's (1986) framework is general and simple, and can be applied in a variety of situations. However, the framework's simplicity restricts the incorporation of specific aspects of varied natural resource use situations. Edwards and Steins (1998) extended this framework for the analysis of resource systems, which support multiple types of uses by multiple types of communities/groups. The main additions incorporated by Edwards and Steins are: i) inclusion of an additional first-order attribute – the social characteristics of the user community; ii) addition of a new component – contextual factors, iii) analysis of the physical and technological characteristics and the decision making arrangements with respect to multiple uses; and iv) incorporation of multi-level analysis of institutions. The revised framework is sensitive to many features of agroforestry systems such as the possibility of multiple uses and multiple agents involved in agroforestry systems, the importance of the social characteristics of the user community and the relevance of contextual factors to agroforestry. However, this framework is unable to address some other main features of agroforestry systems. For example, both the Oakerson (1986) and Edwards and Steins (1998) frameworks, are focused on the management of existing natural resources, while in many agroforestry situations, decision makers may start with a situation where there are no existing trees or crops. Even when there are existing trees, decision makers will be involved in an annual or semi-annual production process of agricultural crops. In these situations, it is important to examine the institutional arrangements related to three components of a production process, namely the factors (inputs), the management of the production process, and the outputs[4]. In addition, these three components may occur in serial stages and not simultaneously, therefore it will also be useful to examine the possible relations and interactions between the institutions related to these three components. There is also a need to distinguish between the users, stakeholders, and external agents in agroforestry systems. Likewise, since interactions among users, stakeholders, and external agents will vary across the stages of production, there is a need to examine these actions and interactions at different stages.

We propose an institutional analysis framework for agroforestry systems (Figure 1) as a conceptual tool for organizing institutional information about agroforestry systems. The framework is set out to raise a series of questions, and not to provide answers, related to institutional arrangements for the system. The relationships identified between resource characteristics, users, stakeholders, external agents, situational factors, and existing institutional arrangements can then be used in diagnosing and understanding problems associated with particular agroforestry systems.

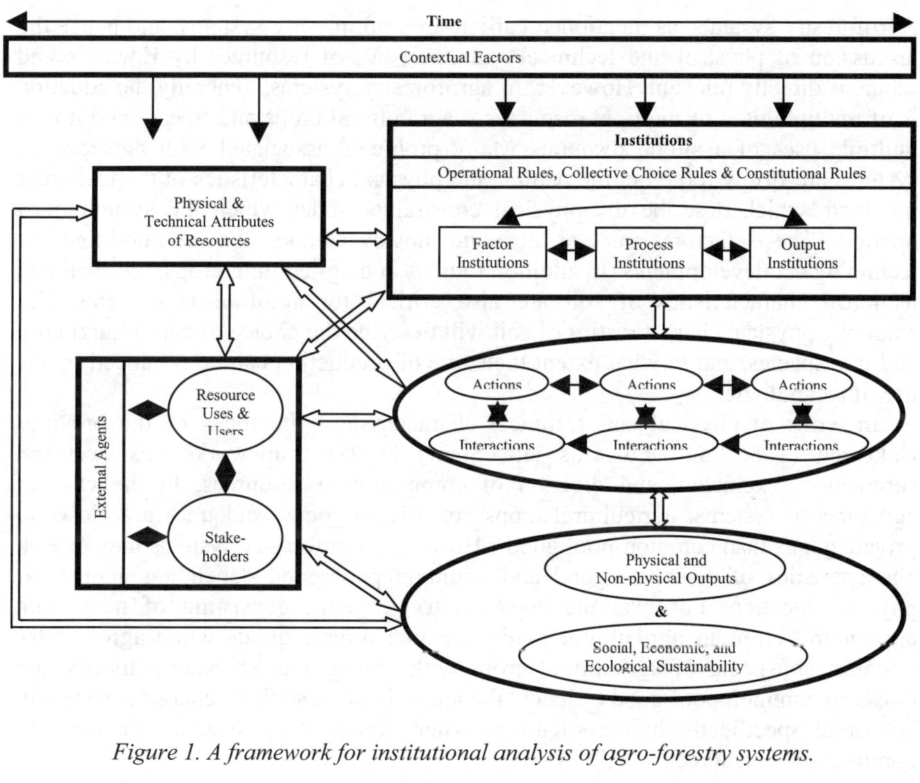

Figure 1. A framework for institutional analysis of agro-forestry systems.

The proposed framework distinguishes six sets of attributes: a) the physical and technical attributes of the resources of an agroforestry system; b) the social, economic, and other attributes of the agents (users, stakeholders, and external agents) associated with an agroforestry system; c) the contextual factors of an agroforestry system; d) the institutional (decision-making) arrangements that govern actions and interactions among the agents associated with an agroforestry system; e) the actions and interactions among the agents associated with an agroforestry system; and f) the outcomes or consequences of an agroforestry system. The first three attributes are focused on agents and resources, which can be termed as the environment for institutions. The fourth attribute is focused on institutions, and the last two attributes on the influence of institutions on agents' behavior and the resulting outcomes. Each of these six sets of attributes and their member attributes are discussed next. Interactions between institutions and the environment for institutions (resources, agents, and contextual factors) are discussed in the subsection 3.5 on interactions.

3.1. Physical and technical attributes of the resources of an agroforestry system

Agroforestry systems, as mentioned earlier, are multiple-use systems, and hence the discussion of physical and technical characteristics of resources by Edwards and Steins is directly relevant. However, in agroforestry systems, generally the situation is of multiple uses of multiple resources – agricultural crops and trees – and not of multiple uses of a single resource. Many problems associated with agroforestry systems are firmly rooted in the natural and physical characteristics of the resources involved, which describe the physical constraints under which the system must operate. These factors are not absolute however, and can be modified by technological developments. In addition to trees and agricultural crops, physical and technical characteristics of soil are also critical for agroforestry systems. For example, physical characteristics of soil will describe the choice of agricultural crop and tree species, and to what extent their rate of production can be enhanced by the use of technologies.

In terms of physical and technical characteristics, the focus of the previous Oakerson (1986) and Edwards and Steins (1998) frameworks has been on subtraction, exclusion, and division of common-pool resources. In the case of agroforestry systems, agricultural crops are private goods, and trees are closer to private goods than common-pool goods. However, agroforestry systems may exhibit characteristics of a private good and common pool good depending upon their physical location. For example, agroforestry systems, consisting of trees with agricultural crops in agricultural fields, are like private goods while agroforestry systems, consisting of agricultural crops with young trees in natural forests, are closer to common pool goods. Hence, the analysis of these three characteristics will be useful specifically in the situations where agroforestry systems are close to common pool resources.

The distinction between 'resource system' and 'resource units', another feature of the previous frameworks, is also critical for the analysis of agroforestry systems. Any agroforestry system is capable of producing a wide-variety of resource units including direct and indirect benefits, and extractive and non-extractive uses. For example, an agroforestry system can produce agricultural resource units of grains, pulses, vegetables, and wood resource units of fuel and timber. In addition, it can also produce non-extractive resource units for soil conservation, enhanced soil-fertility, and windbreaks. Any two resource units will exhibit either competitiveness or complementarities among themselves. Based on the nature of association between the two resource units, these can be termed either as associate products (production complementarities between the two units) or rival products (production competition between the two units). The nature of association between any two resource units may even vary over the complete production cycle. Hence, an agroforestry system will need to be evaluated with respect to: i) each separate extractive and non-extractive use of the system; ii) the nature and dynamics of interactions, competitive or complementary, between the different resource units; and iii) the agroforestry system's productive capacity to support multiple uses.

The productive capacity of the agroforestry system to support multiple uses will depend upon: i) the physical, including biological, characteristics of agricultural crops, tree species, and soils; ii) the compatibility of agricultural crops and tree species in the agroforestry system; iii) the impact of each type of use on soil, agricultural crops, trees, and on the system as a whole; iv) the permutations of each type of use; and v) the extent to which technological aspects might be used to improve production and management of different resource units. The technological aspect includes chemical fertilizers, harvesting equipments, and seed varieties, among others, and the use of technological practices, such as crop rotations, crop mixtures, fallow cycle, spacing and thinning of trees, and local knowledge about crops, trees, and their management and use. Hence, technological aspects will involve improvement in the existing technology and technological practices; use of the existing technology, technological practices, and local knowledge; and development of new technology and technological practices to address new problems.

A specific feature of agroforestry systems, different from the discussions of common pool systems, is that agricultural resource units are drawn continuously on annual or semi-annual cycles while wood units are drawn on much longer cycles depending upon the choice of tree species. In some cases, such as taungya systems, the supply of agricultural resource units may terminate once a forestry crop is established. In other cases, the supply of agricultural resource units may decline significantly after the establishment of a tree crop. The nature of interactions between agricultural crops and tree species will evolve with the age of tree species and may require a continuous attention to technological aspects. Therefore, the dynamic nature of interactions and technological requirements should be an integral part of analysis of agroforestry institutions.

3.2. Social, economic, and other attributes of agents associated with an agroforestry system

The inclusion of social and economic attributes of agents stems from the recognition that the attributes of resources and institutions, alone, do not generate individual or group actions. Rather, individual characteristics also determine how individuals will respond, according to various incentives. The separation of individuals from groups accounts for the different incentives governing action strategies of similar individuals in addition to the different pressures faced by the similar individuals. These attributes were missing from the Oakerson's framework but Blaikie and Brookfield (1987), Tang (1992), and Singh (1994) included 'the social characteristics of the user community' in their modified versions. However, these authors and many subsequent studies used a narrow concept of the user community – the appropriators of the resource units. Edwards and Steins (1998) extended the definition of "user community" by including the presence of occasional users and other stakeholders. However, the inclusion of all the users and stakeholders in one group does not provide clarity about the distinctive roles of the users and stakeholders in resource management and specifically in the institutional analysis of agroforestry systems. In addition to the users and stakeholders, some other agents who are external to agroforestry systems also influence the interactions among the users and stakeholders, and between the users and stakeholders. Hence, in the proposed framework we discuss the attributes of the users, stakeholders, and external agents separately. In our framework, users and stakeholders are internal to an agroforestry system while some other agents are external to the system. These external agents are different from the "contextual factors", or other first order attributes introduced by Edwards and Steins (1998), and are included as a separate attribute in our framework. In addition, the two previous frameworks also do not distinguish between the physical characteristics and uses, or between potential uses and actual uses of resources, which seems necessary for the analysis of agroforestry systems and we establish that distinction.

3.2.1. Resource uses and users' attributes

Physical characteristics of a resource determine in what capacity and to what extent a resource can be used, but do not prescribe how it is used. This recognizes that resources can exist independently from how they are utilized, until users and stakeholders determine how the resource should be used. A tree, for example, has many properties and characteristics that simply exist, and are not considered 'uses' until their function is determined. A tree produces fruits, which can be harvested and used to feed a family, to extract oil or some other by-products, or to gain money, or fruits may not be harvested at all. Similarly, a tree produces branches, which can be used for fuelwood, for fencing, or as a broom for cleaning purposes. Thus, the physical characteristics of a resource, alone, are adequate to determine its potential uses but not sufficient to determine its real uses, which depend upon the social and economic attributes of the users. However, the real uses will also be influenced by the characteristics of stakeholders and external agents, and institutional arrangements, and these issues are discussed in the following sections.

Users are defined as the direct appropriators of the resource units either continuously or occasionally. The relationship between users and uses is reciprocal, in that user groups are also influenced by the choice of uses of a resource. For example, an agroforestry system aimed at commercial production may provide the owner with marketable products and employment opportunities for some local people. Each of these different users will exert different types of pressure on the agroforestry system and thus a clear understanding of user groups is necessary for proper management of agroforestry systems. Hence, all the real uses of all resources in a given agroforestry system should be recorded and for each real use, user groups should be identified. The main social and economic characteristics - such as main occupation, land-ownership, income, and social category – of each group should be examined.

3.2.2. Stakeholders' attributes

Resource users are only a part of the human-side, and invariably in most of the situations, institutional arrangements and resource management in agroforestry systems are affected by many other agents and/or organizations, such as government agencies responsible for agriculture and forest management, forest-based industries, banking institutions, research organizations, and local governments. All of these organizations/agencies have some stake in a given agroforestry system but they may not have direct claim over the resource units. However, some of these organizations may extend direct support to the resource users in the appropriation of resource units. We will call such agents and organizations as stakeholders of an agroforestry system, and these are defined as *"agents and/or organizations that either have direct influence or are directly influenced by the provision of factors, management practices, appropriation and distribution of resource units, outlets for appropriated resource units, and other institutional arrangements and associated practices but are not the direct appropriators of resource units"*. The stakeholders can be individuals, communities, government organizations, industrial organizations, non-government organizations, and other national and international bodies.

Stakeholders and resource user groups may share common actors, but the groups themselves are not interchangeable. For instance, stakeholders such as research organizations can facilitate the implementation of agroforestry projects in selected communities, without being users themselves. However, as a research organization it will have a stake in the success of the project. Similarly, banking institutions may facilitate the implementation of agroforestry projects by providing capital inputs without being resource users, but they will also have some stakes in the success of the project so that they are able to recover their investment.

Stakeholders play an important role in determining who will use the agroforestry system and what the rules surrounding their use will be. Stakeholder characteristics include factors related to their relative affluence, power, education, knowledge, cultures and belief systems. These factors will determine the potential target groups and their likelihood of agroforestry adoption. Hence, interests (stakes) and socio-economic characteristics of each stakeholder in an agroforestry system should be examined to facilitate institutional analysis.

3.2.3 External agent's attributes

In our definition of resource users and stakeholders, we have included only those agents and organizations that either have direct impacts or are directly influenced by an agroforestry system. However, in almost all agroforestry systems, there are also agents and organizations which either have indirect impacts or are indirectly influenced by an agroforestry system. We call such agents and organizations as external agents to the agroforestry system, and there can be numerous categories of such external agents. For example, assume there are two forest companies (FC1 and FC2), in a geographical area, which are promoting their own agroforestry systems (schemes) with local farmers, and both the companies have decided to work in different villages, called VFC1 and VFC2, respectively. Now, according to our classification, participating farmers of VFC1 villages will be the users and FC1 will be one of the stakeholders of the agroforestry system (say AFS1) developed by the FC1 and the farmers of the VFC1. Similarly, the participating farmers of the VFC2 villages will be the users and the FC2 will be one of the stakeholders of the agroforestry system (AFS2) developed by the FC2 and the farmers of the VFC2. However, AFS1 will also be indirectly influenced by the FC2 and the farmers of the VFC2, and similarly the AFS2 will be indirectly influenced by the FC1 and the farmers of the VFC1. Hence, as per our terminology, we will call the FC2 and the farmers of the VFC2 as external agents to the AFS2 system, and the FC1 and the farmers of the VFC1 as external agents to the AFS1 system.

External agents play an important role in the choice set of factors, management practices, outlets for products and services, and what the rules surrounding these components will be. Similar to stakeholders, external agents characteristics include factors related to their relative affluence, power, knowledge, technical know-how, cultures, and belief systems. Hence, all the categories of external agents' main characteristics of each category, possible indirect affect of each category on the system, and the possible indirect effect of the system on each category of external agents should be the part of institutional analysis of agroforestry systems.

3.3. The contextual factors of an agroforestry system

Contextual factors can play an important role in establishing the choice sets from which users and stakeholders can select strategies. These factors include dynamic forces coming from outside the agroforestry system. They are constituted in the social, cultural, political, economic, technological, institutional, and natural environment in which the agroforestry system is embedded but are usually beyond the direct control of the users and the stakeholders of the system. Some of the simple examples are global warming, forest certification, organic farming, and national and international regulations. Even those contextual factors remote from the system may affect its factors, management, and the markets for the products and services from the system. For example, a change in the government policy of industrial raw material supply – from the contracted supply, to industrial units, to the sale in open market – will influence the demand for wood produced from agroforestry systems. Similarly, a change in the interest rate by the central bank or a change in agriculture finance policy by the central government or by the central agriculture bank will influence the availability of capital to agroforestry systems. On the same lines, establishment of a new wood-based production unit in an area, demographic changes in local population, increased personal disposable income, dynamics of consumer preferences for agriculture and wood products will affect the resource users' and the stakeholders' choices for factors, management process, and outputs of the agroforestry system. Other natural contextual factors, such as global warming, prolonged drought, regular floods, can also affect the physical characteristics of the resources of agroforestry systems.

In summary, contextual factors define: i) what is socially, economically, culturally, and legally feasible in terms of supply of factors to the agroforestry system, and products and services from the system; ii) what is socially, economically, culturally, and legally desirable, by establishing the demand factors for factors, products and services of the system; and iii) what are socially, economically, culturally, and legally acceptable management practices. Contextual factors also affect the choice sets of factors, management practices, and outputs of agroforestry systems, and lack of knowledge, or exclusion of contextual factors may lead to simplified judgments of institutional arrangements of agroforestry systems. However, in reality, the inclusion of contextual factors will always have its own boundaries. Nevertheless, the discussion of the choice sets available – in terms of factors, management processes, products, services, and decision-making arrangements – to the users and the stakeholders of the system, and tracing back the source of these choice sets to contextual factors will enhance the analyst's appreciation of the origin of the different strategies of users and stakeholders (Edwards and Steins, 1998).

3.4. Institutional (decision-making) arrangements that govern actions and interactions among the agents of an agroforestry system

Institutions provide a set of rules, either formal or informal, that can be used to regulate different aspects of agroforestry systems. Institutions, exclusively, cannot determine outcomes. Instead, rules can influence an individual's actions and patterns of behavior by altering the costs and benefits (both market and non-market) of each action/strategy. Compliance with these rules is dependent on whether the benefits outweigh the costs, which is most likely to occur if they are reflective of the values of the society in which they are to be applied (Oakerson, 1986). The rigor of a rule is determined not by traditions or laws alone, but by how these institutions are interpreted by individuals and the society as a whole.

The institutional component in the Oakerson (1986) and Edwards and Steins (1998) frameworks focuses on resource appropriation and not on resource provision These frameworks also emphasize that the rules are 'soft' constraints, as opposed to the physical attributes which are 'hard' constraints, because they can only have an indirect impact on outcomes.

In agroforestry systems, since resource provision is as important or even more important than resource appropriation, it requires a similar emphasis on resource provision institutions. In addition, the absence of appropriate institutions for the provision of resources, for example inappropriate land tenure or lack of institutions for technical inputs, may directly affect the outcomes. An equal emphasis on resource provision and appropriation will require sequential analysis in addition to hierarchal (three levels) analysis. Therefore, we propose categorization of institutions associated with agroforestry systems into three groups: i) institutions for the provision of the factors (inputs) required by the system, ii) institutions for the management of production process of the system; and iii) institutions for the appropriation of the outputs of the system. These three categories ensure that all stages of an agroforestry system are considered to avoid unfavorable situations, such as where inputs are wasted on systems with no management. This structure also helps to ensure that agroforestry is evaluated holistically and that institutions associated with different stages are compatible and complete. An examination of all the three categories of institutions on six dimensions – i) short-term and long-term incentive-compatibility to the resource users; ii) short-term and long-term interest-compatibility with the interests of stakeholders; iii) institutional-compatibility among the three categories of institutions – institutions related to the factors, production process and the outputs of an agroforestry system; iv) institutional-compatibility with local traditions and norms; v) the degree of divergence with the institutions in other areas; and vi) institutional flexibilities to meet crisis situations – will assist the institutional analysts of agroforestry systems. The specific features of the three categories of institutions are discussed next.

3.4.1. Institutions for the provision of factors (inputs)

An agroforestry system requires many factors (inputs) such as land, labor, capital, seeds and seedlings, technological knowledge, and other technical inputs. Institutions for the provision of factors, factor institutions are determined by the combination of 'users and uses' of agroforestry systems, the stakeholders and the external agents associated with the system, local traditions and customs, and the existing institutions for the same or similar factors used in other production systems. As discussed previously, the 'users and uses' component unites the producers with their products to identify what inputs are required to implement an agroforestry project, such as which crops should be planted and how land is obtained. The influence of this component on factor institutions is extremely important because it is based realistically on both what the user society wants and what the physical system is capable of producing. These institutions are more likely to be adhered to because they are commonly based on local experience and expertise. However, some stakeholders can have equally strong effects on factor institutions. For example, in the case of agroforestry programs sponsored by forest-based industries, the sponsoring forest-based industry has high stakes in wood supply from the agroforestry system, and is normally a key player in the institutions related to the provision of capital, high quality seedlings, and technical knowledge. External agents can influence factor institutions indirectly by having a different set of institutions for the same factors, which may force the stakeholders and the users to evaluate the existing institutions. Local traditions and cultural norms will generally affect the institutions, specifically the appropriateness of institutions, related to almost all the factors. For example, private ownership based land tenure may not provide the desired results in communities where communal ownership has been practiced for centuries. Similarly, superimposition of formal institutions, for labor and capital, over existing traditional and informal institutions may create some inefficiencies in the system.

An example of this conflict can be seen in Nepal, where traditional institutions are incompatible with government institutions with regard to the planning and design of agroforestry projects. Traditional institutions regarding gender roles determine that women, who play an important role in the cultivation of agroforestry crops, are prohibited from direct communication with men from outside of their community (Kidd & Pimentel, 1992). Government institutions, on the other hand, require that representatives from forestry departments (of which the majority are male) consult with potential users of agroforestry systems during the planning process. Consequently, male forestry representatives have been unable to engage in a meaningful dialogue with this important user group, and efforts to communicate through male members of the community have proven unsuccessful (Kidd & Pimentel, 1992). Hence, the analysis of factor institutions on the six dimensions discussed previously is critical. In addition, the factor institutions should also be examined for i) timely supply of all the required factors during the life-cycle of an agroforestry system; and ii) the security of tenure over immovable factors such as land or natural forests.

3.4.2. Institutions for the management of production process of the system

The production process of agroforestry systems is a composite and complex process system comprising of many sub-processes and interactions between of agricultural crops and trees. Agriculture production processes may be different during different years based on the choices of agricultural crops. Hence, institutions for the management of production processes are crucial components of an agroforestry system and must be very well established and not merely assumed or over looked, as has often been the case in the past. The process institutions may require some adjustments over time based on the variations in agricultural crops, performance of the previous institutions, natural events, the emergence of new technologies, and changes in the composition and socio-economic characteristics of user groups, stakeholders, and external agents. Hence, a continuous evaluation of the process institutions should be an integral component of the institutional analysis of agroforestry systems.

Process institutions, which are frequently neglected during the planning and implementation phases of a project, should be expressly determined to ensure that consecutive sets of institutions are compatible. There exist multiple examples where institutions governing agroforestry systems are established at the beginning of a project but are not re-evaluated during the process resulting in the failure of projects. In certain Costa Rican agroforestry initiatives, the lack of continuity between government-initiated tree planting projects, and proper management following project implementation, was directly responsible for the project's failure (Current, 1995). Negligence of the process institutions has also been the main cause of failure of some agroforestry projects in West Africa (Schlauderer, 1997).

Supply of many factors, such as capital and labor, is an essential ingredient of management of the production process. Hence, the institutions related to the provision of these factors will directly affect the management of the production process, and therefore the complementary interactions between factor and process institutions are essential. For example, in Tanzania, women, who are responsible for collecting and transporting water, were not consulted with regards to crop selection during the initial planning phases, nor did they directly benefit from the chosen crops of agroforestry project. Therefore, during periods when the women were busy cultivating more beneficial crops, such as maize, women refused to carry water for the unwanted agroforestry crops (Kidd & Pimentel, 1992), and the project suffered due to lack of integration of process institutions and factor institutions. Similarly, institutions related to technical knowledge and technical inputs will also have continuous influence over management of production process, and complementary interactions between these institutions and process institutions are essential.

The production processes, of agriculture crops as well as of trees, is comprised of many serial stages such as initial establishment, growth promotion, crop protection from pests, insects, and animals, theft control, and management of matured crops. In addition to the six dimensions discussed earlier for all three categories of institutions, process institutions should be examined on i) comprehensiveness of institutions for all the stages of production; ii) complementarity between the institutions for different stages of production; iii) institutional-compatibility between the institutions for agriculture and tree production processes; and iv) institutional-compatibility with other production processes common in the community.

3.4.3. Institutions for the appropriation of the outputs of the system

Ironically, although outputs are the motivating factors for promotion of agroforestry systems, there has been little attention given to the institutions related to outputs. The two important aspects of outputs, which require appropriate institutional arrangements, are access to and extraction of particular resource units and economic returns from resource units. The first aspect – rights for access and extraction of particular resource units – has been the focus of the previous two frameworks, and we will not elaborate on this except to note that resource unit extraction rights can be either allocated to private individuals (farmers), organizations (forest industries), government agencies (forest departments), and communities, or left unallocated (open access). However, allocation of rights to resource units need not necessarily mean that the right holder can draw optimal economic returns from the resource units. In fact, in the absence of appropriate institutional arrangements, economic returns may be negligible and the right holder may not be able to recover the cost of production. In such cases, an agroforestry system may provide optimal physical output, but the lack of appropriate institutions for marketing of outputs may convert the agroforestry system into a financially non-viable system.

The design of institutions for optimal economic returns, or marketing institutions, is a complex and multi-faceted process which is affected by an array of factors such as local traditions, interests of stakeholders and external agents, government policies about procurement prices of agriculture and forest products, general market conditions in the area, dynamics of the factors affecting demand and supply of the products, and many contextual factors. For example, in West Africa, women act as purchasers of products in rural areas and vendors in urban areas, and play a key role in trade at the local, regional, and interregional levels (Schlauderer, 1997). A similar case is that of the Luo women in Kenya who acquire full ownership of the crop once harvested and full control over the profits from its sale (Rocheleau, 1985). Hence, marketing institutions should be sensitive to local traditions and norms. In some agroforestry schemes, such as the Western India Match Company (WIMCO) in India, forest industries offer marketing institutions through timber purchasing agreements with farmers. In such agreements, the required minimum size (diameter) of a tree, prices offered for different species and sizes of trees (either per tree or per cubic meter), and other components of the agreements such as requirement of insurance and use of technical services of the company are the main provisions. In the case of marketing institutions, the role of external agents is also very critical. For example, establishment of a pulp manufacturing unit in a geographical area rich in agroforestry systems, which have pulpwood as its tree component, will offer immense opportunities for marketing of pulpwood. Similarly, adoption of agroforestry systems by a large number of farmers, based on the success of pioneer farmers in agroforestry systems, will reduce either the marketing opportunities or economic returns to the pioneer farmers. Hence, the examination of marketing institutions is a complex process which will vary across situations. Therefore, in addition to the six dimensions suggested for all three categories of institutions, the institutions for marketing arrangements of outputs should be examined for i) assurance of optimal economic returns and sustainability of institutions in the long-term; ii) sensitivity of institutions to the factors external to an agroforestry system; iii) market risk diversion and risk hedging; and flexibility of institutions to meet demand for economic resources during emergencies and special events, such as marriages, sickness, and the education of children.

Finally, the analytical outcomes of the sequential institutional analyses for factors, process, and output can be improved by using three hierarchal levels –

operational, collective choice, and constitutional rules or external arrangements – for the analysis of each stage.

3.5. Actions and interactions among the agents associated with an agroforestry system

The fifth set of attributes consists of actions of and interactions among the users, the stakeholders, and the external agents of an agroforestry system. Institutions, both formal and informal, provide mechanisms for actions and interactions but do not guarantee the emergence of particular patterns of actions and interactions. Instead, individuals actively choose how they will respond to existing rules in order to maximize their net benefits (benefit-cost). However, these benefits and costs are not restricted to monetary units, rather a "cost" is any perceived obstacle to the choice of some alternative while a "benefit" is any perceived incentive to choose one alternative over another (Buchanan, 1969). Hence, individual choices are conditioned by perceived obstacles and incentives in a relevant institutional structure. In West Africa, for instance, government institutions administer the distribution and supply of crop varieties and fertilizers for agroforestry projects. The government controls the price of all similar products, and maintains it at a certain level. Although this scheme allows inputs to be purchased on credit, the prices are often seen as prohibitive to farmers, who, instead, choose to smuggle inputs on the black market (Schlauderer, 1997). This example illustrates how institutions cannot be relied upon to predict individual behavior, and shows how institutions can be more effective if they accurately reflect the interests of the users.

In any action situation where more than one individual is involved, an important element of individual behavior is inter-dependence. In the case of a two-member action situation one to one direct inter-dependence, or in the case of a more than two-member situation one to many direct inter-dependence and many-to-many indirect inter-dependence, will influence the interactions between different members. Hence, interactions between different members have to be separated from the actions of individuals with respect to the resource. The interactions between members can take different forms such as free-rider, cooperative, exchange and reciprocity. In a free-rider situation, every agent behaves independent of other agents. In a cooperative situation, an individual contributes to a joint undertaking as long as others also contribute. In an exchange situation, *quid pro quo* regulates individuals' behavior and individuals exchange their actions. In reciprocity, members of the group contribute to one another's welfare without an immediate *quid pro quo* as in exchange situations (Boulding, 1972). Hence, exchange is based on *ex ante* conditions while reciprocity is based on *ex post* conditions[5].

The outcomes of any action situation will be determined by the combined effects of the individual's actions and the interactions among individuals. In the case of agroforestry systems, the actions of all the agents – the users, the stakeholders, and the external agents – and interactions between different combinations of agents – among users, stakeholders, and external agents; and between stakeholders and users, external agents and stakeholders, and external agents and users – will influence the outcomes, and therefore should all be examined. In addition, actions and interactions between agents may have different features during the three stages of production – inputs, process, and outputs – and also during the production cycle of agricultural crops and the production cycle of forest crops. The nature of actions and interactions may also evolve over different production cycles of agricultural crops. Generally, all new actions and interactions are dependent on the past actions and interactions; and, therefore, the actions and interactions during the initial phase of the agroforestry system will influence the actions and interactions during later phases. Hence, the analysis of actions and interactions should be divided at least into i) two product classes – agriculture and forestry crops; and ii) three production phases – factors, process, and outputs. In addition, relationships between "actions and interactions" during all the six classes (product and production) should be examined.

3.6. The outcomes of an agroforestry system

Actions and interactions produce outcomes that may be physical or non-physical. Some of these outcomes will be the intended products, such as agricultural and tree products, while others will be by-products such as the technical expertise learned by agroforestry workers and soil conservation. Evaluation of the outcomes against some standard evaluative criteria, rather than in terms of quantities of physical products and some quantitative or qualitative measures of by-products, will be more meaningful and desirable. In addition to standard evaluative criteria, a comparative examination of the two situations, with and without agroforestry systems, for different categories of the users and the stakeholders will also provide useful insights about the outcomes.

The previous two frameworks have proposed the use of economic efficiency, using the test of Pareto-optimality and equity as two evaluative criteria. While these two measures are useful, the measure of equity is not straight forward and is often dependent on questionable indicators such as the satisfaction level of all the users from the agroforestry system. Equity in terms of level of satisfaction or even in terms of direct and indirect returns to all the users as well as all the stakeholders is also problematic. Hence, analysts will have to find an acceptable measure of equity for the given situation, and the measure of equity need not to be the same for all the situations.

Similarly, these two evaluative criteria do not provide any measure of sustainability; the main issue raised in the introduction of this paper, and the main focus of current discussions among scholars, managers, and policy makers on land-use systems. In contrast, we propose to use the evaluative criteria of sustainability, instead of economic efficiency and equity. The sustainability criteria would include economic, social, and ecological sustainability of agroforestry systems; economic efficiency will be encompassed in economic sustainability and equity will be a part of social sustainability criteria.

3.7. Relationships between the sets of attributes

The six sets of attributes and the various intra-set and inter-set relationships are shown in Figure 1. The proposed framework is dynamic in nature and we need not distinguish between the static and dynamic models because time is an influential component of the model. Likewise, in this model almost all the relationships are reciprocal in nature. Intra-set relationships are shown by single lines and inter-set relationships by double lines. While some of these relationships are already mentioned in previous sections we will briefly mention all the relationships here for clarity.

The physical and technical attributes of the resources will determine, at least to some extent, who are resource users, stakeholders, and external agents in a given agroforestry system, and reciprocally these three agents will influence the use of technology and therefore technical attributes of the resources. The three categories of agents will affect the choices of each other, and hence the socio-economic characteristics of the three groups of agents. In the case of institutions, there will be reciprocal interactions among three categories of institutions – factors, process, and outcomes, and overall institutions will be affected by the attributes of resources and agents, and over a period of time, institutions will also affect the attributes of resources and agents. Agents' actions and interactions among agents, at any given time, will be influenced by their previous actions and interactions, and the whole set of actions and interactions will be influenced by institutions and attributes of resources and agents. In a dynamic context, a set of actions and interactions may also influence resource characteristics and attributes of agents. Generally, physical and non-physical outcomes will be the result of actions and interactions and resource attributes, but values of evaluative criteria such as economic and social sustainability will also depend upon the attributes of agents. Similarly, outcomes will also influence the social and economic features of agents and physical and technical attributes of resources. The contextual factors will influence all other attributes continuously, and hence, these are shown at the top covering all other attributes at all times. In the total situation, the influence of the other five sets of attributes of agroforestry systems on contextual factors will be minimal, and therefore the relationship of contextual factors to other attributes is shown as unidirectional.

4. DISCUSSION AND CONCLUSIONS

The initial intent of this chapter was to evaluate different institutional arrangements for agroforestry system. Unfortunately, we could not find a single paper discussing all the institutions related to an agroforestry system. Although many studies revealed an impressive account of the institutions governing some factors (inputs), specifically land tenure, they totally neglected institutions related to other factors, process, and outputs. In addition, none of the papers followed the total sequence of production beginning from resources and agents to the outcomes of the system. Hence, we moved from and evaluation of institutions to a framework for institutional analysis of agroforestry systems.

We have proposed a framework which incorporates the integrated nature of agroforestry systems, the sequential nature of production process, micro-level as well as macro-level and formal, as well as informal institutions, sensitivity of institutions to social, cultural, and ecological diversity, and social, economic, and ecological sustainability as evaluative criteria. The framework treats the different socio-economic aspects of agroforestry systems as a sub-system, and identifies the different elements, physical (natural resource and agents) as well as non-physical (institutions and contextual factors), of the sub-system and the connectedness between these elements. The purpose of the framework, presented in this chapter, is two fold: i) to provide an analytical tool to institutional analysts associated with agroforestry systems; and ii) to help research institutions involved in agroforestry research in the collection and assimilation of system by system analysis of agroforestry systems. Institutional analysts can use this framework at any stage, for in-depth analyses of institutional arrangements of a given agroforestry system and can provide useful inputs to system managers, users, and stakeholders. Research institutions and policy makers can use the framework to collect system-by-system analysis of multiple agroforestry systems across different regions. Using this framework, a consistent and comprehensive method will enhance the comparability of different systems. Comparative outcomes from different agroforestry systems can be used by policy makers for more effective policy interventions.

The proposed framework is a first step in the comprehensive and consistent institutional analysis of agroforestry systems. A great deal of work has yet to be done. There is lot to be learned from the experiences of farmers and multiple agencies, users and stakeholders, associated with agroforestry systems about the varieties of institutional arrangements, and about how these arrangements are nested within the larger set of institutions for forestry and agriculture, and other social and political arrangements found in diverse societies. We hope that some research organizations will take a lead in this direction, and as more and more scholars use and apply this framework, and share ideas, the framework, too, will become the subject of modification and elaboration.

5. NOTES

[1] Although Nair (1989) included socio-economic aspects in the context of land-use systems he included only different combinations of forestry and agriculture crops – the taungya system, tree gardens, and alley cropping – as subsystems of the prominent agroforestry systems in the tropics and sub-tropics; and there is no mention of social systems or economic systems as sub-systems of agroforestry systems. Hence, in this definition of land-use systems, the treatment of socio-economic aspects has been peripheral, and not inclusive. Similarly, in the definitions of agroforestry systems, given by K. F. S. King, M. T. Chandler, P. A. Huxley, and J. B. Raintree (all quoted in Nair, 1989), social and economics systems are not treated as components and/or sub-systems of agroforestry systems. In the Global Inventory of Agroforestry

Systems, maintained by World Agroforestry Centre (ICRAF), socio-economic description and socio-economic characteristics are included, but these are not treated as components of a system.

[2] Oakerson's framework was adopted by the Panel on Common Property Resource Management at the United States National Research Council and was used for the analysis of twenty case studies presented at the International Conference of Common Property Resource Management held in 1986.

[3] In the Oakerson's framework, the third level of rules was "external arrangements" which was replaced by "the constitutional level rules" in Edwards and Steins (1998) framework.

[4] In these situations, institutional analysis using the three-levels of institutions – operational rules, collective choice rules, and constitutional rules – is useful, but not enough. The institutions for each component of agro-forest production process should be analyzed separately. However, the three-levels of institutions can be used to analyze each component of the production process separately.

[5] Please refer to Oakerson (1986) and Edwards and Steins (1998) for further discussion of these situations.

6. REFERENCES

Andreatta, S. (1998). Transformations of the agro-food sector: lessons from the Caribbean. *Human Organization, 57*(4), 414-429.

Arnold, G. W., & de Wit, C. T. (Eds). (1976). *Critical evaluation of systems analysis in ecosystem research and management.* Wageningeng, Netherlands: Centre for Agricultural Publications Documentation.

Bhati, U. N., Klijn, N., Curtotti, R., Dean, M., & Stephen, M. (1992). Impediments to the development of commercial plantations in Australia. In N. Wallace, (Ed.), *Natural resource management: An economic perspective* (pp. 195-204). Canberra, Australia: ABARE.

Blaikie, P., & Brookfield, H. (1987). *Land degradation and society.* London: Methuen.

Boulding, K. E. (1972). The household as Achilles Heel. *Journal of Consumer Affairs, 6,* 111-110.

Bromley, D. W. (1991). *Environment and economy: Property rights and public policy.* Oxford: Blackwell.

Buchanan, J. M (1969). *Cost and choice.* Chicago: Markham.

Current, D. 1995. Economic and institutional analysis of projects promoting on-farm tree planting in Costa Rica. In Current, D., Lutz, E., and Scherr, S. (Eds.). *Costs, benefits and farmer adoption of agroforestry: project experience in Central America and the Caribbean* (pp. 45-70). Washington, D. C: World Bank.

Current, D., Lutz, E., & Scherr, S. (Eds.). (1995). *Costs, benefits and farmer adoption of agroforestry: Project experience in Central America and the Caribbean.* Washington, D. C.: World Bank.

de Wit, C. T., & Goudrian, J. (1974). *Simulation of ecological processes.* Wageningen, Netherlands: Centre for Agricultural Publications Documentation.

Dwivedi, A. P. (1992). *Agroforestry principles and practices.* Delhi: Oxford and IBH Publishing.

Edwards, V. M., & Steins, N. A. (1998). Developing an analytical framework for multiple-use commons. *Journal of Theoretical Politics, 103,* 347-383.

Feder, G., & Onchan, T. (1987). Land ownership security and farm investment in Thailand. *American Journal of Agricultural Economics, 69,* 311-320.

Fujisaka, S. (1994). Learning from six reasons why farmers do not adopt innovations intended to improve sustainability of upland agriculture. *Agricultural Systems, 46,* 409-425.

Gilmour, D. A., & Nurse, M. C. (1995). Farmer initiatives in increasing tree cover in central Nepal. In N. C. Saxena, & V. Ballabh (Eds.). *Farm forestry in south Asia* (pp. 87-103). New Delhi: Sage Publications.

Jackson, W., & Bender, M. (1984). Investigations into perennial poly-culture. In W. Jackson, W. Berry, & B. Coleman, (Eds.), *Meeting the expectations of the land: Essays in sustainable agriculture and stewardship,* (pp. 183-194). San Francisco: North Point Press.

Kidd, C. V., & Pimentel, D. (1992). Social aspects of resource management systems. In C. V. Kidd, & D. Pimentel (Eds.). *Integrated resource management: Agroforestry for development* (pp.176-187). California: Academic Press.

King, K. F. S. (1989). The history of agroforestry. In Nair, P. K. R. Nair. (Ed.). *Agroforestry systems in the tropics* (pp. 3-12). Amsterdam, The Netherlands: Kluwer Academic Publishers.

Lundgren, B. O. (1987). Institutional aspects of agroforestry research and development. In H. A. Steppler & P. K. R. Nair (Eds.), *Agroforestry a decade of development* (pp. 43-51). Nairobi, Kenya: World Agroforestry Centre (ICRAF).

Lundgren, B. O. (1989). Institutional and policy aspects of agroforestry. In P. K. R. Nair (Ed.), *Agroforestry Systems in the tropics* (pp. 610-607). Amsterdam, The Netherlands: Kluwer Academic Publishers.

Nair, P.K.R. (1989). Agroforestry defined. In P. K. R. Nair (Ed.), *Agroforestry Systems in the tropics* (pp. 13-20). Amsterdam, The Netherlands: Kluwer Academic Publishers.

Norgaard, R. B. (1981). Socio-system and ecosystem co-evolution in the Amazon. *Journal of Environmental Economics and Management, 8,* 238-254.

North, D. C. (1981). *Structure and change in economic history*. New York: W. W. Norton and Company.

Oakerson, R. J. (1986). A model for the analysis of common property problems. In National Research Council, *Proceedings of the Conference on Common Property Resource Management* (pp. 13-20). Washington, D. C.: National Academy Press.

Okoth-Ogendo, H. W. O. (1985). Tenure of trees or tenure of lands? In Raintree, J. B. (Ed.), *Land trees and tenure: proceedings* (pp.225-229). Nairobi, Kenya: ICRAF and the Land Tenure Centre.

Ostrom, E. (1990). *Governing the commons: The evolution of institutions for collective action.* Cambridge: Cambridge University Press.

Otsuka, K., Suyanto, S., Sonobe , T., & Tomich, T.P. (2001). Evolution of land tenure institutions and development of agroforestry: evidence from customary land areas of Sumatra. *Agricultural Economics,,* 85-101.

Pannell, D. J. (1999). Social and economic challenges in the development of complex farming systems. *Agroforestry Systems, 45,* 393-409.

Riddell, J. C. (1987). Land tenure and agroforestry: A regional overview. In J. Raintree (Ed.). *Land, trees, and tenure.* Nairobi, Kenya: ICRAF.

Rocheleau, D. E. (1985). Women, trees and tenure: Implications for agroforestry. In L. Fortmann & J. W. Bruce (Eds.), *Whose trees? Proprietary dimensions of forestry* (pp.254-272). London: Westview Press.

Rogers, E. M. (1995). *Diffusion of innovations.* New York: Free Press.

Sanchez, P. A. (1999). Delivering on the promise of agroforestry. *Environment, Development, and Sustainability, 1,* 275-284.

Schlauderer, R. (1997). Socio-economics of the introduction of alley cropping systems in traditional farming. Doctoral thesis. *Farming Systems and Resource Economics in the Tropics, 29.* Center for Agriculture in the Tropics and Subtropics, University of Hohenheim, Germany.

Scherr, S. J., & Hazell, P. B. H. (1994). *Sustainable agricultural development strategies in fragile lands. Environment and production* [Technology Division Discussion Paper No. 1] Washington, D. C.: International Food Policy Research Institute.

Senanayake, R., & Jack, J. (1998). Analogue forestry: An introduction. Monash Publications in Geography Number 49, Department of Geography and Environmental Science, Monash University, Melbourne, Australia.

Singh, K. (1994). *Managing common pool resources: Principles and case Studies.* Delhi: Oxford University Press.

Tang, S. Y. (1992). *Institutions and collective action: Self-Governance in irrigation.* San Francisco: Institute for Contemporary Studies Press.

Urrea, O. S. (1995).Economic and institutional analysis of agroforestry projects in Guatemala. In D. Current, E. Lutz, & S. Scherr (Eds.), *Costs, benefits and farmer adoption of agroforestry: Project experience in Central America and the Caribbean* (pp. 98-113). Washington, D.C.: the World Bank.
Wojtkowski, P. A. (1998). The theory and practice of agroforestry design. A comprehensive study of theories, concepts, and conventions that underlie the successful use of agroforestry. Enfield, New Hampshire: Science Publishers.

7. AUTHOR'S NOTE

We gratefully acknowledge generous financial support from the Sustainable Forest Management Network, Edmonton, and CFS/NSERC/ SSHRC partnership program.

D. EVAN MERCER AND JANAKI R.R. ALAVALAPATI

SUMMARY AND FUTURE DIRECTIONS

1. INTRODUCTION

This chapter summarizes the main results from the preceding chapters, identifies gaps, and provides direction for future economics research on agroforestry systems. Although a common theme throughout the 1990s was that economic research on agroforestry continued to lag the advances made in the bio-physical sciences, the wide range of systems, regions, and techniques presented in this book suggests that economists have taken up the challenges of Scherr (1992), Sanchez (1995), and Mercer and Miller (1998) and are pushing forward the frontiers of economic analysis of agroforestry.

2. SUMMARY OF FINDINGS

2.1. Economic analysis

The first section of the book consists of five chapters (Chapters 2-6) on general economic analysis of the relative profitability of agroforestry compared to alternative land use systems. In Chapter 2, Franzel tackles the problem of assessing financial returns to farmers for three agroforestry systems in Africa (fodder shrubs for milk production in Kenya, rotational woodlots for firewood in Tanzania, and improved fallows for enhancing soil fertility in Zambia). In contrast to much of the agroforestry literature (e.g. Adesina & Chianu, 2002; Bannister & Nair, 2003; Lapar & Pandey, 1999; Sanchez, 1995), Franzel shows why these three systems have been adopted by a large number of farmers (23,000 adoptees of fodder shrubs in Kenya, 961 farmers planting rotational woodlots in Tanzania, and over 20,000 improved fallows planted in Zambia). The financial benefit-cost analyses were all based on comparisons of results on treatment and control plots in on-farm trials designed by researchers (in consultation with farmers) and managed by the farmers. Depending on the system, enterprise and partial budget analyses calculated net discounted returns to land, labor, and capital, and annual net farm income and maize production. The widespread adoption is not surprising since annual net benefits from adoption ranged from $US 68 to US$ 212 (not including non-market environmental

J. R. R. Alavalapati & D. E. Mercer, Valuing Agroforestry Systems, 303–310.

benefits such as reduced soil erosion and deforestation) greater than non-agroforestry alternatives. Given these returns, low establishment costs, and short payback periods (2-5 years) Franzel concludes that credit is not a constraint to adoption.

In Chapter 3, Grado and Husak compare the profitability of cattle-loblolly pine silvopasture systems to four traditional, single-use agricultural and forestland management systems in the southern United States (soybeans, rice, cattle, and loblolly pine plantations). Variations in the silvopastoral system include the impacts of hunting leases and pine straw production on profitability. In addition to the standard net present value (NPV), rates of return (ROR), and equivalent annual income (EAI) analyses, Grado and Husak also examine the impacts on landowners' willingness to pay for land or other assets using the land equivalent value (LEV) under the different scenarios. They find that although pine plantations yield the highest returns, land expectation values and EAI's were only slightly lower for silvopasture, which dominated all other alternatives (LEV for silvopasture was US$ 1253/acre compared to US$ 1284/acre for pine plantations and EAI for silvopasture was US$ 63/acre compared to $US 64/acre for pine plantations). However, risk reduction associated with the more even flow of revenues over the rotation period and the increased diversity of silvopasture may make it more desirable to some landowners.

Drew, Alavalapati, and Nair, use the Policy Analysis Matrix (PAM) method in Chapter 4 to extend profitability analysis by considering both private and social profitability of three agroforestry systems in Pohnpei, Federated States of Micronesia. This type of analysis is crucial because agroforestry is often promoted as a sustainable land use system that produces both private and public goods (e.g., environmental improvements that benefit the wider society such as reduction of soil erosion and carbon sequestration). If private profitability is not sufficient to encourage adoption of agroforestry systems that produce large amounts of public goods, analyses like these are required to determine whether government incentives are appropriate and if so the size and composition of incentives required to encourage socially efficient rates of agroforestry adoption. In addition, the PAM methodology provides a means to quantify the impacts of policy distortions and market failures on farmer decision-making and how potential government strategies may impact different sectors of the economy. Although the PAM has been applied to analyzing single crop systems (including agroforestry) this represents the first time that PAM has been used to examine social and private profitability of multiple crops in complex agroforestry systems and to internalize non-market externalities such as carbon sequestration and soil erosion for agroforestry policy analysis. The case study in Pohnpei provides a striking example of the necessity of policies such as taxes on soil erosion and subsidies for carbon sequestration to correct policy distortions and market failures to promote socially efficient land use with agroforestry.

Bright (Chapter 5) shifts the emphasis from comparing the profitability of agroforestry systems to alternatives to applying production function theory and analysis to determine optimal combinations of inputs and outputs for a given agroforestry system. The basic economic model for using production functions to examine input-output relationships is extended theoretically to account for the more complicated multi-input and multi-output nature of agroforestry systems. He then applies the resulting model of the agroforestry production possibility frontier (PPF) to a case study of a *leucaena*-wheat-maize agroforestry system in India. He shows that using a continuous production function approach that specifically models the competitive interactions and relationships between the multiple inputs and outputs of agroforestry systems allows for more accurate predictions of the benefits, costs, and returns to different factor mixes. Although data intensive, this type of analysis should result in developing more optimal and profitable agroforestry combinations and increase the probability of long-term adoption.

Blandon (Chapter 6) applies portfolio theory to the problem of analyzing the impacts of risk on agroforestry system design, evaluation, and potential adoption. Following an explication of portfolio theory as applied to agroforestry, he demonstrates how it would be applied to the problem of designing the optimal mix of timber trees (ash for joinery), pasture (rye-grass), and sheep in a silvopastoral system in North Wales, United Kingdom. For long rotation, temperate zone agroforestry systems like the sheep-ash mixture, portfolio analysis suggests that the economic risk reduction from diversifying production with silvopasture is quite small. Thus the biological and physical interactions are more important than price diversification for temperate agroforestry. However, for shorter rotation systems (as in most tropical agroforestry systems) Blandon shows that the risk reduction benefits from agroforestry are potentially much larger and critical for the design and adoption of agroforestry.

2.2. Environmental economic analysis

As shown in Chapter 4, the environmental benefits associated with agroforestry will usually be the motivating factor for governments to encourage agroforestry through various policy and programmatic efforts. The environmental economics section (Chapters 7-10) takes a closer look at a variety of tools for addressing environmental services associated with agroforestry systems using case studies from Mexico, Australia, Indonesia, and the United States.

In Chapter 7, De Jong, Ochoa-Gaona, Quechulpa-Montalvo, Esquivel-Bazán, and Pérez-Hernández provide a cost accounting approach to examine the economics of projects and policies to encourage agroforestry for the production of environmental services. Using the Scolel Té project in Chiapas, they estimate a variety of the costs (fixed, variable, opportunity, and monitoring and verification costs) associated with using agroforestry systems for carbon sequestration to mitigate global climate change. They show that the development of cost curves are essential for estimating the carbon sequestration potential of agroforestry for different regions and for determining the efficiency of various programs, policies, and land-use systems.

Cacho and Hean demonstrate, in Chapter 8, the use of dynamic optimization modeling to estimate the optimal combination of trees and crops and rotation period to reduce the costs of land degradation (salinization) and increase the benefits associated with carbon sequestration. Using a dynamic programming approach, where current decisions influence the range of future options, they estimate the direct (timber) and indirect (land degradation and carbon sequestration) benefits and costs of planting trees in association with agricultural crops compared with those of pure agricultural crops. They find that the optimal decision rules for tree stocking and rotation depend on initial land quality, with more trees on shorter rotations being optimal on poorer lands. Land degradation impacts are shown to be more important than carbon sequestration benefits for determining optimal tree stocking and rotation cycles.

In Chapter 9, Pattanayak and Depro demonstrate the use of household production theory for modeling the production of environmental services such as soil erosion and reduced deforestation from fuelwood collection in agroforestry systems. Following the development of the theoretical model they show how it can be used to generate hypotheses and test them econometrically using data from Mangarrai, Indonesia. The model predicts that optimal production of environmental services from the household's perspective is determined by prices of consumption goods and production inputs, and factors influencing production possibilities and adoption choices. Both parametric (multivariate ordered probit) and non-parametric analyses of cross-sectional data support the hypotheses that agroforestry (depending on the specific system) reduces fuelwood harvesting from forests and reduces costs of on-site erosion.

The final chapter (10) in the environmental economic analysis section by Shrestha and Alavalapati, applies contingent valuation to the problem of estimating the amount of compensation Florida ranchers would be willing to accept to adopt silvopastoral systems,that produce environmental services (improved water quality, soil conservation, wildlife habitat, and carbon sequestration) for the citizens of Florida. This innovative application to producer decision-making (as opposed to the more common consumer focused contingent valuation) based on a survey of 421 ranchers found that an average of US$ 0.19 per pound price premium would be required to encourage Florida ranchers to adopt silvopasture. The authors have demonstrated the applicability of survey-based non-market valuation methods to evaluate the adoption potential of agroforestry in the US.

2.3. Household constraints and adoption

It has long been recognized that achieving the full promise of agroforestry requires a fundamental understanding of how and why farmers make long-term land-use decisions and applying this knowledge to the design, development, and "marketing" of agroforestry innovations to improve adoption rates. During the 1990s, a relatively large literature on adoption of agroforestry was produced to facilitate this understanding. Most of this literature is based on *ex-post* binary choice regression analysis (using probit or logit) models to analyze how past adoption decisions are correlated with variables describing farmers, their farms, and demographic and socio-economic variables. This literature, summarized by Pattanayak, Mercer, Sills, and Yang (2003) and Mercer (in press), has made valuable contributions to understanding the characteristics of early adopters, targeting communities and households for projects, and developing policies to promote agroforestry. Yet little work has been done to address the larger question of understanding the decision-making process or on developing rigorous, quantitative *ex-ante* methods for analyzing agroforestry adoption potential. In this section, Chapters 11-13 present alternative methods for analyzing the adoption process.

Mudhara and Hildebrand (Chapter 11) demonstrate how linear programming (LP) models and simulations can be used to analyze the constraints inhibiting households from adopting agroforestry, determine *ex-ante* adoption potential, assess the potential impact of adoption on household welfare, and estimate the viability of various polices for promoting agroforestry adoption. Using a case study of improved fallows in Malawi, Thangata, Alavalapati, and Hildebrand then show in Chapter 12 how the results from a number of LP models (like those described in Chapter 11) can be used to create datasets for meta-regression analysis as a post LP model validation tool and as an alternative means to assess the factors influencing the adoption potential of agroforestry when available survey data is limited. In Chapter 13, Mercer and Snook demonstrate the use of a third alternative for *ex-ante* assessment of agroforestry adoption, attribute-based choice experiments (ACE). Derived from the long history of applying conjoint analysis to analyze the potential demand for new multi-attribute goods and services, this is the first known application of ACE to examine how farmers value a variety of attributes and combinations of attributes of different agroforestry systems. Using a case study in southeastern Mexcio, they apply ACE to the problem of designing new agroforestry systems and projects based on quantifiable farmer preferences for alternatives.

2.4. Macroeconomic and institutional analysis

Regional planners and policy makers are often interested in knowing how and to what extent agroforestry programs and policies impact rural economies and environments. The final section of this book precisely takes a broader perspective by assessing agroforestry from a regional and national perspective. In Chapter 14, Yin uses macro level econometrics to quantify the impact of agroforestry on agricultural productivity and rural development at a regional level in Northern China. Applying time series analysis to a data set on agricultural production from 1978-1990, Yin shows that land tenure reform was the main driver for the increase in agroforestry investment. He also shows that the planting of trees during this period resulted in the increases in both market and non-market (environmental) benefits that generated the observed increase in agricultural productivity in China during the 1980s.

The final chapter (15) develops a framework for a system level institutional analysis of agroforestry systems. The framework incorporates the integrated, sequential nature of agroforestry production processes (physical and socio-economic) with micro, macro, formal, and informal institutions and requirements for social, economic and ecologic sustainability as evaluative criteria. Kant and Lehrer conclude that applying this framework is a first step in a comprehensive and consistent institutional analysis of agroforestry systems to inform effective policy interventions.

3. GAPS AND FUTURE DIRECTIONS

The research reported in this book demonstrates the rapid advances being made in applying economic theory, modeling, and empirical analysis to the study of agroforestry. Although the authors have made significant strides in improving the quantitative rigor of economics and policy research in agroforestry, there is a large potential to take the work presented in this book to a higher level. Areas for future research include:

- Economic analysis can be extended and strengthened by explicitly accounting for variation in future prices of inputs and outputs and risk and uncertainty associated with production processes. Extensive sensitivity analyses and probabilistic modeling might improve the accuracy and credibility of the information.
- Future research should emphasize more whole-farm analyses rather than enterprise-specific budgets to examine potentially crucial interactions between enterprises on a farm and a broader examination of returns beyond profitability to include the impacts of cultural taboos, farmer preferences, resource bottlenecks, policy constraints and market failures.
- Greater emphasis needs to be placed on the role that markets play in determining the profitability and adoptability of agroforestry. Particularly important is analyzing the role of agroforestry production as substitutes for purchased inputs and how prices and markets for the substitutes will influence farmer adoption and vice versa.
- Research on dynamic optimization of agroforestry systems should concentrate more on reducing assumptions of land homogeneity under single management and to include stochastic processes for prices, yields, and weather in the models.
- Econometric analyses could be improved by developing time series or panel data sets and apply the instrumental variables to account for potential endogeneity associated with agroforestry adoption.
- Extensive research and diverse applications of contingent valuation have enabled resource economists to identify various biases and account for many of those anomalies. For example, a variety of elicitation formats (open-ended questions, iterative bidding games, and payment cards) that were explored in resource economics can be introduced to agroforestry. Also, the methods of combining the stated preference data with revealed preference data in estimating environmental values can be applied in agroforestry context.

– One of the major omissions of the book is economy-wide impact analysis of agroforestry systems using applied general equilibrium approaches. Input-output (I-O) models, social accounting matrix (SAM) approaches, and computable general equilibrium (CGE) models are models and approaches provide practical means to conduct economy-wide impact analysis. By incorporating intersectoral linkages, these models capture the multiplying and trade-off effects of agroforestry programs and policies at a regional and/or national level. Although these approaches have been extensively applied in agricultural and forest economics (Alavalapati, Adamowicz, & White, 1998, Cattaneo, 2001; Wong & Alavalapati, 2003), they have yet to make their way into agroforestry. Future research is sorely needed to fill this gap.
– With biophysical, social, and economic objectives, agroforestry management decision criteria become more complex. Many of these criteria are value-laden and cannot be easily expressed in monetary terms. The analytical hierarchy process (AHP), which has been extensively applied to address this issue in natural resources (Schmoldt, Kangas, Mendoza, & Pesonen, 2001), is not discussed in the book. Although Shrestha, Alavalapati, and Kalmbacher (in press) have recently applied this to agroforestry decision-making, more research is needed in this area.

With an objective of providing coverage of a wide range of applied economic and policy analysis methodologies for agroforestry professionals, this book offers practical means for advancing economic analysis of agroforestry. These methodologies assist researchers in conducting rigorous economic and policy research in agroforestry and produce credible information. The diversity of methodologies, issues, and agroforestry practices for decision-making in the book suggest that "one-size-fits-all" does not apply in assessing economic and policy issues of agroforestry systems. Similarly, a single book cannot address all possible cases and appropriate analyses. We hope that the omissions and gaps in this publication serve to motivate agroforestry professionals to explore, investigate, and expand the frontiers of economic and policy research.

4. REFERENCES

Adesina, A. A., & Chianu, J. (2002). Determinants of farmers' adoption and adaptation of alley farming technology in Nigeria. *Agroforestry Systems, 55,* 99-112.

Alavalapati, J. R. R., Adamowicz, W. L., & White, W. (1998). A comparison of economic impact assessment methods: The case of forestry developments in Alberta. *Canadian Journal of Forest Research, 28*(5), 711-719.

Bannister, M. E., & Nair, P. K. R. (2003). Agroforestry adoption in Haiti: The importance of household and farm characteristics. *Agroforestry Systems, 57,* 149-157.

Cattaneo, A. (2001). Deforestation in the Brazilian Amazon: Comparing the impacts of macroeconomic shocks, land tenure, and technological change. *Land Economics, 77,* (2), 219-240.

Lapar, M. L., & Pandey, S. (1999). Adoption of soil conservation: the case of the Philippines uplands. *Agricultural Economics, 21,* 241-256.

Mercer, D. E. (in press). Adoption of agroforestry innovations in the tropics: A review. *Agroforestry Systems.*

Mercer, D. E., & Miller, R. P. (1998). Socioeconomic research in agroforestry: progress, prospects, priorities. *Agroforestry Systems, 38*, 177-193.

Pattanayak, S.K., Mercer, D. E., Sills, E., & Yang, J. (2003). Taking stock of agroforestry adoption studies. *Agroforestry Systems, 57*, 173-186.

Sanchez, P. A. (1995). Science in agroforestry. *Agroforestry Systems, 30,* 5-55.

Schmoldt, D. L, Kangas, J., Mendoza, G.A., & Pesonen, M. (2001). The analytical hierarchy process in natural resource and environmental decision making. Dordrecht, The Netherlands: Kluwer Academic Publishers.

Scherr, S. J. (1992). Not out of the woods yet: Challenges for economics research on agroforestry. *American Journal of Agricultural Economics, 74*, 802-808.

Shrestha, R. K., Alavalapati, J. R. R., & Kalmbacher, R. S. (in press) Exploring the potential for silvopasture adoption in south-central Florida: An application of SWOT-AHP method. *Agricultural Systems*.

Wong, G., & Alavalapati, J.R.R. (2003). The land-use effects of a forest carbon policy in the U.S. *Forest Policy and Economics, 5*, 249-263.

INDEX

J. R. R. Alavalapati & D. E. Mercer, Valuing Agroforestry Systems, 311–314.